Another tour de force by Alex and Dave Bennet, in collaboration with their colleague, Bob Turner. Very few people have the gift to integrate such complex ideas, especially those about learning. We, as humans, will never completely understand ourselves, as we are so very many, so different, and so complex. However, we should keep on trying; and this work can be likened to the Webb Telescope, which gives us more clarity into our mysteries. Well worth the viewing! **– Michael Stankosky, DSc, Author, Philosopher, Professor, Editor-Emeritus, Member of the Academy of Scholars, USA**

Even when one doesn't know the first thing about learning, learning happens anyway. Learning is innate to everything in the universe as it should be seen as the way evolution works. Usually, time is not a big factor in the process of learning—13 billion years and counting—there might, however, come a point at which it might just make the difference for a certain species. The difference here would be between being able to continue keeping a balance, finding a new one (Homeorhesis), or even by implementation of new operational unanticipated and possibly undesired outcomes. As Desire should be seen as residing in Self or even be equated with Self, and the fact that emotions—as the "differences that make a difference"—are instrumental in the balance of Desire/Self and all subsystems in both mind and body, the work presented here allows for many new and valuable insights as well as tools pertaining to the now pressing subject of learning. **– R. D. Heijnen, Independent Researcher, The Netherlands**

With today's obsessive focus on technology, even to the point of building direct brain-computer interfaces (BCI), the concept of evolution has taken on the flavor of humans ultimately becoming blended with technology, much like the Borg as portrayed in science fiction. Thankfully, the Bennets, Robert Turner, and the attuned souls at the deeply remote, RFI-minimal, Mountain Quest Institute are looking at a totally different evolutionary path, that of the human neurophysiology and the spiritual self.

This most magnificent of all organisms has the capacity to sense and act at a quantum level, glimpses into which have only been experienced by sages and seekers on rare occasions throughout history. Let technology continue on its rapid course, by all means, but we must also push the envelope of human potential. Technology is computational, humans are one with nature—the two working together will be able to accomplish far more than either alone. As the Vedic sage Vyasa wrote in the Bhagavad Gita almost 2,000 years ago: "Prakṛitim svaam avastabhya visrijami punaḥ punaḥ" [or] "Curving back onto myself, I create again and again …." **– Dr. Arthur J. Murray, CEO, Applied Knowledge Sciences, Inc.; First Fellow and Director, Enterprise of the Future Program, International Institute for Knowledge and Innovation, USA**

The existential growth risks of the Anthropocene call for a reinvention of learning as the primary means of advancing human development. It could not be timelier for this book. While the relentless economic is proving physically and socially unsustainable, the path towards knowledge-based consilience appears to be both urgent and potentially endless.

– Dr. Francisco Javier Carrillo Gamboa, President, World Capital Institute, Brazil

Another impressive book from the Mountain Quest Institute's thought leaders' team. In our time of fast societal, technological and environmental changes and disruption, experiential learning becomes one of the few solutions to quickly adapt and survive. Through this book, the concept of human learning is deeply analyzed through a neuroscience lens, taking into consideration other connected scientific and human fields. Among the various novel contributions, a new theory of experiential learning is presented based on Complex Adaptive Systems theory. Such a rich and insightful book that will make you discover individual learning in a completely new way.

– Dr. Vincent Ribiere, Managing Director of the Institute for Knowledge and Innovation Southeast Asia (IKI-SEA), Bangkok University, Thailand

As a technologist and a seeker of higher education and self awareness, I am blessed to be constantly exposed to great works from many wise and experienced teachers. Once in a while, I am exposed to a work so profound that it literally causes a massive shift in my own thinking and beliefs. To this end, I deem myself fortunate enough to have been asked to review the pre-published work of Mountain Quest Institute's newly created written work by David Bennet, Alex Bennet and Robert Turner entitled "Unleashing the Human Mind, A Consilience Approach to Managing Self". This book delves into one of the most difficult factions of a lifelong quest to understand one's self and experiences throughout life. In my opinion, thinking about how one "thinks" or learns is a very difficult discipline as it requires an existentialist approach while at the same time being aware of one's emotions and academic inclinations.

I highly recommend this work to anyone seeking to further their ontological awareness, simply stated, the study of "what it is to be". This work explores both the abstract and practical nature of the learning process and how absorbing information can sometimes be counterproductive to the higher learning one seeks. I found myself riveted to each paragraph as I embarked on a journey that vastly deepened my understanding of the learning process. The exploration of different facets and influences that can impact perceptions is a must read for anyone determined to understand the true nature of how we learn. **– Duane Nickull, Author, Technologist, and Seeker of Higher Truth, Canada**

You are in an old library and have walked here along the street past all the trashy street magazine stands and Internet cafes. Something inside you has unconsciously led you here, to a single bookcase. You look up high to see a book that someone has left hanging halfway out from the shelf. At this very moment it is falling and now lands on your head, bouncing to the floor and opening at the very page and subject you currently need to know the answer to, which is this page, miraculously.

This synchronistic event has happened to you several times before, but this time it seems obvious that something is trying to tell you something. The book says it is a guidebook to navigate a path through a mountain of knowledge, helping you move through an uncertain future, as well as providing tools and techniques that will get you there with understanding and science. It's also an enabling guide in itself such that you will no longer need to go to libraries any more to receive knowledge. The exact books you need will come to you wherever you are, whenever you need them. You won't even need to read them, because when the information arrives you will know and understand all that they have to say in that moment.

It appears to be a sort of Knowledge guide to unleashing learning and understanding. It looks quite long and complicated though, and we all know how you prefer quick and easy. There, there, there is that guilty feeling again. Why does learning and understanding have to be so difficult? It does though look interesting, even in itself offering free membership to some sort of global virtual dynamic interconnected knowledge book club—a club that makes the Internet seem like a slow chaotic naive dinosaur. A system where you don't need to use a search engine; it already knows what you want and presents it to you instantly when you need it. All in real time, and sometimes even spookily using your own words. All free to use for everyone, and constantly updated and refined by people like you. If you can be bothered though, that is.

It's probably worth a read then... especially if someone—probably yourself—is taking the time to make you see, and learn, and open your mind to itself and our shared mindscape. Or, maybe you can just put it back, wander down the road again, and buy a trashy magazine, or sit on your own in the Internet cafe drinking your overpriced latte—after all, who will care and what difference does it make anyway, you tell yourself.

You are probably right; you aren't quite ready for this yet. And besides, if I showed you evidence of what I can do with this knowledge, I wouldn't want to be responsible for you choking on your coffee.

– Nick Sambrook, Author of The IT Trilogy and *A Brief Guide to the Collective Unconscious Overmind*, UK

Thank you for the opportunity to read through this Masterpiece. This book is thought provoking in a way that provides practical ontologies, taxonomies, epistemologies, examples and fusions of greater concepts that are meaningful and important for us all. In his book "Mind: A Journey to the Heart of Being Human", Dr. Dan Siegel defines the "mind," as an "emergent, self-organizing, embodied, and relational process that regulates the flow of energy and information." In "Unleashing", David Bennet, Alex Bennet and Robert Turner expand our collective understanding of the Human Mind by embedding timeless wisdoms, transliterations, and their collective expansive knowledge and experiences that enrich our lives from cover to cover.

Beyond this book, I have had the pure pleasure of sitting at Alex and David's feet and listening to fascinating stories, experiences, and conversations that show what exceptional minds and lives they have created together, laying a foundation that expands prior understandings of Human Thought, and now appears in many of these writings. I am excited that they and their colleague Robert Turner are sharing these successes, methodologies, frameworks and inspirations with each of us.

In meeting Alex and David, I continue to be amazed by their positive intentions and successes that previously led the way for the U.S. Department of the Navy and the Mountain Quest Institute, and are now made available to you and me. This comprehensive desktop reference provides applicable science and physics, and is essentially a step-by-step guide moving from deep learning into action. The author's contributions to learning, consciousness and managing knowledge and Self continue to enhance our lives today in beneficial, applicable, and needful ways for the improvement of humanity and our conscious ability to upgrade in each new moment.

– Bob Beringer, CEO of The Electronic On-Ramp, Inc. (EOR), Intelligence Professional, Author of *Linux Clustering with CSM and GPFS* and other books on SuperComputing, IT Security, Biohacking, Health, and Consciousness, USA

For once it's allowed for Alex to put humbleness aside, which she is always advocating! To describe this masterpiece of accumulated knowledge about knowledge, I could use various superlatives! This publication is the brilliant sublimation of a life-long accumulation of knowledge about the potential of our brain to learn, adapt and evolve to the best version of ourselves. David, Alex, and Robert are the living proof of what you can accomplish if you apply all the concepts and knowledge that are condensed in this magnificent book. This is turning knowledge into personal power by taking effective and efficient action, and to the degree that is possible, taking that action with passion and dedication. I'm delighted, inspired and grateful, and would advise everybody who has an interest in the field of knowledge, learning and the brain to enjoy the rich content and eye-opening insights of this book.

– Johan Cools, Higher Architecture Institute of Saint-Lucas Ghent, Belgium

Unleashing the Human Mind

A Consilience Approach to Managing Self

David Bennet, Alex Bennet, Robert Turner
Mountain Quest Institute

MQIPress (2022)

Frost, West Virginia

ISBN 978-1-949829-63-1

Learning is the personification of change, the gateway to knowledge,
the enlightening experience of education.

FIRST EDITION

MQIPress
Frost, West Virginia
303 Mountain Quest Lane, Marlinton, WV 24954
United States of America
Telephone: 304-799-7267
eMail: alex@mountainquestinstitute.com
www.mountainquestinstitute.com
www.mountainquestinn.com
www.MQIPress.com

ISBN 978-1-949829-63-1
Cover drawing by MQI Resident Artist Cindy Taylor

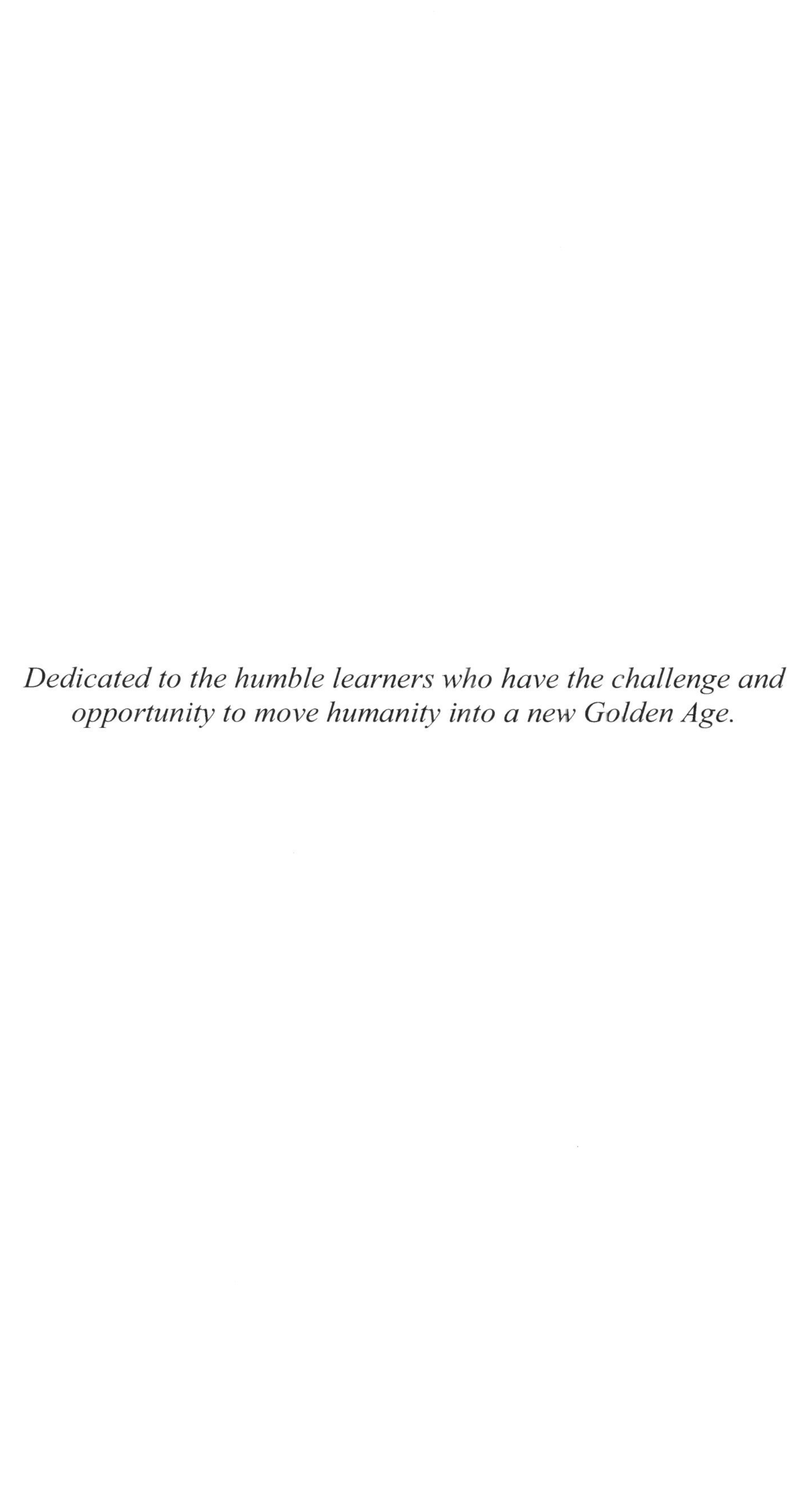

Dedicated to the humble learners who have the challenge and opportunity to move humanity into a new Golden Age.

Prologue

Perhaps one of the most striking characteristics of contemporary man is his internal fragmentation: generally, in the concreteness of everyday life, the instruments that contemporary humanity has to overcome daily difficulties are separated from each other. And so, the breakdown of cognitive, emotional, and spiritual unity is inevitable. But this is an apparent phenomenon, and the approach to experience based on a principle of unity, on the interconnectivity of knowledge and instruments for analyzing the reality that surrounds us and within which we are placed, is more than necessary.

Music, Science, Mathematics, Philosophy and the immediate perception of reality, although conceptually different from each other, are not *different things*; they are linked in the unity of humanity's "total" experience of reality, and this unity is clearly decanted in the concept of consilience, a powerful concept that leads the Self, by means of a creative rational thinking, to manage itself from its unity with other human beings and with reality, since knowing, above all, is a relationship between a Self with a multiplicity of factors that must be united in intelligence.

Professor Dr. Walter Gomide,
Universidade Federal de Mato Grosso, Brazil

Preface

Thank you for your interest in our latest work illumining advances in understanding learning through the lens of neuroscience. May we provide some background as introduction, touch on the beginnings of this research, the structure of this book, and then suggest a strategy to help you dive into this text.

Unleashing Background

Fundamentally, what is learning in the human experience? How is learning evolving in this accelerating age of technological and knowledge advancement? As educators who have worked in government, industry, and academia through the years with senior leaders—and as senior leaders who have spoken around the world on learning and education, knowledge management, change, organizational development, and a myriad of related topics—these questions repeatedly pop up. And, in the context of keynote addresses, lectures, or workshops and webinars, whether face-to-face or virtual, a constant inquiry is: "What learning is needed to prevail in the unprecedented future?"

While there are various answers, and certainly differences implied in terms of individual, organizational and societal approaches, for one of the authors it was a professional focus on knowledge as the Chief Knowledge Officer of the U.S. Department of the Navy that brought this core insight into sharper focus. *We are change, ever growing or declining, moving forward or fading away, living or dying.* Our bodies have some 50 trillion cells continuously changing. Nothing about us is static, and it never has been. We pulse with change: absorbing food, processing air, dividing cells, making and pushing out fluids, transmitting heat and energy, and on and on. At the microscopic level, everything about us teems with change, creating patterns of activity linked with energy from our environment.[1] So, forget what we learned from early language lessons. We are—quite literally—a proverbial *verb*, not a common noun.

As complex adaptive systems—created to adapt and evolve in a changing environment—we cannot stay in stasis, that is, things cannot stay the same. There is no constant. We often think of our heartbeat as our constant rhythm of life; but even our hearts must skip here and there. If the heart stayed in the same rut it we would wear out by the time we were 30 years old! Change is a natural part of our systems, whether we define those systems as physical, mental, emotional, or spiritual, or whether our focus is on the individual, family, community, nation, or humanity as a whole. Everywhere we look, everywhere we go, change is at play, and play is a natural way we change. It is synonymous with the act of living, and life could not exist without change.

The 20th century multiplied human powers—military, economic, social interaction, technological—are all compelling us to closely examine our beliefs, values, ethics, and way of life. And larger changes are upon us, challenging us—and blessing us—as individuals and as a humanity. These changes also offer us new opportunities for learning. *Learning*. At the core of human life, *learning is the personification of change, the gateway to knowledge, the enlightening experience of education*. Across time, for millennia, our freedom and our ability to reason, and our capacities and resources for learning have increased. Today we are engaged in learning more broadly and more deeply across humanity and we *rise on the minds of brilliant thinkers*. Among them, René Descartes, who boldly proclaimed as a scientist and philosopher in the Age of Enlightenment, *Cogito ergo sum*, "I think, therefore I am." With gratitude, we now understand *Disco, ergo fio*, "I learn, therefore I become."

Nevertheless, a perspective of caution as we quest for learning. In our schools and universities, we've focused on the experience of education in terms of the practice of teaching, with teachers giving and students receiving instruction in the prescribed curriculum that needs to be provided in each specific domain of knowledge. While all this certainly is of import to providing education, somehow, we've often deviated far from learning based on self-individuation to "learning" based on required "knowledge", with that knowledge directly connected to carefully constructed outcomes tied to test scores and grades which can demonstrate an institution's worth. In other words, the "system" has memetically taken on a life of its own, and while not forgotten, relegated the learning experience to a potential emergent outcome.

Today that learning is increasingly learning in place by means of various technologies and networking platforms, with value built on respect for and resonance with ideas. This has not always been the case. In the early days of bureaucracy, relationship-focused value was built on trust and respect of people with whom you had personal relationships. As businesses expanded, relationship and idea-focused value was built on trust of structure (the workplace) and people in the structure (partners and friends of partners). In today's virtual world network connections are idea focused, that is, trusting in ideas that resonate with us. With the larger-than-life events of today coupled with global conditions beyond belief—all escalated through chaotic arousal challenging the very foundations upon which our lives are grounded—humans are moved to reach out for help, searching for a "savior" of sorts who is perceived as having the power to break through these threats, and on whose coattails one can ride, a persona becoming the "idea". This is the phenomenon of "idea locking", trust and commitment to ONE BIG IDEA or person, with value based on identity, the need to define and establish "I am" in a turbulent world. All of which offer huge challenges—and opportunities—to the future of learning and education.

And then came the pandemic, and the continuous waning of truth and trust at the individual, organizational, national, and global levels, and even as we achieve greater technological connection, many have divagated back to separation, only trusting those we perceive as like us … the same color, the same religion, the same political persuasion, which is accompanied by an arrogance: "I'm right, you're wrong, and I'm not listening". Yet this response—as only one potential in a quantum reality—closes off learning and—as complex adaptive systems—sets us on the decline toward death. Death of the individual, death of democracy, death of humanity.

Unleashing Beginnings

It is with these thoughts in mind, that this book came into being, although it began with the completion of a ten-year research study by Dr. David Bennet exploring neuroscience findings in terms of experiential learning. Taking a living system's perspective, Bennet recognized that limitations and challenging problems often derive from our inability to rise above the tendency to categorize and specialize in separate disciplines. While disciplines are clearly convenient, they are filled with artificial constructs that may be effective within their boundaries, but may also be limiting by their frame of reference and accepted procedures and practices. This can happen within a discipline as well as across disciplines. An example of such a limitation within a discipline was the dominance of behaviorism on learning research in the 20th century. By looking across fields it is possible to see interactions and possibilities that are not obvious within individual disciplines. This is the fodder for synthesis, the ability to gather information from disparate sources and knit it into a coherent whole. With this in mind—and acknowledging that the human is an integrated, biological, and complex system entangled across the physical, mental, emotional, and spiritual dimensions—Bennet began a consilient scientific analysis of thinking and learning in the human brain through the lens of neuroscience. This enabled a new perspective which coalesced a kaleidoscopic view of wisdom about our mind/brain at work.

Bennet's ten-year research study conducted in the MQI environment resulted in publication of an in-depth text titled *Expanding the Self: The Intelligent Complex Adaptive Learning System* (ICALS), of which parts are excerpted for this book. Then, in 2021, David's work and life partner, Dr. Alex Bennet—whose research focus is knowledge, consciousness, and human and organizational systems—was asked to keynote at the United Professionals for Development and Advancement of Teacher Education webinar titled "Rehumanizing Education in the Time of Technological Advancement", which began her mind churning (and re-churning) on the important topic of human learning, asking how this body of work could be used to address today's large issues of life and learning and, specifically, the challenge of managing the self.

Robert Turner, an associate of the Institute since its inception and a long-time partner in this work, joined in this quest as well, adding a diversity of insight through a rich background in adult learning, change management, systems thinking, modeling, and accelerated collaboration and decision-making as we bring this learning into the times of today. Our intent was that together we could develop a deeper understanding of the human mind in terms of the opportunities and challenges offered in the global and local current and future environments.

In practical terms, and building on the consilience imperative, this book takes flight from so much of the work on knowledge, learning, change, physics, philosophy, decision-making, leadership and management, systems and complexity, neuroscience, music, cognitive psychology, sociology, strategy and planning, mathematics, human and organizational development, sustainability, consciousness, and spirituality—and touching a myriad of other fields—published through the Mountain Quest Institute (MQI). Using the ICALS theory as the baseline and focusing on the neuroscience findings from that research, this book is about the success factors and the learning and skill sets needed for the unknown future in the midst of which we find ourselves. See the ICALS Consilience Framework in the Afterword for a detailed accounting of the expansive research approach that set ICALS apart from the beginning. The value of the Consilience Framework begs consideration for large-scale applications of this bold strategy.

Unleashing Structure

This book can be thought about in terms of four distinct chapter groupings. Chapters 1, 2 and 3 provide a foundation: Chapter 1 begins with a look at exactly what is meant by being human, then shifting in Chapter 2 to a look at today's environmental currency, and in Chapter 3 taking a glimpse at our potential future. Given our mind/brain propensity to anticipate and predict the future—envision what is coming—it's certainly intriguing to consider our future.

A significant factor is that the current capacity of our human brain exceeds our heretofore needs and uses. The potential is exemplified in the human population by the abilities of savants and geniuses. Moreover, the capabilities of the general population are especially encouraging. When we look at a wide range of human endeavors, we find evidence of this. For example, think about human transportation not so many generations ago. Early on, most humans traveled about by foot, then we used animals, followed by a combination of animals and wheels. Then came motorized modes—trains, cars, trucks, etc. Before long, we took to the sky, traveling vast distances, even around the planet, in a matter of hours. And how did we go from our first successful plane flights to a six-day trip to the moon in less than one average lifetime? Now, think about a category of human activity where you hope for breakthrough progress. What do you think is possible? How long could it take? What are we capable of achieving as we unleash the human mind?

The second grouping—Chapters 4, 5 and 6—begins with an introduction to the Intelligent Complex Adaptive Learning System (ICALS), that is, the human learning system, which provides the foundational system model for the rest of the book. Chapter 5 dives into diverse modalities of thoughts and thinking related to the modes of learning, and Chapter 6 focuses on the self—which is central to the ICALS model—as the ground of learning. The third grouping—Chapters 7, 8, 9, 10 and 11—addresses aspects of the self. *The importance of the self cannot be overstated.* Thus, at the core of this book and the number one skill set needed for a successful future (encompassing all other skills) is managing the self, which is therefore addressed from the viewpoint of the mind/body (Chapter 7); heart-mind entrainment (Chapter 8); the extended reach of self in terms of social engagement (Chapter 9); the expanded self in terms of inner connections to larger fields—whether you consider that in terms of an energy field, a consciousness field, a quantum field, or a God field (Chapter 10); and the human gift of humility (Chapter 11).

Figure 1 shows related areas addressed in this book that are correlated to helping unleash the human mind and move beyond the perceived limitations of self. These are success factors, skill sets, and the emergent areas of neuroscience findings, all grounded in self as the foundation of the Intelligent Complex Adaptive Learning System, expanding beyond the experiential to include existential experiential learning as humanity becomes fully participative in the consciousness shift underway. The numbers in parentheses represent the chapters where each specific success factor and skill set is addressed.

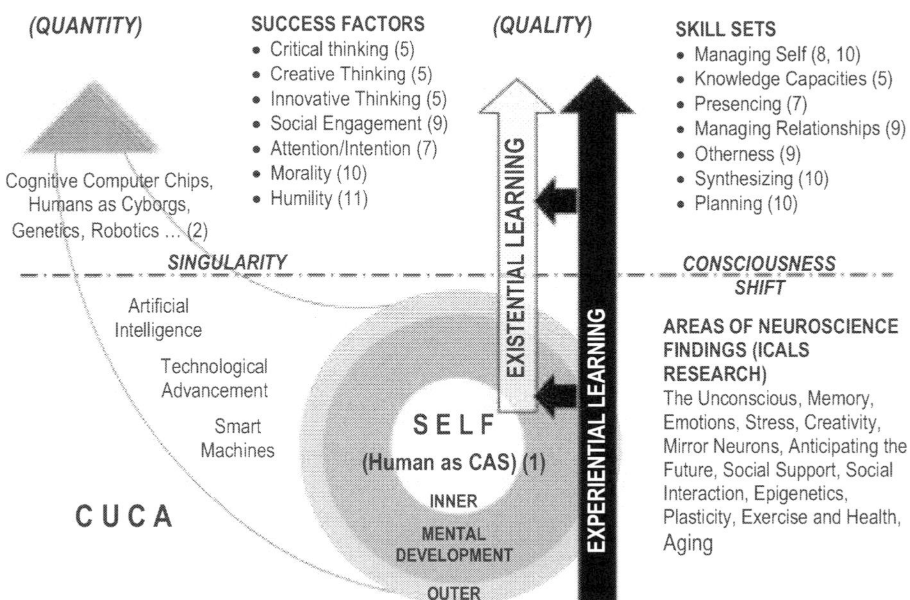

Figure 1. Unleashing the human mind as we move into a new Golden Age.

The neuroscience findings from the ICALS research study are spread throughout the text as we lay the foundation for this work, introduce the expanded experiential learning model, and address managing self from the perspective of the physical, mental, emotional, and spiritual dimensions. For example, in the discussion of Knowledge Capacities in Chapter 5, it was forwarded that as we mature and age—and as our senses lose some of their effectiveness—the ways we perceive and operate in the world change, and with them a preference for higher engagement of the intuitive emerges. Thus, the Knowledge Capacity of "Knowing and Sensing" would dominate decision-making. This is supported by the research finding: **Despite certain cognitive losses, the engaged, mature brain can make effective decisions at more intuitive levels**. Note that Knowledge Capacities, identified as a skill set, support Critical Thinking, identified as a success factor (see Figure 1). As this book unfolds, you will discover for yourself the strong relationships among the findings, the skill sets, and the success factors, all in service of the self as we learn to navigate an uncertain future.

We do not propose to have the answers for life in the future, but rather attempt to expand the understanding of human learning from a neuroscience perspective, and suggest some very specific human characteristics, potentials, capabilities, and actions. When these are consciously developed, they will enable you as you move through the Intelligent Social Change Journey of which we are all a part.

As we move through this text, we will interject select tools developed in our change series, *The Profundity and Bifurcation of Change*, as well as where it makes sense sharing pieces of the large body of research emerging from MQI throughout the years.

And in the final grouping—Chapters 12 and 13—we offer ideas and thoughts emerging from this learning, potential actions, and ways to unleash the human mind.

Unleashing Strategy

This second volume from MQI focused on learning in the 21st Century—following an academic treatment of the findings—is truly a breakthrough. If you choose, it can help you connect the dots in powerful ways to expand your inherent learning capacities and to increase your abilities to guide others in unleashing the human mind.

To that end, we wholeheartedly recommend you:

1. Read quickly through the range of topics under the chapter headings in the Contents.

2. Scan the entire book for an overview.

3. Notice the illustrated figures for concepts and models.

4. Become aware of the range of available tools.

5. Go back and read from the beginning, spending time on what peaks your attention. *Remember*: revisit topics to probe for insight.

6. Be alert for tools that are new to you and ones you can readily use.

7. Most importantly, create a broad set of tools for yourself by reviewing and regularly managing your learning over time.

You will note that there is a fair amount of repetition throughout, specifically related to the neuroscience findings and related quotes by leading researchers. The intent of this is (1) because repetition aids learning and memory recall, and (2) to show the entanglement of the various systems connected to what it is to be human, that is, an intelligent complex adaptive learning system (ICALS). In many cases—but not all—we will indicate this repetition. As a result of this approach, for those who read this book all the way through, you will discover that the language and thought becomes easier and easier to comprehend and reflect upon. And, no doubt, you will begin connecting previous learning—and your own experiences—to what is being read. In this manner you expand the learning, and perhaps prepare to write a follow-on book on managing self, perhaps a "how to" reflecting your personal learning journey, in today's world a topic sorely needed in every field of endeavor.

Along with building on generally accepted theory, newer ideas are distributed throughout the text and connected to one or more chapter focus areas as appropriate. For example, ideas that you may find as exciting and thought-provoking as we do include:

- The recognition of Self as the underlying foundation to experiential learning.

- Our advanced capacity as humans to engage and manage our brains for adaptation and change, for anticipating the future, for enhancing learning, and for development of the soul and spirituality.

- The fifth mode of Social Engagement added to the expanded experiential learning model (ICALS), which opens up a dialogue with the environment, brings social support and social interaction into the sphere of the self as learner.

- Acknowledgement of the neocortex as the "organ of intelligence" and a deeper understanding of what that means, bringing in brilliant research focused on cortical columns and reference frames.

- Recognition that the human mind/brain is primed for critical, creative, and innovative thinking in general—and paradoxical thinking in particular—and exploring how these can be engaged in learning.

- The discovery through mathematics—specifically, transreal numbers—that knowledge by sentience gives us the possibility of evaluating the truth value of non-propositional objects such as sensations or feelings.

- The learning potential offered by *psyche*cology educational games (PEGs), virtual games that cross the boundaries between the physical "real" and "virtual", fully engaging the entangled mental, emotional, and spiritual aspects of the human.

- The collision of experiential learning and existential learning as humans move through a turbulent environment toward a new Golden Age.

Incrementally (after chapters 3, 6, 9 and 12), you will find an UNLEASHING Litmus Test, which includes reflective questions related to the previous material. It is suggested that you reflect on each question for one minute prior to answering it, which may be done silently, verbally or in writing. The questions are based on the Bennet Individual Change Model.[2] CHANGE comes from within, that is, unleashing your mind is YOUR choice. It does not just happen because someone says you should. However, before you can choose to change, you must first be aware of the desired change, understanding the current situation along with the context and *need* for change. But that's not enough. Along with awareness and understanding, you must believe in the need for change, feel good about it, and have ownership of making the change. Even then, you will be unable to change unless you possess the knowledge of *how* to change and the *courage to act*. See Appendix H for an expanded change model in the context of experiential learning.

Here is a short description for each of those stages:

1. **AWARENESS** means something has come to your attention, it is perceived, it has been mentally engaged.

2. **UNDERSTANDING** includes your perception of the situation—the who, what, where, when, and why, and the anticipated results. As the situation becomes more complex, you need to re-create your understanding.

3. **BELIEF** means you accept what you are aware of as true and understand it really exists. Beliefs which dominate other patterns are prominent. Strong patterns are created by experiences and are closely related to emotions.

4. **FEELINGS** are the foundation of learning—positive feelings make actions important to you and worthy of your efforts. Reason cannot operate without emotions.

5. **OWNERSHIP** implies a personal commitment for you to take responsibility and act.

6. **EMPOWERMENT** refers to self-empowerment, that is, having the *knowledge* to make the necessary change and the *courage to act* on what you have learned.

There are no right or wrong answers to the reflective questions offered in the UNLEASHING Litmus Tests; rather, the intention is to engage YOUR thinking about what is being shared.

This book has been carefully designed on multiple levels to facilitate your use. In addition to development of a comprehensive set of content resources and the flow of the subject matter interjected with a variety of tools, there is an extensive set of Endnotes (page 341) and a supporting Bibliography (page 371), and you will enjoy the quality of the Subject Index (page 391), which is both detailed and highlights key learning points, and the convenience of the Index by ICALS Findings (page 405) with page number references.

As you move through this text—and as you sense positive shifts in your personal learning—*you will experience moving into the new Golden Age* we envision. It is within reach, just beyond the quantum door to all possibilities. Yes, there is that much potential in each of us, and in our shared humanity, as we unleash the human mind.

In learning, David, Alex, and Bob

Contents

Chapter 13: **New Beginnings** | 287

A Nexus of Choice … Moving Toward AGI … Human Identity … The Intelligent Complex Adaptive Learning System

Afterword: The ICALS Consilience Framework | 297

Appendices

Figures

Tables

Table 4. (Appendix F). The Relationship Network Management Assessment Chart | 334

Tools

Tool 1: Situational Backtalk | 93-94

Tool 2: Activating New Resources | 96-98

Tool 3: Self-Belief Assessment | 116-117

Tool 4: Sleep on It | 123-124

Tool 5: Letting Go | 137

Tool 6: Holding Neurovascular Reflect Points | 152-153

Tool 7: Connecting through the Heart | 169-171

Tool 8: Releasing Emotions Technique | 186

Tool 9: Relationship Network Management | 203-204

Tool 10: Quieting the Mind | 217-218

Tool 11: Thinking Patterns | 224-225

Tool 12: Scenario Building | 230-231

Tool 13: Grounding through Nature | 243-244

Tool 14: The Choice of Humility | 251-252

Consilience
Systems
ICALS
Constructivism
Epistemology

Foreword

Living is a large story of learning. The life evolution itself is also a large story of learning processes. Caught up in the tumult of daily activities and focused on comprehending/understanding and the solving of life problems—be they born from the interaction with nature/surrounding environment or with society—most of the time we look for solutions outside, although *the solutions are actually found within ourselves.* We forget to address ourselves, our power, our eternal root: the power to learn.

Yes, life itself is a long race of learning and adaptation, as science begins to discover and show us through recent results not only from biology, neuroscience, psychology and philosophy, but also from the perspective of information science in this information age, and as this book reveals us. And not just reveals us, but teaches us about the science of learning, scientifically based, and concretized by progressive techniques of connection and self-control of the mind. The central objective is not only knowledge itself, but this is oriented towards personal progress, towards the way through which this knowledge serves our understanding the world, the surrounding reality, the personal relationships with it and with society, allowing the inner enlightenment/widening of our understanding horizon in an 'open-mind'/'open-soul' fashion, and on this basis the rediscovery of the self in new dimensions and the restructuration/renewal of our judgment and decision criteria, thus enabling us to advance at the same pace as the increasingly rapid change of society.

This book is therefore not an inert multipage picture found or exhibited in an exhibition or in an archaic library, *it is a dynamic moving 'living' guide,* stimulating us to follow the path of change and personal development not only as an active/essential factor induced by the change of society itself, but as well a promoter of change through our co-participation in this evolutionary/revolutionary process. What this volume definitively shows us is that we are, and should be therefore, not only the product—sometimes becoming obsolete—of a society in the full process of change, but an active and pro-active factor such that we initiate and follow the path of the full affirmation of the human as an individual and also as a species, of the deep human essence that unites us and ensures our existence as a community.

This volume is not a textbook of psychology or neuroscience, although it is based on their most recent discoveries, it is an active and contributory expression of a 'vivant'/living pulse throughout the progressive and interactive chapters, in a logical and sentient living 'heart'-like rhythm, leaving to the reader the satisfaction of discovering one by one and in an alert but not rushed or precipitated way—methodical but not nagging, perseveringly but not in a 'mantra'-like style, didactically but masterful, persuasive but argued and reasoned. This unfolding follows only the way of the true truth, based on science but also on practice and experience, connecting these wonderful internal reserves that each of us dispose of and can be used with skill, attention and thoughtfulness to achieve our best performance in attaining our proposed objectives and sought solutions, in agreement not only with ourselves, but also with the members of society and the society itself, and with all the other co-participating members within and without. Going through the pages of this volume one by one, the reader him/her self starts to feel an invited/involved welcome, becoming a co-participant in this beautiful and magical—and necessary and useful—incursion of self-progress toward knowledge and wisdom.

Life itself is a whirlwind of events in which we integrate with or without our will. Do we let ourselves be dragged by these, often thinking that it is the will of destiny? Or do we try to become our own authors and co-participants in drawing/designing the destiny of our lives? These are questions that each of us reflect on sometime, at least in crossroads moments, or if we haven't done this yet, it is time to reflect on them now. Can we be the masters of our own destiny? Can we adjust our abilities to shape it, or will we model ourselves to fit a perceived reality? This volume helps us answer these key questions for ourselves. The discovery/rediscovery of these sources/resources, of our internal reserves, of confidence in our own strength, all put at the service of our evolution as human beings and as members of the society in which we live today and will live in the future, is not only a noble—but also a necessary—

objective, available to the reader of any age, training and profession, for whom this book opens generously.

It is the mastery of the authors of this book to open the reader's mind and soul, thus offering the opportunity for the content of this live transmission to be discovered and interpreted in the most appropriate way by each reader, with his/her way of looking at the world but also wanting change, in step with the change of the world itself, so that only the reader's desire is needed to let him/her self be seduced by this wealth of wisdom, generously placed at the reader's disposal.

Dr. Florin Gaiseanu, Research Professor,
Science and Technology of Information Bucharest
(Romania) and Barcelona (Spain), Honor Member
of *NeuroQuantology* (Europe) and *International
Journal of Neuropsychology and Behavioural
Sciences* (USA)

Chapter 1

Being Human

The Human as a Complex Adaptive System ... The Inner and Outer Self ... The Plasticity of the Brain ... Our Emoting Guidance System ... The Intelligent Social Change Journey ... The ISCJ and Human Learning ... The Human Search for Wiscom ... So, What Does It Mean to be Human?

TABLE: (1) Comparison of phases of the ISCJ with levels of learning.

What does it mean to be human? Increasingly, we recognize that we are infinitely complex beings with immense physical, mental, emotional and spiritual capacities. Presiding over our human systems, *our human brains are fully integrated, biological, and extraordinary organs that are preeminent in the known Universe.*

As our understanding moves into a quantum frame of reference, recognizing that energy and matter are indefinite and that thought affects energy, more and more individuals are realizing that the limitations and boundaries we create close us off from fields of possibilities. By looking across disciplines and sciences, observing from multiple points of reference, we see interaction and potential that is not obvious within single perspectives. Thus, in our brief presentation here of what it means to be human, multiple viewpoints are engaged to explore human learning from both the inside and outside, opening a larger conversation of how we can unleash our minds in a continuing journey of expansion.

Since the Intelligent Complex Adaptive Learning System (ICALS)[3] embraces *all* it is to be human, we offer introductory definitions to focus this discussion. First, the brain consists of an atomic and molecular structure and the fluids that flow through this structure. The mind is the totality of the patterns created in the brain's 86 billion neurons and their firings across neurotransmitters and trillions of synaptic inter-connections. These patterns encompass all our thoughts. The term mind/brain refers to both the structure and the patterns emerging within the structure.[4] Spiritual is taken to mean pertaining to the soul, or "standing in relationship to another based on matters of the soul."[5] In this context, "soul" represents the animating principle of human life in terms of *thought and action,* specifically focused on its moral aspects, the emotional part of human nature, and higher development of the mental faculties.

The Human as a Complex Adaptive System

It was near the end of the 20th century before we began to clearly recognize that people are complex adaptive systems, and that the entangled physical, mental, emotional, and spiritual systems *cannot be separated from each other*. Reality System Theory says just that, comingling the dynamically interacting influences of the intellectual, emotional, and spiritual. As Stebbins describes, "The human reality is a dynamic holistic system subject to the continuous ebb and flow of intellectual, emotional, and spiritual influences."[6]

A system is a group of elements or objects, the relationships among them, their attributes, and some boundary that allows one to distinguish whether an element is inside or outside the system. Complex systems consist of a large number of interrelated elements (parts) that may or may not have nonlinear relationships, feedback loops, and dynamic uncertainties very difficult to understand and predict. A complex adaptive system (CAS) co-evolves with the environment through adaptation, the process by which a system improves its ability to survive and grow through internal adjustments. Adaptation may be responsive, internally adjusting to external forces, or it may be proactive, internally changing so that it can influence the external environment. A CAS has partially ordered subsystems—with various levels of self-organization—that evolve over time.[7]

As complex adaptive systems continuously interact with their environment and adapt, they operate at some level of perpetual disequilibrium, which contributes to their unpredictable behavior.[8] Having nonlinear relationships, a CAS creates global properties that are called emergent because they emerge from the multitude of elements and their relationships and actions, making it difficult to trace these emergent characteristics back to their origins. Thus, emergent properties cannot typically be understood through analysis and logic because of the large number of elements and relationships. As Johnson points out, "It wouldn't truly be considered emergent until those local interactions resulted in some kind of discernible macro-behavior."[9] An example is life, along with many of the behaviors we each exhibit.

As can be seen, emergence is a property of a system that its separate parts do not have. A good example is consciousness. "No single neuron has consciousness, but the human brain does have consciousness as an emergent property."[10] In the midst of all this change—and despite the need for the disequilibrium to adapt—most individuals have a tendency to seek stability. Yet this very disequilibrium provides the uncertainty that challenges stability.

Reality System Theory looks through the lens of quantum physics, building on Heisenberg's uncertainty principle[11] and the contextualism of the quantum field.[12] The now-classic uncertainty principle refers to the particle/wave fluctuation and contextualism refers to context sensitivity, that is, change as a function of time and surroundings. Of contextualism, Wilber reminds us that, "Meaning is context-dependent, and contexts are boundless."[13] As we now recognize through the focus on knowledge management since the turn of the century, all knowledge is context sensitive and situation dependent.[14]

The ICALS research which resulted in an expanded experiential learning model (see Chapter 4) refers to the human as an Intelligent Complex Adaptive Learning System. When the term "intelligent" is added to the concept of a complex adaptive system, it infers a capacity for reasoning and understanding, or an aptitude for grasping truths.[15] Wiig broadens this view of intelligence and considers it the ability of a person to think, reason, understand, and *act*,[16] which relates the concepts of knowledge (tied to effective action) and intelligence.

Hawkins says that "We are intelligent not because we can do one thing particularly well, but because we can learn to do practically anything."[17] He calls out four attributes of the human brain which provide the flexibility needed to form a baseline for intelligence. These are learning continuously, learning via movement, creating many models, and using reference frames to store knowledge.[18] It is important to recognize that humans are verbs, not nouns, continuously learning and changing, forming new connections and patterns with new synapses. Learning via movement refers to both thought and the active human body, which through sensing provides us the data to develop a model of the world in which we live. This relates to the ICALS finding that **the mind/brain creates an internal representation of the world**. We now know that this modeling occurs in the cortical columns in the neocortex, which receives input from movement and generates behaviors, and predicts the next input. The ability of the mind to predict is addressed in Chapter 3. "Many models" refers to the tens of thousands of cortical columns in the neocortex, which provide the flexibility to process and navigate a changing, uncertain and complex environment. Reference frames, which are established in each cortical column, are a fairly new concept to the field of neuroscience. These reference frames enable the ability to perceive shape, changes and locations relative to each other. As Hawkins describes: "Thinking occurs as the brain activates one location at a time in a reference frame and the associated piece of knowledge is retrieved."[19] Reference frames are found in most systems that exhibit planning and complex, goal-oriented behaviors.

As is becoming clear, humans are integral parts of a larger whole, *part of a holistic human* potentially capable of intelligent decisions and actions. As Zohar and Marshall state,

> *Neither IQ nor EQ [nor SQ][20], separately or in combination, is enough to explain the full complexity of human intelligence nor the vast richness of the human soul and imagination.[21]*

The Inner and Outer Self

It was true yesterday, it is true today, and it will be true tomorrow—which in the turbulent environment that we live in can't be said about very many things! Each and every human being is unique, whether considered from the frame of reference of DNA, experience, culture, family, thought patterns, beliefs, values, or emotions and feelings. And at the core of all this difference is the self, with a subjective mind, exploring the world from the inside out, a protagonist ready for action. Yet in the magical environment of the womb, the self exists in embryo. The thinker and feeler of the future that we become (the self) is a rather fresh slate, even as energies entangle to create the unique web of associations and responses that will help ensure survival of the budding human, what we describe as the "personality".[22]

Self is an emergent quality of the human which moves beyond biological drives and cultural habits. As a working definition, self is considered as the totality of the conscious and unconscious mind, the brain and the body. While there is a close relationship of self to our understanding of consciousness, we take an expanded view, with self-inclusive of the personality, and—as our consciousness expands and we become a co-creator of our reality—*inclusive of every aspect of what it is to be human.* Csikszentmihalyi frames this well. As he says, "Inside each person there is a wonderful capacity to reflect on the information that the various sense organs register, and to direct and control these experiences." This is the self, somewhat like a figment of our imagination, "something we create to account for the multiplicity of impressions, emotions, thoughts, and feelings that the brain records in consciousness."[23]

Every self has a mind and, from Csikszentmihalyi's viewpoint, "Nowadays learning to control the mind may have become a greater priority for survival than seeking any further advantages the hard sciences could bring."[24] We agree. Recall that the mind is the totality of the *patterns* in the brain and throughout the body created by neurons and their firings and connections.

From a quantum field theory perspective, the mind can be considered a field of potential—of possibilities—where at least half of the field has chosen to head the same direction. This field could also be thought of in terms of focused consciousness or will. The mind is the seat of consciousness, enabling awareness of our self as a knower, an observer and a learner, and as one who takes action, with our neurons forming a continuous memory of our thoughts and actions. *We can identify self as a critical node—and learning as a primary catalyst—for state change (quantum change) in humans and living systems.*

Self—including self-referential memory, self-description, self-awareness and the personality—coevolves with, and is the creator of, its environment. Life is a process between the individual and the environment. Our self-awareness—which represents the unique ability to reflect on the past and potential future of ourselves, our world and the Universe—is possible because of the *relationship between* ourselves and our environment.[25]

Self has many facets. As humans, we often perceive ourselves as the roles we play, the jobs we have, the identities we take on—the outer self—but those are just the *way we choose* to manifest self. Self is so much more. American psychologist William James argues,

> *Although the self might feel like a unitary thing, it has many facets— from awareness of one's own body to memories of one's self to the sense of where one fits into society.*[26]

Thus, there is no single point within the mind/brain/body complex where we could situate self. It is the interactions among all these neuronal patterns, firings and connections that create the self.

Similarly, contemporary neuroscience does not identify a separate neurological function or structure where self—or consciousness, which enables the recognition of self—exists. Self is the quantum leap that occurred after the emergence of self-reflective consciousness, a distinct self that could take charge of the domain of consciousness, and determine which feelings or ideas take precedence. "Having had this experience of something inside us directing consciousness we gave it a name—the self—and took its reality for granted. And the self became an increasingly important part of human beings."[27]

One of the main roles of consciousness is to tie our life together into a coherent story, a concept of self.[28] Moving through various life experiences, the individual singles out and accentuates what is significant and connects these events to historic events to create a narrative unity, what can be described as a fictionalized history. This autobiographical self—the idea of who we are, the image we build up of ourselves and where we fit socially—is built up over years of experience and constantly being remodeled, a product of continuous learning in an experiential life.[29] Damasio believes that much of this model is created by the unconscious.[30] While this is undoubtedly true, *it is the conscious mind that perceives the idea of self* and through active experimentation with objects and the external world is typically very aware of the perceived boundaries between the individual and the external world.

Thus, *the story of you lives in the mind*—the what, why and how of your life, a summary of the conclusions you have about yourself and life, which often include the harshest judgments you have about yourself. As Walsh reminds us, this is *not* the place to search for the answers to life.

> *Burrowing deep within that mess rarely produced clarity. In fact, I will say that it never does, because your story is not real. It exists only within your mind. It may seem very real to you, but it is not reality.*[31]

The Plasticity of the Brain

The brain maintains a high degree of plasticity, changing itself in response to experience and learning. It has been shaped by evolution to adapt to the changes in its external environment through changes in the brain's chemistry and architecture.[32] This process of neural plasticity comes from the ability of neurons to change their structure and relationships according to environmental demands or personal decisions and action. Plasticity is increased through the production of neurotransmitters and the role of growth hormones, which facilitate neural connections and cortical organization.[33]

In 2000, Eric Kandel was awarded the Nobel Prize for showing that when individuals learn, the wiring (neuronal patterns, connections, and synapse strengths) in the brain changes. He showed that when even simple information comes into the brain it creates a physical alteration of the structure of neurons that participate in the process. Thus, we are all continuously altering the patterns and structure of the connections in our brains. The conclusion is significant. Thoughts change the physiological structure of our brains. What and how we think and believe impacts our physical bodies.[34]

The following neuroscience finding emerged from the ICALS research: **Plasticity is a result of the connection between neural patterns in the mind and the physical world—what we think and believe impacts our physical bodies**. Evolution has created a brain that can adapt and readapt to a changing world. [35] This finding is relative to concrete experience; such plasticity can broaden the scope of sensing as well as feelings.

A second related finding is **learning depends on modification of the brain's chemistry and architecture**. From a learning viewpoint, brain plasticity opens the door to the possibility of continuously learning and adapting to external environments and internal needs. This phenomenon, applying to all physiologically healthy individuals, means that *everyone has the potential to improve themselves by learning and modifying the structure of their brains*.

Our Emoting Guidance System

Emotions—present in all sentient complex adaptive systems—play a powerful role in influencing our perception of reality and how we respond to that perception. They assign values to options and alternatives, often without our even knowing it! There is growing evidence that fundamental ethical stances in life stem from underlying emotional capacities, with those stances creating the basic belief system, the values, and often the underlying assumptions that we use to see the world. In other words, our mental models. In short, we can say that emotions enable us to manage our lives.

As working definitions, while we agree the early description of emotions as "a mental state that arises spontaneously rather than through conscious effort and is often accompanied by physiological changes"[36], we also honor the more-recent description of emotions as skills,

> *... organized patterns of thoughts and behaviors that we actively construct in the moment and across our life spans to adaptively accommodate to various kinds of circumstances ...*[37]

Focusing on the biochemical substrate of emotion, Pert explains that "Neuropeptides and their receptors thus join the brain, glands, and immune system in a network of communication between brain and body, probably representing the biochemical substrate of emotion."[38] Neuroscientist Antonia Damasio separates emotions and feelings, and this is an important distinction. As Damasio says,

> *The term feeling should be reserved for the private, mental experience of an emotion, while the term emotion should be used to designate the collection of responses, many of which are publicly observable.*[39]

Thus, "feelings" represent the private mental experience of emotion, and the use of the term "emotions" represents those public expressions of response.[40] But accepting this distinction, there is an even larger difference. "Feelings" do not necessarily spontaneously emerge. For example, consider *desire* or *passion*. While certainly these feelings can emerge when triggered, that is, come into conscious awareness in the instant at hand, they are largely residing in the unconscious, perhaps building up over time in response to our experiences and imaginings, and—consciously or unconsciously—serving as powerful motivators for our choices and actions.

Our emotions *are* a building block of consciousness,[41] with both emotions and feelings serving as a guidance system for survival and the pain and pleasure portals of personality. Indeed, emotional content is almost always present in verbal and non-verbal communication, with the strength of that emotion determined by the personal preferences and relevance reflective of our inner map of the external world developed and adapted as we experience and contemplate our lives. As Henry Plotkin— who brings together evolutionary biology, psychology and philosophy in his writing—describes, "Normal human life is lived within a sea of experienced and expressed emotions."[42]

Drawing on neuroscience findings, we know that emotions play a strong role in learning. All incoming signals and information are immediately passed to the amygdala, where they are assessed for potential harm to the individual. The amygdala places a tag on the signal that gives it a level of emotional importance. If the incoming information is considered dangerous to the individual, the amygdala immediately starts the body's response, such as pulling a hand away from a hot stove. In parallel, but slower than the amygdala's quick response, the incoming information is processed and cognitively interpreted.

Quite literally, **emotions influence all incoming information**. This is another research finding that could potentially influence all aspects in the adult experiential learning model. However, there are three specific aspects that will most likely be subjected to emotional influence: understanding, meaning and truth (how things work). As Mulvihill notes:

> *During both the initial processing and linking with information from the different senses, it becomes clear that there is no thought, memory, or knowledge which is "objective", or "detached" from the personal experience of knowing.*[43]

See Chapter 8 on heart-mind entrainment for a deeper discussion of emotions in relationship to understanding, meaning and truth.

Another aspect of the emotional system is the role it plays in individual memory. Emotions assign values to options or alternatives, and, as noted above, often without our knowing it! Situations that have a high emotional impact are much easier to recall, sometimes remembered throughout life, and hard to lay aside even when we desire to do so. From a learning perspective, this means that (consciously or unconsciously) the learner is always evaluating the importance of incoming information, and this process helps the individual to remember the information.[44]

A related finding is **emotional tags influence memory recall**. We remember things that are emotionally significant, those things that are personally relevant. This helps our understanding and the creation of meaning, as well as our ability to anticipate the outcomes of actions. Thus, this item supports both reflective observation and abstract conceptualization. The mind solves many complex problems by recalling the core (or meaning) of past solutions. The most significant past problems have emotional tags that aid in recall. This finding will emerge several times in this book.

The interplay of the physical and mental with the emotional is going on throughout the body; these are interdependent. Candice Pert's study of information-processing receptors on nerve cell membranes led her to discover the presence of neural receptors on most of the body's cells.[45] This "established that the 'mind' was not focused in the head, but was distributed via signal molecules to the whole body,"[46] with the mind, spirit, and emotions unified with the physical body as *part of a single intelligent system*. In a sense, we have "feelings" associated with all our other senses, and our whole body has sensory capability![47]

A related finding is **the entire body is involved in emotions and the body drives the emotions**. LeDoux offers that emotions are a slave to physiology, not vice versa.[48] This aspect of the emotions may singularly impact how we act on the environment and how much control, rigor, and/or discipline we can exert. It would also impact on how well we can focus attention and maintain awareness.

Desire, courage and drive represent different combinations of emotional and mental activity which we call cognitive conveyors, a subset of words that represent concepts filled with a combination of thought, emotion and feelings, most of which are developed over time and some of which become a permanent part of the individual. The mental activity of "wonder" would also fit into this set. While emotional arousal plays an important role in these concepts, affecting mental activity and having a physiological effect on the body, these are often not identified as emotions, but they certainly are feelings, proving the usefulness of Damasio's differentiation between emotions and feelings. Damasio goes even further, suggesting that emotions such as fear and anger, happiness and sadness, are both cognitive and physiological processes involving the body and the mind,[49] which is supportive of the cognitive conveyor model.

Cognitive conveyors weave their way throughout various modalities of change. For example, experiential learning covers much territory from living in a certain environment to direct interaction with another person, to a frightening event, to the internal experiences of dreaming, meditation, reading, or reflection on action. In early human development this experiential learning loop was intertwined with fear driven by the need/desire to survive. While survival today is rarely a matter of being eaten by a tiger, nonetheless, fear and desire often remain in the learning loop related to financial success, saving face, bullying, divisive politics, personal desires and selfishness, and even an arrogance of personal entitlement.

Desire as used here is the expression of a feeling, to want or wish for something. As such, it is closely related to intention,[50] which is discussed in Chapter 7. Desire is a sustaining life force, indeed, an animating, continuously *expanding* life force which is unstoppable as long as we live and breathe. It often occurs without our conscious awareness—either rational or non-rational—a bedfellow driving our thoughts and actions. As Irvine forwards,

> ... *many of our most profound, life-affecting desires are not rational, in the sense that we don't use rational thought processes to form them. Indeed, we don't' form them; they form themselves within us. They simply pop into our heads, uninvited and unannounced. While they reside there, they take control of our lives. A single rogue desire can trample the plans we had for our lives and thereby alter our destinies.*[51]

Heijnen—who sees life as a game to be played in our current universe which is operating between its last reboot and a possible next one—says desire is the reason that started creation, and which is—in its current sentient form—still driving it. Further,

> ... *the notion of emotions as the later stage version of motion—with pain and pleasure from which we either run away from or get drawn to—can be seen as the sources of information by which sentient beings not only operate but also learn, and by which desire finds new direction.*[52]

Desire, then, has a potential unlike other emotions. As Alexander Faulkner Shand, an English writer and barrister, recognized in 1920, "Every emotion, when its end is obstructed, tends to develop its impulse into desire, and so give rise to the prospective emotions" and thus "the system of every emotion potentially contains desire with its prospective emotions."[53] This stance has been forwarded through various other theories. For example, the Humean Theory of Motivation, which forwards there is a belief/desire pair behind all motivation, and the philosophical thesis that "all normative reasons must be grounded in desires."[54]

The unknown is the foremost object of human fear. **Courage** is about choices made and actions taken in an uncertain environment, the courage to think and act. This does *not* mean there is a lack of fear. As psychologist Rollo May says, "Courage is not the absence of despair; it is, rather, the capacity to move ahead *in spite of despair.*"[55] In the latest *Green Lantern* movie,[56] the hero is a test pilot. Following a near fatality, the hero's young nephew asks, "Were you afraid?" The hero's answer is a response heard earlier in the movie from his father. "It's my business not to be." He sits with his nephew and explains that while fear is always close at hand, *courage is a choice.*

Drive is to move toward or through something with mental or physical force, to cause something to happen. While it is often urged forward by our desires, it can also become an automated process or an unconscious emergent force, sometimes beyond our control. The drive focused on here is a *choice* connected to desire. Every individual has energy levels that vary with physical health, interest, experiences, age, and so forth. How we focus this energy and the level of energy focused in a particular direction can be a choice. For example, if we get joy out of learning, we can choose to put ourselves in learning situations ranging from the classroom—virtual or otherwise—to dialogues with respected and knowledgeable others. In this example, the concept of "drive" would insinuate that we had a goal to achieve and would continue that learning until the goal is achieved, is no longer of interest, or is replaced by another goal.

Note that the same concept can act as a cognitive conveyor or a cognitive impeder. Drive is a good example of that possibility. There is no doubt that the way drive is described above is a cognitive conveyor, accelerating movement toward a defined goal. However, drive could also very much get in the way, that is, it might be so strong that an individual fails to take the necessary time to think through actions. In this case, drive would serve as a cognitive impeder.

Through this brief treatment, it can be seen that cognitive conveyors such as desire, courage and drive (as emotions and feelings) strongly affect our thoughts and actions and weave in and out of the changing dramas of our lives. By understanding their importance and impact—which also applies to other cognitive conveyors and impeders—we can consciously harness their energy to bring about the change we choose.[57] Managing emotions (as part of managing self) is addressed in Chapter 8.

The Intelligent Social Change Journey

As a human we are on a journey of learning and expansion, what we call the Intelligent Social Change Journey (ISCJ). This is a developmental journey of the body, mind and heart, moving from the heaviness of cause-and-effect linear extrapolations, to the fluidity of co-evolving with our environment, to the lightness of breathing our thought and feelings into reality. These are phase changes grounded in development of our mental faculties, each building on and expanding previous learning in our movement toward intelligent activity.

This is very much a social journey, for change does not happen in isolation. Thus, intelligent activity has to do with interaction with others, a state where intent, purpose, direction, values and expected outcomes are clearly understood and communicated among all parties, reflecting wisdom and achieving a higher truth.[58] By definition, that means cooperating and collaborating with others. It also brings in the value of wisdom and truth.

Wisdom does not convey directly from knowledge. Back in the early days of knowledge management (KM), there was an early model referred to as the DIKW (Data-Information-Knowledge-Wisdom) continuum. This was the beginning of a longer conversation. And while some KM advocates still support this relationship, as another says, "DIKW is ontologically and epistemologically flawed; it is not consistent with modern cognitive neuroscience or epistemology."[59]

During the 90's, Tom Stonier, a theoretical biologist, developed a workable theory of information, discovering new relationships between information and the physical Universe as fundamental as matter and energy.[60] Building on his foundational work, we take information to be a measure of the degree of organization expressed by any non-random pattern or set of patterns. Data (a form of information) would then be simple patterns, and while data and information are both patterns, they have no meaning until some organism recognizes and interprets the patterns.[61] Thus, information exists in the human brain in the form of stored or expressed neuronal patterns that may be activated and reflected upon through conscious thought.

When that information is effectively acted upon—whether consciously or unconsciously—it is considered knowledge, which is applied information to create value. This is consistent with more formal definitions such as "justified true belief" (Plato) or, a bit more recent, "the capacity (potential or actual) to take effective action".[62] Thus, "knowledge" is directly tied to action, situation-dependent and context-sensitive, and at the very core of what it is to be human.

In Phase 1 of the ISCJ—learning from the past—we act on the physical; we "see" changes happening, cause-and-effect relationships, from our physical form. Actions have consequences, both directly and indirectly, and sometimes delayed, but visible in simple situations. Thus, Phase 1 reinforces the characteristics of how we interact with the simplest aspects of our world, with elements that are predictable and repeatable and make us feel comfortable because we know what to expect and how to prepare for them. The challenge, of course, is that they only remain predictable if the causes remain constant, and, frankly, that just doesn't happen in today's world! In this phase we may emotionally develop sympathy for others as we realize that supporting and caring for the people involved in the change helps mitigate the force of resistance, which improves the opportunity for successful outcomes.

As we expand into Phase 2 of the ISCJ—learning in the present—we begin to recognize patterns from past experiences that repeat themselves over and over. This recognition enables us to "see" (in the mind's eye) the relationships among events in terms of time and space. Thus, we are no longer in an action and reaction mode, but in a position to co-evolve with our environment in the present (the NOW), enabling us to better navigate a world full of diverse challenges and opportunities. We have moved into a higher level of thinking and feeling about how we interact with the world, including the interesting area of human social interactions. Although complex, recognizable patterns enable us to explore and progress through uncertainty, making life more interesting and enjoyable. Patterns grow into concepts (higher mental thought), and we begin to see larger connections among "things" as we search for a higher level of truth. To co-evolve with our environment requires us to develop a larger understanding of our self (values, beliefs, emotions, desires, etc.) and of others, developing empathy, which provides a direct understanding of another individual and a heightened awareness of the context of their lives and their desires and needs in the instant at hand.

Development of higher mental faculties comes with instinctive knowledge of the workings of the Universe, which helps cultivate intuition and develop insights in service to our self and society. For example, building on the recognition of patterns is the idea of infinite symmetry. Symmetry is proportional or balanced harmony, with the exact correspondent of form on opposite sides of a centerline or point, a harmony of proportions.[63] Since nature is fond of doing things in the most economical and efficient way, symmetry plays an important role in patterns in the physical world. All the mineral substances that are part of the Earth's crust can be described by the Platonic solids, five shapes that each have equal faces, lines and angles. These are the tetrahedron (4 triangles), the cube (6 squares), the octahedron (8 triangles), the dodecahedron (12 pentagons) and the icosahedron (20 triangles). These five simple shapes are a template for all three-dimensional forms in the

Universe. The idea of *infinite symmetry* insinuates that we know much more about the Universe than we know that we know. If we can understand the models of life within our context, we then have the keys to understanding higher order patterns beyond our cognizance. And it follows that the more we discover about the Universe, the more we understand ourselves.

Throughout all our experiences, humans are always looking for relationships among things—thoughts, events, happenings, and people. We are on a grand search for patterns, for *creative association*.[64] Consider the example of symbiotic thinking, which shows us that the very concept of "cause" *cannot exist* with the concept of "effect". *This deep relationship is not from causality, but from existence.* The very existence of a thing or idea requires the existence of something else. The concept of "day" would not exist if not for the concept of "night". In the physical Universe, there is "matter" and "antimatter". This same pattern plays out in the discussion of time and space, noting that space cannot exist without objects, and objects exist because they are surrounded by space. And with symbiotic thinking, there is reason to expect that our individual ideas cannot exist without *a larger consciousness which seeks to incorporate these ideas.*

There is a freedom that occurs as we leave behind the thinking patterns of Phase 2 of the ISCJ and open to the choices and discoveries of Phase 3. For many, this is where the world is poised at this instant in time. As we enter this phase—co-creating the future—we are acquiring the ability to tap into the larger intuitional field that energetically connects all people, whether you call that an energy field, consciousness field, quantum field, or God field. This can only be accomplished when energy is focused outward in service to the larger whole, not constrained within, thus deepening our connection to others. Compassion deepens that connection. Thus, each phase of the ISCJ calls for an increasing depth of connection to others, moving from sympathy to empathy to compassion, as we become fully engaged in co-creating our reality.

Time and space play a significant role in the phase changes. Using Jung's psychological type classifications, feelings come from the past, sensations occur in the present, intuition is oriented to the future, and thinking embraces the past, present *and* future. Forecasting and visioning work is done at a point of change[65] when a balance is struck continuously between short-term and long-term survival. Salk describes this as a shift from "Epoch A", dominated by ego and short-term considerations, to

"Epoch B", where both being and ego co-exist.[66] In the ISCJ, this shift occurs somewhere in Phase 2, with beingness advancing as we journey toward Phase 3, which is where we can fully engage the existential experience (see Chapter 10). Appendix A shows the three phases of the ISCJ from the viewpoints of the nature of knowledge, points of reflection and cognitive shifts.

As we expand through the phases of the Intelligent Social Change Journey, our idea of "self" shifts and changes. We have a higher awareness of being aware, and with consciousness comes responsibility. We move into a state of self-governance, anchoring and putting into action every level of our multidimensional self in our continuous search for higher truth. We move beyond the mental ego into the heart, embracing diversity and expanding our individuation, while simultaneously connecting more deeply with others through conscious compassion. And, as those connections deepen, we move ever closer to intelligent activity, and that unconditional love that once appeared illusive and, even, unobtainable for humanity.

The ISCJ and Human Learning

As a cognitive-based ordering of change based on Bertrand Russell's work in logic and mathematics, we forward the concept of logical levels of learning consistent with levels of change developed by anthropologist Gregory Bateson. This logical typing was both a mathematical theory and a law of nature, recognizing long before neuroscience research findings confirmed the relationship of the mind/brain, that we literally create our reality, with thought affecting the physical structure of the brain and the physical structure of the brain affecting thought.[67]

Bateson's levels of change range from simplistic habit formation (Learning I) to large-scale change in the evolutionary process of the human (Learning IV), with each higher level synthesizing and organizing the levels below it, and thus creating a greater impact on people and organizations.[68] This is a hierarchy of logical levels, ordered groupings within a system, with the implication that as the levels reach toward the source or beginning **there is a sacredness of power or importance informing this hierarchy of values**.[69] Similar to Bateson's levels of change, each higher phase of the ISCJ synthesizes and organizes the levels below it, thus creating a great impact in interacting with the world.

With *Learning 0* representing the status quo, a particular behavioral response to a specific situation, *Learning I* (first-order change) is stimulus-response conditioning (cause-and-effect change), which includes learning simple skills such as walking, eating, and driving. These basic skills are pattern forming, becoming habits, which occur through repetitiveness without conceptualizing the content. For example, we don't have to understand concepts of motion and movement in order to learn to walk. Animals engage in Learning I. Because it is not necessary to understand the concepts, or underlying theories, no questions of reality are raised. Learning I occurs in Phase 1 of the ISCJ.

Learning II (second-order change) is deuteron learning and includes creation, or a change of context inclusive of new images or concepts, or shifts the understanding of, and connections among, existing concepts such that meaning may be interpreted. These changes are based on mental constructs that *depend on a sense of reality*.[70] While these concepts may represent real things, relations or qualities, they also may be symbolic, specifically created for the situation at hand. They provide the means for reconstructing existing concepts, using one reality to modify another, from which new ways of thinking and behaviors emerge. Argyris and Schon's concept of double loop learning reflects Level II change.[71] Learning II occurs in Phase 2 of the ISCJ.

Learning III (third-order change) requires thinking beyond our current logic, calling us to change our system of beliefs and values, and offering different sets of alternatives from which choices can be made. Suggesting that Learning III is learning *about* the concepts used in Learning II, Bateson says,

> *In transcending the promises and habits of Learning II, one will gain 'a freedom from its bondages', bondages we characterize, for example, as 'drive', 'dependency', 'pride', and 'fatalism'. One might learn to change the premises acquired by Learning II and to readily choose among the roles through which we express concepts and thus the 'self'.*[72]

Similarly, Berman defines Learning III as, "an experience in which a person suddenly realizes the arbitrary nature of his or her own paradigm."[73] This is the breaking open of our personal mental models, our current logic, losing the differential of subject/object, blending into connection while simultaneously following pathways of diverse belief systems. Learning III occurs as we move into Phase 3 of the ISCJ.

Learning IV deals with revolutionary change, getting outside the system to look at the larger system of systems, awakening to something completely new, different, unique and transformative. This is the space of *incluessence*,[74] a future state far beyond that which we know to dream. As Bateson described this highest level of change:

> *The individual mind is immanent but not only in the body. It is immanent in pathways and messages outside the body; and there is a larger Mind of which the individual mind is only a sub-system. This larger Mind is comparable to God and is perhaps what people mean by 'God,' but it is still immanent in the total interconnected social system and planetary ecology.*[75]

Table 1 below is a comparison of the phases of the ISCJ and the four levels of learning espoused by Bateson based on the work in logic and mathematics of Bertrand Russell.

An example of Learning IV is Buddha's use of intuitional thought to understand others. He used his ability to think in greater and greater ways to help people cooperate and share together, and think better. Learning IV is descriptive of controlled intuition in support of the creative (quantum) leap in Phase 3 of the ISCJ, moving beyond what we can comprehend at this point in time, and deepening the connections of sympathy, empathy and compassion.

A deeper treatment of the ISCJ is available in the five-book series titled *The Profundity and Bifurcation of Change* and in the little conscious look book titled *The Intelligent Social Change Journey*, which is the foundational book for the 22-book *Possibilities that are YOU!* series.[76]

Phase of the Intelligent Social Change Journey	Level of Learning [NOTE: LEARNING 0 represents the status quo; a behavioral response to a specific situation.]
PHASE 1: Cause and Effect (Requires sympathy) • Linear, and sequential • Repeatable • Engaging past learning • Starting from current state • Cause and effect relationships	**LEARNING 1: (First order change)** • Stimulus-response conditioning • Includes learning simple skills such as walking, eating and driving • Basic skills are pattern forming, becoming habits occurring through repetitiveness without conceptualizing the content • No questions of reality
PHASE 2: Co-Evolving (Requires empathy) • Recognition of patterns	**LEARNING II: (Deutero Learning)** (Second order change)

• Social interaction • Co-evolving with environment through continuous learning, quick response, robustness, flexibility, adaptability, alignment.	• Includes creation or change of context inclusive of new images or concepts • Shifts the understanding of, and connections among, existing concepts such that meaning may be interpreted • Based on mental constructions that depend on a sense of reality
[Moving into Phase 3] **PHASE 3: Creative Leap** (Requires compassion) • Creative imagination • Recognition of global oneness • Mental in service to the intuitive • Balancing senses • Bringing together past, present and future • Knowing; beauty; wisdom	**LEARNING III: (Third order change)** • Thinking beyond current logic • Changing our system of beliefs and values • Different sets of alternatives from which choices can be made • Freedom from bondages **LEARNING IV:** • Intelligent Purpose • Quantum change • Getting outside the system to look at the larger system of systems • Awakening to something completely new, different, unique and transformative • Tapping into the larger mind of which the individual mind is a sub-system

Table 1. Comparison of phases of the ISCJ with levels of learning.

The Human Search for Wisdom

The highest part of mental thought is wisdom, yet it is *more* than mental thought. Representing completeness and wholeness of thought, wisdom is universally a lofty consideration, and too often it eludes us. The more we seek it—and it is in our human nature to do so—the more we understand that it comes through experiencing and learning, and brings with it the desire to learn more. It also comes with an ever-deepening connection to others.

Wisdom occurs when activity matches the choices that are made and structured concepts are intelligently acted upon, thus **directly connecting wisdom to intelligent action**, which is a goal of the developmental Intelligent Social Change Journey. Action occurs in our perceived physical reality, with intelligent activity representing a state of interaction where intent, purpose, direction, values and expected outcomes are clearly understood and communicated among all parties, reflecting wisdom and achieving a higher truth, a definition that will be repeated throughout the text. More briefly stated, MacDonald says intelligent activity is "acting with the well-being of the whole in mind."[77]

Knowledge and wisdom both have an information component and a process component dealing with the nature and structure of information, with *nature* representing the quality or constitution of information, and *structure* representing the process of building new information, or learning. Csikszentmihalyi and Nakamura refer to these two components as the *content* of wisdom (information) and the *capacity to think or act wisely*. Content would be context-sensitive and situation dependent,[78] which is consistent with the position that wisdom is grounded in life's rich experiences.[79] However, wisdom moves above and beyond specific circumstances as patterns develop. Wisdom "therefore is developed through the process of aging … [and] seems to consist of the ability to move away from absolute truths, to be reflective, to make sound judgments related to our daily existence, whatever our circumstances."[80] This is consistent with the ICALS finding that **the engaged, mature brain can make effective decisions at more intuitive levels**. It is also consistent with the recognition that memory is stored in invariant form, as well as our understanding of the plasticity of the brain and the finding that **the mind/brain unconsciously tailors internal knowledge to the situation at hand.** Thus, patterns become more important than exact actions.

In a beautiful flow of words and concepts, artist Joan Erikson says that a sense of the complexity of living is an attribute of wisdom. A wise person embraces the

> *… sense of the complexity of living, of relationships, of all negotiations. There is certainly no immediate, discernible, and absolute right and wrong, just as light and dark are separated by innumerable shadings … [the] interweaving of time and space, light and dark, and the complexity of human nature suggests that … this wholeness of perception to be given particularly and realized, must of necessity be made up of a merging of the sensual, the logical and the aesthetic perceptions of the individual.*[81]

Wisdom appears to deal with the cognitive and emotional, personal and social, as well as the moral and spiritual aspects of life. "In essence, wisdom grows through the learning of more knowledge, and the practiced experience of day-to-day life—both filtered through a code of moral conviction"[82] and, as forwarded above, "acting with the well-being of the whole in mind."[83] Imagine if that was the criteria for each and every decision we each make?

While all too short a discussion on this critical element of what it is to be human, relating this journey to both our knowledge development (the action growth path) and the need for developing deeper connections with others (the connection growth path), we provide a wisdom model as Appendix B for your reflection. You will note that the connection growth path follows the same deepening connections of sympathy, empathy and compassion in our journey toward unconditional love developed as the individual moves through the ISCJ (see Chapter 1). As previously voiced, wisdom is also reflective of Learning IV.

So, What Does It Mean to be Human?

At a high level, we've touched on a few potential answers to this question. We are superior energy beings in a physical body with a mind/brain, emotions, and a spiritual essence. We exist, we perceive our existence, with knowledge at our core providing the ability to think, choose, act, interact, learn, and change. We are complex adaptive systems, forever learning and changing, and cannot exist in stasis. We have emotions and feelings that—both consciously and unconsciously—serve as a personal human guidance system as we move through the phases of the Intelligent Social Change Journey, an experiential journey of learning and expansion.

In this work, it is not a coincidence that the human journey is described as an Intelligent Social Change Journey and that the human is considered an Intelligent Complex Adaptive Learning System. The human movement toward intelligent activity occurs through development of wisdom, a combination of knowledge and experience, following the tracks of both developing the mental faculties **and** increasingly deeper and more humble connections with others and the environment. (See Appendix B.) Yet the human is more than the sum of those parts, beyond reason, beyond emotion, beyond creativity, with an "otherness" about it, a caring for the larger whole. We will have a deeper discussion of intelligent activity and the emerging wisdom of the human self in Chapter 9.

Each of us is incredibly unique, one of a kind, which means that the specificity of what it means to be human—the capacities and capabilities, the potential and possibilities, the needs and wants, the dreams and desires—are entangled with the physical, mental, emotional and spiritual self—you. And our relationship with learning? Quite simply, **learning is living; and living is learning**. The more fully we can engage the human learning processes—with agency, making wise choices in that regard—the richer our lives and the greater potential for unleashing the human mind and contributing to the noble evolution of humanity.

Chapter 2
Environmental Currency

More about CUCA ... The Current Learning Environment ... The Impact of Technologically Enabled Networks on Learning ... Continuing Technological Advancement ... The Potential Loss of Deep Knowledge

FIGURE: (2) Nominal shift in focus of levels of knowledge from 2000 to 2020.

For years we and many others have been writing about our CUCA world, that is, accelerating **Change**, rising **Uncertainty**, and increasing **Complexity** combined with the human response of **Anxiety**. As we wrote at the beginning of this century: "Time accelerates. Distance shrinks. Networks expand. Information over-whelms. Interdependencies grow geometrically. Uncertainty dominates. Complexity boggles the mind."[84] And never has the term "CUCA" been more appropriate than today! Such is the environment and the context within which we must compete, survive, and thrive.

While the elements of CUCA have always been with us, they accelerated at the turn of the century as the world virtually connected, and in today's world these *forces are building*. A force occurs when one source of energy affects (pushes against, interferes with, or influences) another source of energy, whether with positive or negative results. Energy can never be lost—it can only be transformed—and it is through this transformation that light, heat, electricity, and *life and learning* are produced! Advanced sentient beings largely have a choice in the forces they engage; those things in the natural setting of our planet that are less sentient are controlled by the *forces of nature*. And forces bring change.

More About CUCA

The first big **C** of CUCA represents change. And yes, while there's always been change, today it's taking leaps and bounds in every area of life—*the pace of change has changed*. And along with that change, there is so much Uncertainty, even down to the point of survival. Where will my next rent or mortgage payment come from? How do I go safely to the store? And the chatter continues: Life, liberty, and happiness. I'm entitled. If I wear a mask, it takes away my freedom; besides which, they are uncomfortable. If I don't wear a mask, it may take away someone's life—or maybe even my own—that is, if this whole bunch of fake news is anywhere close to reality! Life is illusion anyway, so it's only illusion that's going to die.

Guess it's easy to tell why the second **C** in CUCA stands for complexity. Our systems continue to increase in complexity, whether small or large—technology, medical care, water supply, power grids, the internet, distribution, politics, the economy—and the more they increase in complexity, the more vulnerable to failure, either from natural causes or from intentional sabotage. Why? First, as systems become more complex, they have more internal connections and networks, making them more susceptible to possible failures. Secondly, these same complex systems are—or can easily become—unpredictable because they no longer operate via deterministic cause and effect. While each connection may, or may not, be causal, the number of connections, the possible feedback loops (or sneak circuits), the time delays and nonlinear effects, plus the possibility of sensitivity to input values and the effect of their own environment all create a situation of non-predictability. And that leads to the **A**, which represents the anxiety we all feel in response to the CUC part.[85]

And while it's nice to talk about all this as something separate, something being imposed *upon* humanity, with us in the role of victims, it is humanity that is the cause, and, to a large extent, the accelerated mental development that has accompanied our technological advancements. Our inner and outer worlds are out of balance. As Harvey contends, we are "in a period of evolution where the world has expanded and developed outwardly and left the inner world behind. Our inner selves must catch up and restore the balance."[86] If we pause for a moment and reflect within, we can recognize the truth of this statement.

As far back as we can collectively remember, humans have been primarily focused on development of the mental faculties. In this rush (in evolutionary terms) for mental achievement, we have diminished our spiritual development, which facilitates access to higher intelligence or energy levels. Energy follows thought; everything that physically exists was a thought first (see Chapter 5). When our thought is focused on the perceived objects of our physical reality that surround us, it is difficult to tap into the energy flows within, much less tap into that which is beyond our current understanding. This unbalanced mental growth is what has led to the egotism and arrogance often visible in the global business and political environments.

Let's explore that a bit deeper. The sharp edge of win-lose competition combined with this accelerated mental development has pushed individuals and organizations through ego into arrogance. While egotism says, "I am right", arrogance says, "I am right. You are wrong. And I'm not listening." *Egotism closes the door to learning, and arrogance builds forces that can lock that door.* This is addressed from the viewpoint of humility in Chapter 11.

This surge of arrogance displayed in today's world builds on the modern focus on humanism, that is, faith in our own omnipotence. In the expansion of our mental capabilities, we seem to have forgotten the nobler parts of humanism, individually and collectively seeking the potential value and goodness of humanity through rational thought. As biologist David Ehrenfeld describes, we have "chosen to transform our original faith in a higher authority to faith in the power of reason and human capabilities", and this has proven a misplaced trust. He goes on to point out that the modern world has chosen Humanism as the guiding philosophy of life, and that with that choice we become "responsible for all the consequences that flow from that choice".[87] Only, our "humanism" must balance the physical, mental, emotional and spiritual.

While it may sound like accelerated and unbalance mental growth is a failing in current human history, that is not the intent. Whether we look at accelerated mental development from the viewpoint of the need to learn to survive and thrive in a CUCA environment, or from the viewpoint of the innate human hunger for "more" (which we all can admit to at some level), *it can be a good thing.* Unfortunately, as is wont to happen with good things, it is easy to push ourselves out of balance. For example, as we become more proficient in a domain of knowledge, building confidence in our knowledge and thinking of ourselves as an "expert", it is quite easy for the ego to expand and begin to identify our self with that knowledge. When this happens—when we allow ourselves to become our knowledge—we tie ourselves to a moment in time, limiting the continuous learning and growth necessary to co-evolve with our environment.

This rapid development of the mental *does* offer considerable opportunity for the future of humanity, that is, if we are able to now bring the mental into balance on the physical and emotional planes, with the spiritual counterbalance woven throughout, which means *bringing truth, compassion and unconditional love and purpose into our everyday lives* as we cooperatively and collaboratively work together to achieve intelligent activity.[88] There are several reasons this is true. First, no one can argue the technological advancements over the past thousands of years that have led us to today, nor the incredible potentially positive shifts underway today in nearly every area of human endeavor. When used for the benefit of all, these advances offer the potential to take humanity to the next level in terms of the growth of both individual and collective consciousness.

Second, even when an individual is able to access higher intelligence sources and has brief flashes of intuition or inspiration, without development of the mental faculties in a related domain of knowledge— or the wisdom involved in translating a concept across domains of knowledge—it is difficult, if not impossible, to successfully act on that insightful flash. Further, these flashes happen sporadically, and the large amount of insight gleaned from the experience is difficult to grasp and quickly forgotten. As this is being written, a colleague and friend with whom the authors have had a number of conversations related to the importance of developing the mental faculties in support of the intuitive, called to say that at lunch in a Chinese restaurant she had just received a fortune that read, "Intuition and knowledge walk hand in hand." Well said!

The Current Learning Environment

Built on the deteriorating foundations of democracy, today we live in a society where truth as a commodity has less value and consistency than the markets, catering to the whims of wealth and power. Yet truth is the highest virtue of mental growth. Add in a pandemic—whether facilitated by man or an emerging historic pattern whose time has come—and life and living has changed for everyone. With these unwanted, and often denied, changes has come fear and pain, the loss of freedoms and loved ones, and a virus of anger. And all this is occurring in the wake of over-stimulation and divisiveness emerging from political and economic uncertainties, and unfettered despotism.

Layered with occasional whisps of truth, misinformation, and disinformation, propaganda abounds, washing across our world in a wave of media bantering that is continuously challenging our trust and learning and thinking processes. And during the largest lockdowns in living memory, and the continuing rise of death as an everyday tolling, the young continued to be programmed by specialized computer games that glorify violence, potentially leading to acceptance of the use of violence—as evident in the increasing death toll of the young by the young—and a belief that the end justifies the means. Almost daily in the news we read about mass murders, about random victims, about law officers who kill those they are bound to protect, about parents who kill their young, children who kill their parents, and people of all ages who kill their friends and strangers alike over petty grievances.

Death is a part of life, and as the large population of Baby Boomers continue to age, there is certainly an expected increase in deaths, as has occurred year after year, howbeit with the pandemic daily death became more visible. What is unexpected, keeping up with the pandemic, is the rise in gun-related crime, with one example the surge in homicides. For example, the Council on Criminal Justice found that homicides, which had increased by 42% during the summer and 34% in the fall of 2019, were even 15% higher in the first half of 2020. The Gun Violence Archive shows that 2020 had the highest number of gun violence deaths in the U.S. in more than 20 years. There was also an increase in aggravated assault and gun assault. While there is no consensus on the *why,* there are many theories: a bad economy, new economic inequality with the emergence of the post-industrial economy, a battling political environment, de-policing and distrust in police, an increase in the purchase of guns, overwhelmed hospitals, boredom and, as a result of the pandemic, the eruption of emotions tied to the loss of freedom and perceived rights, and so forth. And then, expected by some and unexpected by many, an aggressive war promulgated by a super power, threatening a sovereign nation, the notion of world peace, and the supposition that humanity had learned the lessons of World Wars I and II.

Now, reflect for a moment on all that we can learn from neuroscience research, which will be scattered throughout this book, that is, *the powerful influence of our environment and social situations on our thought and action.* For example, an emergent neuroscience finding is that **mirror neurons are a form of cognitive mimicry that transfers active behavior and other cultural norms**. This means that when we see something happening, our mind creates the same patterns of neurons that we would use to enact the same thing ourselves. This is directly related to the sub-element of accelerated learning and enhancing understanding, meaning, truth and how things work as well as related to the sensing, feelings, and attention in concrete experience. *Reflect with this new* *understanding:* How is this current world stage affecting learning and behaviors? How much of the repetitive negative activity in our systems is unconsciously being transferred to learners of all ages? How much constructive learning from across time and across humanity is seeping away?

The weakening effects on humans of media violence alone can lead to subtle grades of depression, and *depression is mankind's leading cause of death*, killing more than that all other diseases combined.[89] Or, at least, up until the present. While there are many excellent studies in this domain, let's briefly explore Hawkins' work to better understand the full import of media violence on the physical body. Using Kinesiology testing, Hawkins was able to show that a typical television show **produced weakening about 113 times in a single episode**. As Hawkins explains:

> *Each of these weakening events suppressed the observer's immune system and reflected an insult to the viewer's central as well as autonomic nervous system. Invariably accompanying each of these 113 disruptions of the acupuncture system were suppressions of the thymus gland; each insult also resulted in damage to the brain's delicate neurohormonal and neurotransmitter systems. Each negative input brought the watcher closer to eventual sickness and to imminent depression ....* [90]

Hawkins uses the concept of levels of consciousness to represent calibrated levels correlated with a specific process of consciousness—emotions, perceptions, attitudes, worldviews, and spiritual beliefs. As the culmination of research over a 20-year period involving thousands of people of all ages and personality types, Hawkins mapped the energy field of consciousness, with the levels ranging from 0 to 1,000. The progression is as follows: 20 (Shame); 30 (Guilt); 50 (Apathy); 75 (Grief); 100 (Fear); 125 (Desire); 150 (Anger); 175 (Pride); 200 (Courage); 250 (Neutrality); 310 (Willingness); 350 (Acceptance); 400 (Reason); 500 (Love); 540 (Joy); 600 (Peace); 700-1,000 (Enlightenment). The 200 level, that associated with integrity and courage, serves as a critical response point, "the balance point between weak and strong attractors, between negative and positive influence."[91]

Using Hawkins' levels of consciousness, any media events producing a state of consciousness that calibrates below 200—which represents such states as Shame, Guilt, Apathy, Grief, Fear, Desire, Anger and Pride—are destructive and unsupportive of life. *Reflect with this new understanding:*

What effect is political untruth having on our pandemic-honed population? How does it impact learning?

The Impact of Technology Enabled Networks on Learning

During the last half century, the entire global humanity, either directly or indirectly, has undergone a social transformation beyond comprehension, exceeding the transformation from the Industrial Economy to the Knowledge Economy. At exponentially accelerating rates we have become networked together through technologies that compound their inherent capabilities and functionalities with each new iteration of innovation. The most apparent contributing technologies include transportation, mail, telephone, television, internet, and cellular. When examined carefully, the impacts are almost too new, too different, and too complex to adequately describe. For example, the internet began primarily as an information distribution capability and cell phones as portable telephones. Now, with interlacing layers of technological advancements, the convergence of the internet and cell phones has given rise to a nearly indispensable human tool that serves widely as an augmentation device for the human brain in nearly everything we do, including formal learning. What percentage of its use is now dedicated to phone calls? Currently, the Apple iPhone iOS offers nearly 2 million apps to choose from. What is next? Do we have an inkling of how cell phones will perform with augmented reality in the coming decade? Eventually, new forms and functions will again become unrecognizable in every dimension of human activity, especially learning, self-management, personal performance, and self-development. But, before we venture into the impact of technologies on learning, let's step back a minute.

Manuel Castells is the preeminent harbinger of the impact of networking on global society. As a renown Spanish sociologist, he has been affiliated in various capacities with the Universitat Oberta de Catalunya in Barcelona, the Annenberg School of Communication at the University of Southern California, the University of California at Berkeley, St. John's College at the University of Cambridge, and the College d'Etudes Mondiales in Paris. He is the author or editor/co-editor of 38 books and is one of the foremost quoted scholars internationally in social science and communications.[92] The most noted work is his landmark trilogy, entitled "The Information Age: Economy, Society and Culture", with Volume 1 being "The Rise of the Network Society".[93] This seminal book has been translated into 20 languages. Providing this background is to point out a resource that will help to understand this societal phenomenon that is revolutionizing learning and will change learning across the world.

Castells, in his Volume 1 referenced above, begins his Conclusion section with this statement:

Our exploration of emergent social structures across domains of human activity and experience leads to an overarching conclusion: as an historical trend, dominant functions and processes in the Information Age are increasingly organized around networks. Networks constitute the new social morphology of our societies, and the diffusion of networking logic substantially modifies the operation and outcomes in processes of production, experience, power, and culture. While the networking form of social organization has existed in other times and spaces, the new information technology paradigm provides the material basis for its pervasive expansion throughout the entire social structure. Furthermore, I would argue that this networking logic induces a social determination of a higher level than that of the specific social interests expressed through the networks: the power of flows takes precedence over the flows of power. Presence or absence in the network and the dynamics of each network vis-a-vis others are critical sources of domination and change in our society: a society that, therefore, we may properly call the network society, characterized by the preeminence of social morphology over social action.[94]

In that same volume, Castells begins the "Prologue: The Net and the Self" with a profound quote:

"Do you think me a learned, well-read man?"

"Certainly," replied Zi-gong, "Aren't you?"

"Not at all," said Confucius. "I have simply grasped one thread which links up the rest."[95]

To be sure, every aspect of learning has myriad sources in a networked world. In this hyperconnected environment, the volume of data flows, information generation, and knowledge proliferation mushrooms continuously and exponentially. From a learning viewpoint, there is both the informal and formal. Castell writes of both, and draws upon both. To a great extent, he professionally thrives within a formal multi-university-based world on the one hand, and on the other as he writes in his Volume 1 Acknowledgement, the book required 12 years working across informal international collegial networks, including four "exceptional forums".[96]

Formal vis-à-vis informal learning is an increasing topic of popular interest. For a whole host of reasons, the nature of learning, as impacted by networks, has new unpredictable dynamics. Speed, accessibility, reach, ubiquity, standards, reliability, etc. are evolving, albeit frequently chaotically. How about this comment from a Gen Z 2021 high school graduate in the top three percent? "I would rather learn a trade than spend an insane amount of money for something I can learn online for free."

In 2004, Sal Khan began helping a cousin with mathematics using an internet Yahoo app. Before long, other cousins joined in his tutoring and he transferred his instruction to YouTube. Over the years, the Khan Academy (KA) platform has grown to approximately 8,000 videos that have had nearly two billion viewings. Now, 100 million users are calculated to have used KA worldwide in dozens of languages, and 10 million users subscribe to KA. On the KA website home page, we read "We're a nonprofit with the mission to provide a free, world-class education for anyone, anywhere." We hear from Anjali in India, "I come from a poor family. At home it's one room, just a room we live in. When I was a child, I used to fear mathematics. But now, I am in love with mathematics because of Khan Academy." A Gen Z student, just beginning his first year of college on full tuition scholarship, reported, "Before I started my academic year, I took a Khan Academy introductory course in calculus. Now, I can see that if I had not prepped with Khan, I would have dropped out of my first term calculus course."

This Khan Academy example raises a number of questions about informal and formal learning activities in a networked environment. Some include: How was one person able to launch a global learning service? How was the migration from informal to formal achieved? How was it possible to attract financial support?

Generally speaking, how do networking environments present new learning opportunities? Moreover, what is the impact on populations that have traditionally been restricted from learning opportunities by poverty or gender? The significance of networks in emerging social issues is brilliantly documented in Castell's book *Networks of Outrage and Hope, Social Movements in the Internet Age.* Related to this topic is the increasing interest in social media platforms being used as political forces in national politics internationally. A consideration is for global networking technologies, especially the internet, to provide a world-wide predominant voice for learning about global security challenges. Russia's 2022 invasion of Ukraine, a democratic nation of 43 million, would fit in this category. A global-networked voice with 100-plus nations linked together to express their commitments could be a dominant voice offering invaluable real-time learning capabilities from the collective resources of its contributing nations and could react much faster than independent news services.

Furthermore, in a world where the principal networking modality is the internet and there is increasing use of it, how much concern should there be about the quality and integrity of the flows? The current reality is that fake news, propaganda, bogus conspiracy theories, and misinformation in general are inordinately prevalent. Learning in a networked world inherently needs increased and discerning learning skills for more reliable learning. Chapters 6-11 address skills that serve us in self-managing our learning.

Continuing Technological Advancement

The creation of smart machines that are becoming autonomous—far beyond that which could be conceived a few years ago as we pursued artificial neural networks—is an offshoot of our accelerated intellect. Let's look at a few examples. Describing today's computers as glorified calculators, Dr. Dharmendra Modha, the founder of IBM's Cognitive Computing Group, set out to have computers think more like humans. Modha and his team say that a human brain-inspired cognitive computer chip and software ecosystem would consume much less power and space than a computer of today while simultaneously powering "everything from search and rescue robots in hazardous environments to intelligent buoys which float on ocean waves, predicting tsunamis or warning of oil pollution."[97] We agree. This shift towards brain-based models is underway.

Similarly, in video game development, sentient computer chips are under development (and most likely already on the market) that can analyze a player's personality, changing accordingly. For example, to reach this capability, Black Mirror pauses at key points during its (terrifying) narrative for players to answer a questionnaire based on the "Big Five" personality test, which is a test used for personality research by academic psychologists. People are asked questions such as: "Are you a private person? Do you listen to other people's feelings?" Such answers help build a player's psychological profile. These responses are combined with data picked up during the process of playing the game such as "how long they spent exploring each area before moving on; whether they strayed from clearly marked paths; whether they faced non-player characters while they talked", with every action and response having an effect on the narrative of the game.[98] This leads into the value offered through *Psyche*cology Games (PEGs) which include brain entrainment technologies to enhance heart coherence and the flow state, a rich condition for learning which can circumvent the analytical mind. PEGs are discussed more fully in terms of their educational value in Chapter 9 and are exampled in Chapter 12.

Which brings us back to the learners of today. Indeed, the generation of learners emerging in the workforce today are quite different from the generation of yesterday, and the next generation emerging from the virtual reality of a global pandemic moves us into unknown territory. We live in unprecedented technologically advanced times. Unparalleled economic shifts over the last century have taken us from predominantly agrarian societies to post-industrial and post-truth societies with divisive political and economic models. In terms of the scale of societal change and the need for learning and adaptation, this is the most radical environmental transformation in human history. Every aspect of our existence is impacted. We are quite literally in a consciousness shift, as will be pointed out throughout this text. Amidst the transformational challenges, including the diminished convergence with nature in the environment and the accelerated speed of change, we make some choices that potentially portend dire consequences. Nevertheless, progress and opportunity can carry the day.

At the heart of everything we create now in every field of human endeavor—in health and medicine, safety, transportation, energy, exploration, manufacturing and technological innovation, finance, education—and in whatever work we pursue, in our sociality and in our governance, *our mind/brain is augmented in increasingly unimaginable ways*. That augmentation is computers and advanced communication technologies. These two technologies are increasingly inseparable, and their relationship with us is increasingly synergistic. Already we may say that the interdependence between humans, computers, and communication systems is so rich and dynamic that technologies are shifting from a focus on specialized design and specific functionality to the dynamics and sharing of information, misinformation, disinformation, and propaganda, cracking open the illusion of trust and truth.

Interactive man-machine learning not only applies to how we learn from the technologies, but also *how the technologies learn from us*. Not long ago, we hoped that the technologies would remember our previous settings, requests, and choices. When you use these technologies today, there is increasing likelihood that they will learn from what you're doing and adjust their functionalities to accommodate your mode of use. Further, various software and computer systems not only anticipate what we want, but, when we select a certain type of function or process, the system will suggest next steps or new levels of procedures. The technology gives us choices, supports us in the background, and guides/directs us as we proceed. However, let's be sure and acknowledge the caveat that while these technologies are beginning to exhibit human-like intelligence, these "deep-learning AI systems exhibit almost no flexibility".[99] That is, they are specific to a domain of knowledge or a process, etc., with less flexibility in regards to domain shifting than that of a normal inquisitive five-year-old.

While admittedly limited by domain-specific programming and automated networking capability, the systems of today *simultaneously enable performance while enhancing our learning*. A key factor in this kind of future is the ability of technologies to accommodate our individual needs on multiple levels. Our technologies are increasingly able to reach out to whatever we need and whomever we seek, *increasingly augmenting our brain capacity* and serving as teachers. As Horstman describes:

> *Forget book learning, physical classrooms, and didactic teaching, even physical books themselves. Brains today learn through Internet interaction, wirelessly at lightning speed and all the time, networked globally across social, political, and geographical boundaries. Scientists aren't sure exactly what that's really doing to our brains, but they're sure it's doing something, and that microprocessors that will WiFi our brains directly to the Internet are next up.*[100]

While there are other core technological enablers in our new environment, computer and communication technologies are leverage points. As we continue to interact and learn in this new environment, there will be increases in interaction speed, frequency, and types; and in this environment we can—and will—resolve dire human issues, prevail against our human inadequacies, prosper in new ways, and exponentially accelerate our mind/brain capacity to learn. And whether this "enriched environment" will prove as beneficial as we would like, or, coupled with development of artificial general intelligence (AGI), overpower the mind has yet to be seen.

The Potential Loss of Deep Knowledge

From the viewpoint of one of the authors, as technological advancements moved into everyday life and social media entered the mainstream, he warned of the potential loss of deep knowledge. This refers to the levels of knowledge (surface, shallow and deep) based on a model developed by the U.S. Department of the Navy. *Surface knowledge* is predominantly but not exclusively information, answering the questions of what, when, where and who—and representing visible choices that are generally easily understood. In the form of information, it can be stored in books and computers and shared explicitly. *Shallow knowledge* is information plus some amount of context below the surface in terms of understanding, meaning, and sense-making such that the knowledge maker can identify cohesion and integration of the information in a manner that makes sense. This meaning can be created via logic, analysis, observation, emotion, reflection, and even—to some extent—prediction. Shallow knowledge, which is primarily social knowledge in nature, emerges (and expands) through interactions as people engage in conversations and dialogue.

Deep knowledge occurs when a knowledge worker has developed and integrated the following components in a specific domain of knowledge: understanding, meaning, insight, creativity, judgment, and the ability to anticipate the outcome of one's actions. It represents the ability to shift frames of reference as the context and situation shift, with the unconscious playing a large role. The source of deep knowledge lies in creativity, intuition, forecasting, experience, pattern recognition and the use of concepts and theories (also important to a limited degree in shallow knowledge). Deep knowledge is the realm of the expert, whose unconscious has learned to detect patterns and evaluate their importance in anticipating the behavior of situations that are too complex for the conscious mind to understand. [101]

In a 2008 MQI study, it appeared that expansion of shallow knowledge was an area of strength for the next generation of knowledge workers (the Net Generation). This was a shift from a primary focus on surface knowledge in an earlier 2000 study. Based on this learning, a nominal representation of this shift is shown in Figure 2 below. In that figure, the representation to the left (Figure 2a) is based on studies in education, organizations, and complexity.[102] The representation to the right (Figure 2b) was speculative based on the recognized trend emerging in 2008 and anticipated social aspects of developing shallow knowledge based on the rapid expansion of social networking with an almost instant exchange of thought and the immediate response to questions.

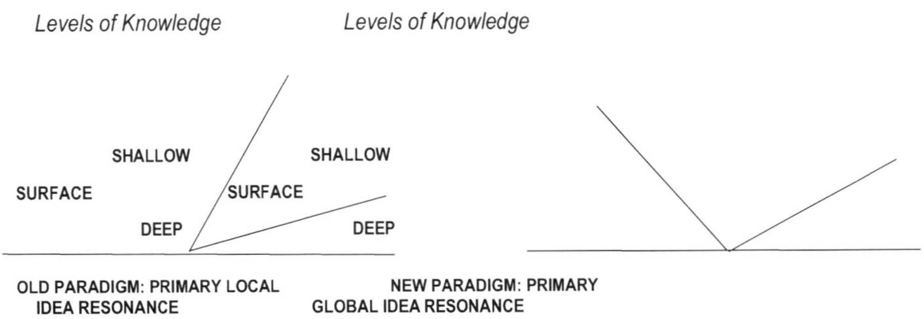

Figure 2a: A nominal graph illustrating the historic (2000) level of knowledge achieved by knowledge workers. Note that these levels are consistent with the level of decisions made in an organization.

Figure 2b: A nominal graph illustrating the 2020 level of knowledge achieved by knowledge workers. The increase in shallow knowledge is a result of consistent expanded interactions via social media.

Figure 2. Nominal shift in focus of levels of knowledge from 2000 to 2020.

For example, consider texting. In a 2008 study, as global connectivity exploded, a shift from a primary focus on surface knowledge was identified. This is where we entered the age of everyday social networking, enabling instant exchanges among individuals versed in diverse domains of knowledge, which allowed for a wider dialogue and sharing of shallow (social) knowledge.

By 2011 in a Pew Research Study, it was reported that youths averaged more than 100 personal texts per day. A 2012 study by Trapnell and Sinclair reported that some youths received or sent upward of 300 personal texts per day.[103] While texting is both interruptive and short, these two traits cause instant attention, and messages right to point which can be—and are—ingested and responded to immediately. By 2016 there were 1,661 billion texts being exchanged in the U.S., and by 2019 that number was up to 2,098 billion. And while the average worldwide user sends or receives 41.5 messages per day, the average American user sends or receives 94 messages per day. To break that down a bit further, according to Nielsen the average teenager sends 3,339 texts per month, which is disproportionately distributed by sex, with the female average 4,050 and the male average 2,539.[104]

At the time of the 2008 MQI study, it was felt that increasing complexity would compel individuals in various domains to develop the deep knowledge necessary to continue technological advancement and co-evolve with an increasingly CUCA environment. Has this occurred? Certainly, for those of us coming from a historic context, complexity continues to expand. However, as will be presented in the Chapter 3 discussion on Singularity, what "we" consider complex is NOT necessarily complex for subsequent generations. Knowledge that was considered "deep" 20 years ago, perhaps resides in the shallows or may have become so well-known that it glides along the surface, visible and available to all.

Chapter 3
Contemplating the Future

The Universal Knowledge Guild ... Transparency, Participation and Collaboration ... Learning and Education in Service to Freedom and Democracy ... The Shift Underway ... The Point of Singularity ... The Future of Work

FIGURES: (3) The shift of humanity underway embracing a new Golden Age; (4) Complexity begets complexity, from which emerges simplicity.

A significant aspect of the mind/brain is its capability to continuously make sense of its environment and anticipate what's coming next. As Buzsaki states,

> *Brains are foretelling devices and their predictive powers emerge from the various rhythms they perpetually generate ... the specific physiological functions of brain rhythms vary from the obvious to the utterly impenetrable.*[105]

In other words, our behavior is closely related to our capacity to form accurate predictions. Anticipating the future is one of the thirteen areas of findings in the ICALS research upon which this book is based. As is forwarded, prediction is seen as a ubiquitous function of the neocortex. **The neocortex constantly tries to predict the next experience.**

This perspective is reinforced by Llinas, who considered predicting the outcome of future events as the most important and common of all global brain functions.[106] The sense of movement of the body provides a simple demonstration of the need—and power—of anticipating the future. Imagine walking down a staircase and accidentally missing a step, recognizing the surprise one has when beginning to fall. Since for thousands of years survival has depended upon humans being capable of anticipating their environment and taking the right actions to survive, perhaps it should be no surprise that this capability has come through the evolution of the brain. As Damasio explains,

> *Survival in a complex environment, that is, efficient management of life regulation, depends on taking the right action, and that, in turn, can be greatly improved by purposeful preview and manipulation of*

images in mind and optimal planning. Consciousness allowed the connection of the two disparate aspects of the process—inner life regulation and image making.[107]

Jeff Hawkins, a computer scientist who was a founder of Palm Computing, a co-creator of the Palm Pilot, and the founder of Redwood Neuroscience Institute that promotes research on memory and cognition, wrote a book called *On Intelligence* that investigates how the mind predicts the future. This is not about long-term forecasting, rather it is about how the brain anticipates the outcomes of its actions. The definition of knowledge—whether you prefer "the capacity to take effective action" or "justified true belief"—clearly requires anticipation of the outcome. As forwarded above, prediction is a ubiquitous function of the neocortex. Hawkins believes that the brain constantly predicts and that what we perceive is not coming only from our senses. Our perceptions are the result of a combination of what we sense, our brain's memory, *and* our capacity to anticipate the outcome of our actions (prediction). As Hawkins explains,

Your brain makes low-level sensory predictions about what it expects to see, hear, and feel at every given moment, and it does so in parallel. All regions of your neocortex are simultaneously trying to predict what their next experience will be.[108]

According to these sources, then, prediction is a primary function of the neocortex and is a part of intelligence, more recently credited to cortical columns.[109] Through experience, the brain creates a rich and complicated model of the world.[110] Hawkins says that intelligence is "measured by the capacity to remember and predict patterns in the world, including language, mathematics, physical properties of objects, and social situations."[111] This conclusion ties in nicely with Kolb's experiential learning concept of comprehension (or abstract conceptualization)[112] and with the definition of knowledge as the capacity to take effective action.

One way the brain anticipates the future is through the process of storing sequences of patterns. Since we never see the same world twice, the brain (as distinct from a computer) does not store exact replicas of past events or memories. Rather, it stores invariant representations. These forms represent the basic source of recognition and understanding of the broader patterns. According to Hawkins, "The problem of understanding how your cortex forms invariant representations remains one of the biggest mysteries in all of science." It is so much so that "no one, not even using the most powerful computers in the world, is able to solve it. And it isn't for a lack of trying."[113] As Kandel explains,

By storing memories in invariant forms, individuals are able to apply memories to situations that are similar but not identical to previous experiences. Cognitive psychologists would describe this as developing an internal representation of the external world, a cognitive map that generates a meaningful image or interpretation of our experience.[114]

While it has historically been believed that the brain stores patterns in a hierarchical and nested fashion—and certainly some patterns are stored in this manner—it is also now recognized that patterns (as information and potential knowledge) are stored at locations relative to reference frames,[115] which are located in the cortical columns in the neocortex. Recall that thoughts are represented by patterns of neuronal firings, their synaptic connections, and the strengths between the synaptic spaces. For example, a single thought could be represented in the brain by a network of a million neurons, with each neuron connecting to anywhere from 1 to 10,000 other neurons.[116] Incoming external information (new information) is mixed, or associated, with internal information, creating new neuronal patterns that may represent understanding, meaning, and/or the anticipation of the consequences of actions, in other words, knowledge.[117] *Associative patterning* describes this continuous process of learning by complexing incoming information with stored core patterns—which are stored at locations relative to reference frames—and creating new patterns in the mind pertinent to the situation at hand.[118] As an example, the mind/brain could not possibly do all the calculations to compute the movements needed for a baseball outfielder to run and catch a baseball. Instead, the sighting of the ball automatically connects to the appropriate memory, which is then adjusted to meet the specific situation at hand.[119] In this case, the outfielder's movements are guided by a simple heuristic: Keep the angle between the ball and the outfielder constant.

Occurring through the neocortex, and specifically in the cortical columns throughout the neocortex, anticipation (prediction) of the future is one aspect of the human operation of mind that enhances the abstract conceptualization learning process. From the viewpoint of the mind/brain, any knowledge that is being "re-used" is actually being "re-created" and, in an area of continuing interest, most likely complexed over and over again as a continuous flow of incoming information is associated with internally stored information. This phenomenon of relating external and internal forms of experience is called "appresentation" and is an example of the mind's search for meaning.[120] Moon, an educator and expert on reflective and experiential learning, explains,

> *Appresentation is the manner in which a part of something that is perceived as an external experience can stimulate a much more complete or richer internal experience of the 'whole' of that thing to be conjured up.*[121]

We address this concept in relationship to mirror neurons in Chapter 9.

The Universal Knowledge Guild

The first gathering of what would laughingly be called the Universal Knowledge Guild began on December 18, 2012, at the Mountain Quest Institute. We were gathered around the conference table in the great room, having spent the early part of the morning exploring the research library. Through the windows—regardless of the direction you looked—could be seen the rounded hills of the Allegheny Mountains of West Virginia.

The gathering had originally been scheduled for two days, but at the last minute, Dr. Franc Calabrisi, one of the participants coming from George Washington University, shot an email to the group suggesting extending the visit an extra day. And here is what another participant, Dr. John Lewis, responded,

> *After some deep thought, I agree with Franc's wisdom on staying over another day. I just reread the Institute's mission statement. If there was ever a time for a meeting of the minds, it would be now. Our ability to articulate the central importance that learning and knowledge and innovation management play in the productivity and survival of an organization, even a nation, as a collective mind share is more important than any one of our individual efforts. I want to have the time to learn, share, and then move from the interesting aspects to the productive aspects.*

The challenge posed to the group: With all this CUCA activity, how are we going to develop learning aids for the future when we don't know what the future looks like? An age-old question, really, one asked by any futurist with our little twist on learning! And from a larger frame of reference, how can we help humanity move to the next level, whatever that means? Are those same questions ever pertinent today!

Those sitting around the table included a scientist, a philosopher, a professor, a mathematician, an engineer, an educator, and a psychologist. Disjointed thoughts emerged from the ensuing conversation, becoming ideas and coalescing into a connected whole. Of course, it was first necessary to agree on assumptions, which in itself was an enlightening experience! Here's what emerged:

1. We were currently operating in a knowledge economy, sharing a deep appreciation for the collaborative entanglement of people and their communities, of organizations and their competitors.

2. We fully appreciate the values of transparency, participation and collaboration rapidly becoming embraced throughout the developed world.

3. We recognize the human as holistic (physical, mental, emotional, spiritual) and the power of the mind/brain; that each person is sovereign and has choice.

Today we might wish to add something further about the state of the world, and it is clear that there are many people who do not fully appreciate global collaborative entanglement. And if we look at number 2 above, there has been a downward spiral in this regard, with truth and trust becoming personal judgments very much linked to political and economic considerations.

Agreeing on these assumptions in 2012, the guild next focused on the critical elements, little pieces of understanding that would drive our modeling of the future. Out of the pursuing conversation seven elements emerged:

First, the sense of, "We are here." I recall standing out under the country sky situated here on the edge of the Milky Way and thinking the thought, "I am here." Other places were no longer important; I didn't require difference to have appreciation, no lessons to dampen my full expression. So, this "We are here" is a *presencing* that does not require the duality of black and white, good, and bad. An awareness, a consciousness of being present. Presencing is a critical capability in a CUCA environment.

Second, that creativity is based on "Unk Unks" (that's short for "unknown unknowns"). This is all the stuff that's going on within us of which most people are unaware! So much knowledge and knowing stored up in our minds and in our bodies. This is an honoring of our *Selves*, what we each know that we don't know we know.

Third, the idea of a memory shift underway. Things are so complex that we no longer can afford to hold on to all the unfruitful thought and emotion from the past. Rather, ideas are floating in and out continuously, and we want to grab hold of the ones that resonate, connecting with others with whom they also resonate, and explore *those* ideas!

Fourth, the concept of simple elegance. We don't have to be burdened down by the things we own, tethered by the weight of overwhelming commitments. Time to simplify, discovering the elegance of a simpler way of living.

Fifth, the value of diversity. We've had our lessons in this as a humanity. Diversity is the spark for ideas, the action power for intelligent activity. This means honoring each and every individual for the unique person they are.

Sixth, the idea of the singularity. We've reached a tipping point, and things are changing. Our job is not to try and hold it back, but to stay open to the amazing energy patterns of thought emerging from our expanding social engagement. We are all connected. We are one world.

Seventh, the human drive for—and deserving of—physical, mental, emotional, and spiritual freedom. Ultimately this brings us full circle to "choice". "We are here and we have choice."

Since leaping from these assumptions and critical elements to the world of the future seemed impossible, it was necessary to take incremental steps. The world of 2010? Still very much caught up in duality, WIFM (What's in it for me?), control and separateness, although at that time it appeared much headway had occurred from the early power domination and bureaucratic models. And, admittedly, at this point there was an emerging focus on participation, collaboration, and transparency, although there's been a waxing and waning of trust in government media since that point occurred, which happens whenever there is a shift of those with power. And there was likely no greater shift than that between the 2008/2012 and 2016 administrations in the United States and, again, between the 2016 and 2020 administrations. In 2009, one of the hallmark achievements aimed at building trust in U.S. government was the Open Government Directive, which was committed to government "for the people", ensuring the engagement of both government and private sector expertise, and the sharing of that expertise among people through the three principles of transparency, participation, and collaboration.[122]

Transparency, Participation and Collaboration

In 2010, Avedisian and Bennet published a paper titled "Values as Knowledge: A New Frame of Reference for a New Generation of Knowledge Workers" which explored fundamental and operational values emerging in the new generation.[123] Eight values were identified, three of which were transparency, participation, and collaboration. The other five values identified as important to the new generation were integrity, empathy, contribution, learning, and creativity.

Transparency, an operational value, is *being candid or open, easily seen through or detected, and free from guile.*[124] In an extensive study of the Net Generation also known as Millennials—the generation rapidly becoming the key decision-makers in today's organizations, who are internet savvy and engage heavily in social media—Don Tapscott, co-founder of the Blockchain Research Institute, identified transparency as a core value which is *critical to establishing trusting, long-term relationships.*[125] Emerging from this research, Tapscott forwards that true transparency "must make the processes, underlying assumptions, and political presuppositions (including supporting research) of policy explicit and subject to criticism."[126] This moves far beyond sharing surface knowledge on websites to including the open sharing of ideas, feelings and personal view points. Now it becomes increasingly clear as we accelerate learning about the learning brain, that the values identified are key enablers of effective learning itself. Indeed, they are essential to unleashing the human mind.

Participation, also an operational value, can be called a keystone for the Net Generation, who reach out worldwide to creatively engage people and ideas, extending to both community service and political engagement. For example, in 2004 in the first U.S. Presidential election where the Net Generation was coming of age, more people cast votes under the age of 30 than over 65, with the largest increase in the 18-24 age group. At that time, a group of researchers forecast, "Signs indicate that Millennials are civic-minded, politically engaged, and hold values long associated with progressives, such as concern about economic inequalities … and a strong belief in government."[127] Similarly, in a 2006 report, the Corporation for

National and Community Service found that teens 16 to 19 years of age were volunteering twice as much as in 1989.[128] Note that these descriptions *reflect a coherence between the heart and the mind*, that is, intelligent caring about others, and acting on that caring (see Chapter 8).

In a 2017 G-LINK conference on "Leading Digital and Cultural Transformation" held in Bangkok, Thailand, a research study was presented which highlighted the diversity of opinion around the unfolding culture of connection.[129] There was a clear distinction in the responses from different generations. From the Baby Boomers—the generation who entered a world throbbing with materialistic potential and promise—there were fears and feelings voiced that global connectivity facilitates invasion of privacy, energizes loss of individuality, reduces deep communication, intensifies paranoia, and enables bully leadership. Conversely, from the Millennials—the generation not constrained by old concepts and prejudices—there was clear excitement about a culture which invigorates creativity, illuminates understanding, promotes mutual influence, expedites problem solving, and eliminates cultural boundaries. Note that for the Centennials, the internet is—and always has been—the norm, thus they have always lived in a culture of connection. This study affirms the findings and forecasting of other researchers and provides a depth of understanding in terms of the voting outcomes of the U.S. 2020 Presidential election in younger and older populations.

Then in 2022, howbeit with perceived cause, Russia, a major world power, invaded Ukraine, a sovereign country, with the people of Ukraine rising against superior forces and weapons to protect their homeland and their freedom. As the western world watched—staying on the fringes of this war, not directly engaging, in an attempt to prevent WW III while supporting the people of Ukraine in terms of weapons, economic and political sanctions, and humanitarian aid—a shift began that potentially has wide ramifications for the future of humanity. The EU, NATO, and UN as individuated organizations aligned their entangled relationships in terms of thoughts and actions. Russia itself—despite attempting to close off all outside social media, and shutting down any news media not regurgitating Putin's words—had a swath of the population that clearly recognized the power play of the aggressive and senseless war being waged on Ukraine, the resulting eradication of personal and national infrastructure, and the annihilation of life. And whether we refer to the divisiveness of war or depraved political ambitions occurring in the U.S. and other countries around the world, a question that continuously emerges pertinent to managing self is: Despite the efforts by despots to control the media and how people think, how long will it take in a connected world for truth to more fully emerge and prevail?

Collaboration means to work together, especially in a joint effort.[130] In a culture of connection, collaboration is not limited to internal groups at the team, unit, or business level, but potentially reaches out to a fluid, ever-changing, interdependent network of diverse experts around the world. This delivers a new type of peer network to decision-makers, one that moves from autonomy to interdependence and from deference to dialogue. And the primary focus moves from doing a job well to making a contribution to collective purposes.[131] In this type of network, the values of collaboration, transparency and contribution make it possible to successfully work in open, changing and diverse environments. Fortunately for the shift underway, collaboration—which involves both engagement and participation—is a core value embraced by the Net Geners. As Tapscott says, "Collaboration as Net Geners know it, is achieving something *with* other people, experiencing power through other people, not by ordering a gaggle of followers to do your bidding."[132] And as Panetta corroborates, "Collaboration and communication are second nature for the Millennial generation." [133]

The U.S. Open Government Directive charged government organizations—and, by extension, all those organizations in the private, nonprofit and education sectors who support the government—with specific directions for achieving behavior changes to support transparency, participation, and collaboration. A starting point was expanding access to information, making it available in open formats online, and developing policy to support the use of emerging technologies. Concurrent with release of the Open Government Directive, the U.S. Attorney General issued new guidelines under the Freedom of Information Act (FOIA), reinforcing openness as the Federal Government's default position. As Departmental directives and guidelines were issued consistent with this higher-level directive, individual behaviors began to change, which over time became part of the way work was done, the culture. And as individuals in resonance with these principles repeated actions over and over again, these organizational values became personal values.[134]

This democratization of information content continued with the launching in 2009 of data.gov, the official U.S. government site aimed at providing increased public access to government datasets, which was managed by the General Services Administration. While the drivers were to meet regulatory compliance and better communicate with citizens and stakeholders, and empower better decision-making, there was also a focus on helping create new businesses with open data. Countries were encouraged to share data to create new economic development and kickstart innovation. The website was created as a data ecosystem, which includes a simple and continuous five-step process: (1) gathering data from multiple sources and sharing it freely; (2) connecting the community, facilitating collaboration through social media, events and platforms; (3) providing an infrastructure built on standards and interoperability; (4) encouraging technology developers to create applications, maps and visualizations to empower people's choices; and (5) gathering more data and connecting more people.[135]

The site took off. By 2013, there were 400,000 datasets in easy-to-use format, with contributions from 180 agencies, the UN, and the World Bank. Open communities were supported on safety, energy, health, law, education, manufacturing, ethics, sustainable supply chains, the ocean, and many more areas, connecting innovators, industry, academia, and government at the federal, state, and local levels. There were recognizable successes. Weather-related businesses were stoked by NOAA's data, and when satellite data was released by the Department of Defense, private industry was able to create affordable GPS devices. With the slogan of "Let's work together to set our data free!", data.gov began a global movement to provide transparency and democratization of data, which by 2013 including 40 U.S. states, 21 U.S. cities and counties, 39 international countries, and 115 international regional areas.

Despite counter forces launched during the Trump era, guidance supporting open government data continued to emerge at the Congressional committee level. The OPEN Government Data Act, sent to the House committee on Oversight and Government Reform in March 2017, remains in committee. However, the Foundations for Evidence-based Policymaking Act of 2018 (the "Evidence Act") was signed into law in January 2019, with an emphasis on coordination and collaboration in order to advance both open government data and Federal evidence-building functions. Meanwhile, while the OPEN Government Data Act was not law, the Office of Management and Budget, the Office of Government Information Services of the National Archives, and the General Services Administration collectively developed and launched in July 2019 a repository of Federal Enterprise Data Resources,

resources.data.gov, to *provide guidance and tools for implementation of OPEN Government.* Thus, these remain important values to achieve in today's world for information and digital literacy, for political freedom, and for effective learning.[136]

Learning and Education in Service to Freedom and Democracy

Now, turn with us briefly to learning and education in the matter of empowering relationships between governments, societies, cultures, and populations, and the function of our human mind/brain use and development. In Plymouth, Massachusetts, on Plymouth Bay where the full-size replica of the Mayflower rests at anchor, you can take a reasonable walk from the Plymouth Rock to a nearby icon of the great American experience. There, in a quiet neighborhood, you will behold an amazing memorial to the founding of the nation. The 81-foot-tall monument resides as the United States' largest freestanding solid granite statute and, in its unique majesty, it presides over what may be one of the most insightful stone engraved messages in human history. What you find before you is *The National Monument to the Forefathers.* The message that is conceived in the Pilgrim's miraculous undertaking resonates across time as the covenant that supported the creation of an unprecedented democratic endeavor.

All of the monument's figures, symbolism, and inscriptions invite careful consideration; nevertheless, here we refer to only one portion of the monument's message. The other portions of the message and their interrelationships with the one we share are indeed interwoven. The central image of the monument is the figure Faith reaching Heavenward, then there are four figures surrounding her which represent the key pillars the pilgrims regarded for governance. They are Morality, Law, Education, and Liberty. Education is our focus here, but you could select any of these to consider their regard for its influence on the creation and preservation of a free society.

The Statute of Education is a celebration of their new found liberty and the opportunity to read, discuss, teach, and learn, especially with their youth, those things which lead to wisdom among the citizens. It is particularly relevant that one of the most valued and prominent artifacts of the Mayflower cargo was the Geneva Bible, which was not only available in English, but also contained a substantial volume of footnotes. This book was a catalyst for education and learning and for thinking, freedom-loving citizenry. What we find with these forefathers is a manifestation of the Reformation. Not only was their desire to know about the world for themselves, they also wanted the ability to learn and think about it deeply, which they knew required education.

This emphasis on education in general as a basis for learning and thinking engendered regard for liberal education in the new world. Over time a liberal education orientation was adopted and championed extensively as fundamental to education curriculums. In general, it provided a well-rounded approach for exploring subjects and disciplines, presenting opportunities to sample potential interests, and fostering a wider range of communication, working, learning, and thinking skills. Liberal education should also be recognized for encouraging lifelong learning, which is increasingly essential for continuous professional development and personal edification.

In the middle of the 20th Century, a landmark undertaking was conducted by a select group of American and international educators. The result was a compilation of *Great Books of the Western World.* The 54-volume collection was introduced by Volume 1 entitled *The Great Conversation.*[137] In this volume, perhaps the deepest and most regarded observations are the many explanations regarding the value of education to a democratic nation and specifically the impact of liberal education on safeguarding and strengthening the thinking of its citizens. A focal point of concern is expressed in this rather strong admonition.

We believe that the reduction of the citizen to an object of propaganda, private and public, is one of the greatest dangers to democracy. A prevalent notion is that the great mass of the people cannot understand and cannot form an independent judgment upon any matter; they cannot be educated, in the sense of developing their intellectual powers, but they can be bamboozled. The reiteration of slogans, the distortions of the news, the great storm of propaganda that beats upon the citizens twenty-four hours a day all his life long mean either that democracy must fall prey to the loudest and most persistent propagandists or that the people must save themselves by strengthening their minds so that they can appraise the issues for themselves.

Not only was education regarded as a guardian of freedom and democracy, it was inherently revered as a pillar of human capacity. Again, in *The Great Conversation,* the value of education for all is extensively discussed and well established. In brief, it is characterized to this degree of worth:

The aim of liberal education is human excellence, both private and public (for man is a political animal). Its object is the excellence of man as man and man as citizen."[138]

On the innovative and economic value of education, a poignant observation was made by Peter F. Drucker, an icon of modern management. He wrote,

The postwar [WWII] GI Bill of Rights—and the enthusiastic response to it on the part of America's veterans—signaled the shift to the knowledge society. Future historians may consider it the most important event of the twentieth century. We are clearly in the midst of this transformation; indeed, if history is any guide, it will not be completed until 2010 or 2020. But already it has changed the political, economic and moral landscape of the world.[139]

We clearly recognize, in this third decade of the 21st Century, prolific gains in so many domains of human civilization, the great advances of our education systems, our learning resources, and our thinking. Nevertheless, there are increasing concerns about the relatively new and pervasive dangerous impact of misinformation, the proliferation of which is accelerated by the powerful reaches of our advanced communication and information technologies. While the internet, cellular, and broadcast technologies change the fiber of human interaction, learning, and thinking they also present distraction, deception, distrust, deviation, and destruction.

Now, in our learning and thinking we have to become vigilant to soundbites and communication patterns that undermine how our brain serves us. Of particular concern in recent decades is the invasion of our networks of communication and interaction by organized operations from foreign governments whose primary purpose is to penetrate and undermine our cultures and create chaos. Tragically, within our own country some key politicians not only avoid addressing the onslaught, but also use the same technologies and processes to wield power, deception, and manipulation.

Some nations have become acutely aware of this and have planned and organized national programs to safeguard against misinformation and fake news and sustain effective thinking and learning. In that regard, we would invite your attention to the extent of research and tracking of trends and impacts. An excellent example is the success in Finland with tackling misinformation, as reported by the World Economic Forum:

> *Finland has an effective weapon to combat fake news: education. The Nordic nation tops a list of (35) European countries deemed the most resilient to disinformation, the Media Literacy Index, compiled by the Open Society Institute in Sofia ... "We need to train a new generation of critical minds", Jean-Pierre Bourguignon, President of the European Research Council told the World Economic Forum's Annual Meeting of the New Champions in September. 'We must tackle this issue through improved literacy, and it is the task of our educators and society at large to teach children how to use doubt intelligently and to understand that uncertainty can be quantified and measured.*
> 140

Jurgen Renn in his work *The Evolution of Knowledge*--looking at the structure and spread of knowledge across time—provides a succinct caution at length that is summarized here:

The future of humanity will depend on its ability to preserve and develop its shared resources: natural resources such as clean air and water, sources of food and energy, and cultural resources like infrastructure for mobility, communication, and the dissemination of knowledge. Many of these are common-pool resources and are thus endangered by the 'tragedy of the commons' which is to say their overuse and potential destruction owing to the tendency of individuals to maximize their own benefits at the cost of the common good.[141]

The Shift Underway

To look toward the future, the Universal Knowledge Guild agreed that it was necessary to let go of the 2010 status quo, breaking the bonds of the past to live in the NOW (a presencing), although there was no doubt that participation, collaboration, and transparency were laying the groundwork for this shift.

The early 2020's have seen a decline in global collaboration, largely due to political wrangling and power plays. Still, as Millennials move into positions of authority in and out of government and private industry, new attitudes are emerging which, in turn, promise new decisions and actions. While Baby Boomers still hold tightly to the higher seats of power in the government—including in the U.S. Congress, which has no age or term limits—there are increasing numbers of Millennials, most with progressive ideas, who are seeking and filling those positions as they become available. In other words, a new generation with new ideas and experiences is coming to the fore, and the potential for large-scale change is afoot. Simultaneously, the breaking open of historical patterns and beliefs facilitated through the pandemic and life-and-death political and geographical power plays offer the opportunity for each and every human to consider their choices in terms of the future they wish to embrace.

The construction of a global culture, a culture of connection, and perhaps of coherence, is still a possibility. In fact, as boundaries are released between nations and connectivity continues to expand—given humanity can move beyond political and economic divisiveness—transparency plays an even larger role. With transparency comes a need for authenticity and global truth, greater freedom of thought for all, and the desire and opportunity for choice that leads to intelligent activity. With a feeling of oneness riding on the coattails of connection, compassion emerges. This is reflective of the model developed by the Universal Knowledge Guild, howbeit without awareness of the backtracking and pitfalls currently presenting to the world. However, while acknowledging a level of raised consciousness occurring in 2012—the year where many of those with a spiritual nature had expectations of an external event causing peace, love, and harmony to enter humanity at an instant in time (supposedly correlated with the end of the Mayan calendar, December 21)—the guild had enough sense not to tie future conscious expansion levels to specific years. See Figure 3 on the following page.

What this graphic depicts—for those who are ready—is that with connectivity and the maturing focus on knowledge in a global world comes the understanding that people are at the core of our organizations and communities, bringing with them the power of diversity and individuation as they join hands to dream the future. A raising of awareness is coupled with knowledge resonance, that is, the free flow of ideas around the world, offering a bouquet of choices for each person.

Coupled with an openness to new ideas and learning is *thought attendance*, that is, a rising awareness that form follows thought, a consciousness of our own consciousness, and the thoughts and feelings emerging from self. With that awareness empathy emerges, a stronger, more intimate form of feeling and connection with the other, while in the larger population a resonance occurs, with new ideas precognating the coming renaissance, a new Golden Age of Humanity.

When we paint the future with our words, beautiful thoughts stream their way into our awareness, popping up again and again, connecting themselves to other thoughts. As we look at this model, it is clear that these ideas are woven throughout this book. Perhaps that is not surprising, although what could not have been anticipated—or at least were not in mind—are the global natural disasters that would tear open the fragile economic, political, and ecological systems within which we live. However, these extremes are emerging leaders around the world who seek coherence.

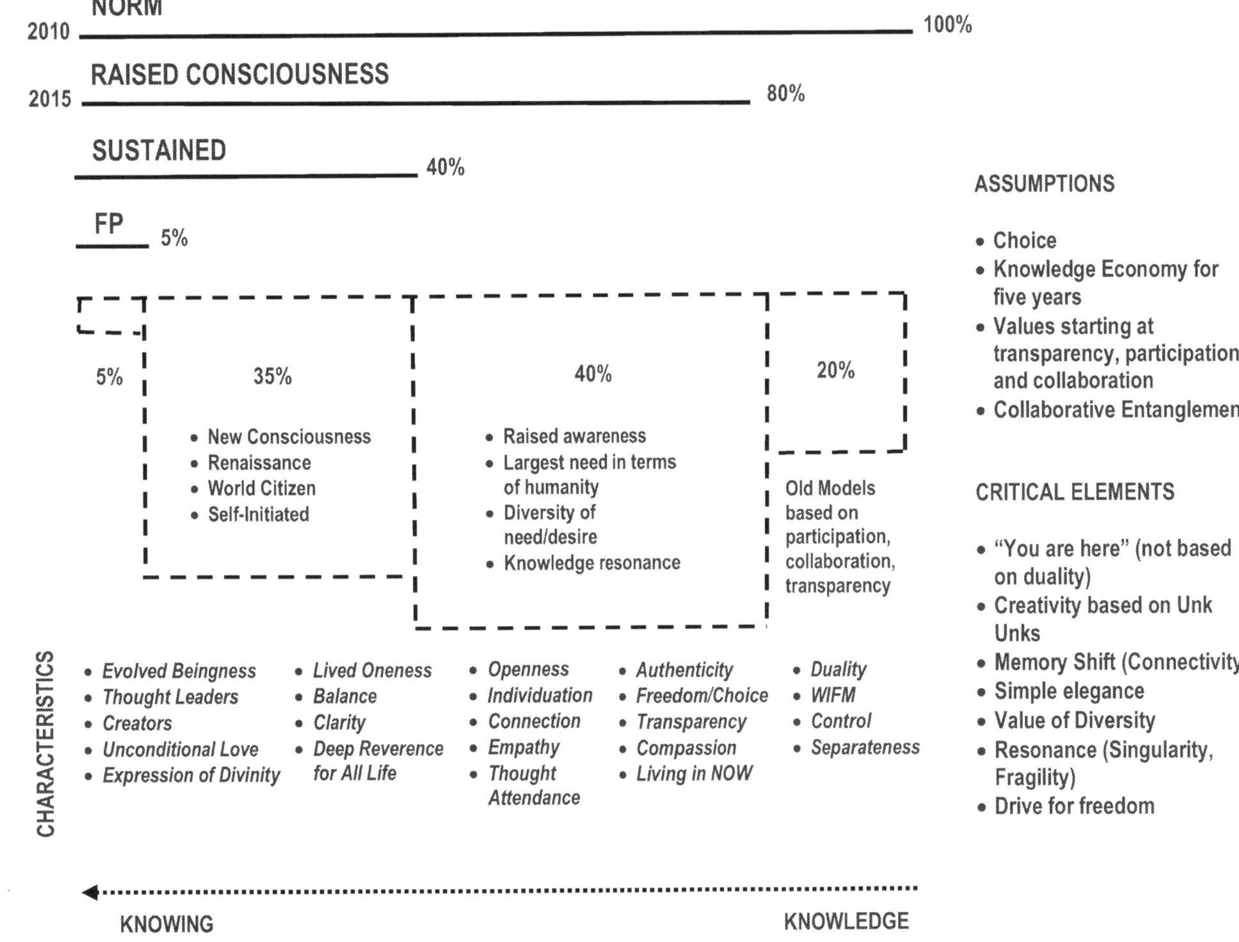

Figure 3. The shift of humanity underway embracing a new Golden Age.

In this new era, knowledge workers are no longer sacred. Today there are so many advanced technologies enabling natural language capacities and voice and facial recognition engaging in pattern recognition and deductive reasoning, and the kinks will inevitably be worked through! The bottom line is that in most fields of endeavor, humans will be less necessary for historical value creation. As smart machines increasingly take over tasks previously considered as handled exclusively by people, there are several areas which demand serious attention. Two of these we will touch on lightly: (1) the Singularity and (2) what the human-AI tradeoff means to the future of work.

The Point of Singularity

In 2006, Ray Kurzweil published his quick-to-become a best-selling book, *The Singularity is Near: When Humans Transcend Biology*.[142] Filled with imaginative speculation, the idea was that Artificial Intelligence (AI) technology would soon outdistance human intelligence, that at some hypothetical point in time technological growth would become uncontrollable (and irreversible), which would result in unforeseeable changes to human civilization. The book goes into detail on genetics (the intersection of information and biology), nanotechnology (the intersection of information and the physical world), robotics (as strong AI), and humans as cyborgs (with technology transforming the body and brain in terms of longevity, warfare, learning, work, and play).

When would this occur? Different futurists thrust out different timelines, some of which have come and gone, and others still lingering in the future. Nonetheless, in 2014, the inevitability of a technological singularity was addressed by Luca Maria Aiello in the *MIT Technology Review*. We agree. From the authors' point of view, it is most likely that the Singularity will occur in concert with a human consciousness shift—which is underway now—as part of our human journey, the instant where we surpass our current level of thinking, whether expressed through technology or through the expansion of our consciousness and the mastering of our mind/brain. Only, perhaps our interpretation of what is happening may play out differently than expected.

Historically, we as humans work and strive to create change with only slightly visible results, then some event occurs that connects all this prior activity, and the understanding of value pushes everyone to a new stratum of recognition, with the entire plane of behavior shifting upward to a new starting point. One of the authors had a related personal experience when taking a six-week summer course in calculus in preparation for a decision-making series that started in the fall semester. No one can learn calculus in six weeks, much less someone returning to school after a ten-year plus absence and never having had pre-calculus or any other similar prep course! After struggling to understand what was being presented—and following some good advice—the author just memorized logical, repetitive relationships. *If this, then do this, etc.* So, lots of memorization was packed into what felt like the very limited space of mind, but no understanding. Still, there was the resultant passing grade, along with some pride for passing! Almost a year later, while in the second decision-making course in the series, fully understanding the decision situation and context, the instructor drew a differential equation to explain the situation. In that instant, an amazing thing happened. The author felt an actual "swoosh"—an expanding of the upper part of the mind/brain—and understood the equation! A good analogy for this experience is the growth of bamboo. For the first four years the young bamboo plant is watered with relatively little visible evidence of growth. But during this time, out of sight the roots are spreading, interconnecting, and growing in strength. Then, during the fifth year, the bamboo plant streaks upward some 20 or more feet.

Humanity is at the sprouting point in terms of our expanding consciousness. We've done a great deal of work growing our mental faculties and learning through our connections to others. We look through the lens of Ken Wilber to further explore this transition. Wilber says that humanity—both as individuals and as members of a historically located culture—is undergoing a process of psychospiritual development.[143] This transition is not much different than the shifting developmental stages each of us experiences in life, that is, moving from a symbiotic state with our mother to separation; from focus on an individual body to membership in a group; and then into development of the mental ego, from where we start the Intelligent Social Change Journey introduced in Chapter 1, moving from the logic of lower mental thinking to the concepts of higher mental thinking (which is further detailed in Chapter 5).

There are two dimensions that Wilber says are necessary to transition from one stage to another: *the creative urge* and *the willingness to let go*, that is, be open to new thoughts and experiences.[144] Note the similarity of these dimensions to those required for innovation! This is not surprising. Certainly, the beginning of this transition is the global network of technology facilitating the movement of information around the world and enriching the creative field from which innovation emerges. As introduced above, coupled to this capacity are the attitudes of a new generation of decision-makers who are growing up green and growing up connected, and who are not satisfied with the world as it is.[145] Jean Houston, scholar, philosopher, and teacher, is more dramatic in her description of the choice upon us. She says that this time in Earth's history addresses the imperative "grow or die". As she quite eloquently describes:

> *We are guests at a wake for a way of being that has been ours for hundreds, even thousands of years .... Our challenge is to cultivate the vision and lay out the practical steps necessary to move through the opening times that follow upon closing times .... the new millennium we have entered is the intersection between worlds, between species, between ourselves and forever .... we as a species stand at a crossroads faced with radical choices, any of which promise to make tomorrow look nothing like yesterday.*[146]

That wake is turbulent, an ending—breaking apart and crumbling—of all that we have known, the future that we expected. We are beyond our comfort level moving toward the unknown. Yet from this very place, when we look inside, we can perceive great potential, unlimited possibilities. We've entered the *Jump Time* that Houston describes as "a whole system transition, a condition of interactive change that affects every aspect of life as we know it."[147] Recognizing that virtually every human system—organizational, economic, environmental, political—is undergoing a state of deconstruction and breakdown, in this journey *the choices of each individual make a profound difference.* "Our individual life is part of the unfinished symphony of the cosmos."[148]

Considering the expansion of consciousness juxtaposed with what we have described as the point of Singularity, let's see what that might look like as a graphic. In Figure 4 (displayed on page 59), we show the relationship of the expanding CUCA environment introduced in Chapter 2 (increasing change, uncertainty, complexity, and anxiety) and the creative leap that represents an *expanded state or level of consciousness.* In the current environment (lower left of figure), we are embedded in ever-increasing complexity, opining about how much simpler things were in he past. The older you are, the more you can identify with that concept. The topics in the text box in the center of the graphic are all explored deeply in *The Profundity and Bifurcation of Change* five-book series, all of which play a role in expanding our consciousness.

An insight is that the worldview at this expanded level of consciousness, represented by the upper pair of glasses, is quite different, such that the current condition of the world—the way things are—is simpler, providing (for a new generation) a new "norm", a new starting point for learning and increasing complexity. From that viewpoint, we experience the total complexity of today as chaste simplicity. *And we begin our journey of expansion anew.* Focusing and creating from this *new place of learning and expansion*, there is a new point of singularity to move towards.

As can be seen, it is insinuated in this graphic that *this shift is coincidental with the point of Singularity*, a time when humans have advanced technology so far and fast that it enables humans to transcend their biological limits. Recognizing the current human limits to fully understanding the capabilities and possibilities of the human mind/brain, we contend that the Singularity, which we agree could happen in the foreseeable future*, is thought of and defined by our **current** limitations*. Thinking from the quantum field perspective, we recognize that behind a closed door "all that is possible exists", and that prior to opening the door, what is behind that door is potential. Once that door is open, our frame of reference has shifted such that what "is possible" *after* the door is opened can be quite different than what "is possible" *before* the door is opened. This is the realm of *incluessence*, that which is beyond our ability to imagine or dream.[149]

What is expressed here is that the Singularity will most likely occur in concert with our consciousness shift, and indeed this is part of our human journey, the instant where we surpass our current level of thinking, whether expressed through technology or through the expansion of our consciousness and mastering of our mind/brain. As forwarded and shown in Figure 4, focusing and creating from this *new place of learning and expansion*, there is a new point of Singularity to move towards.

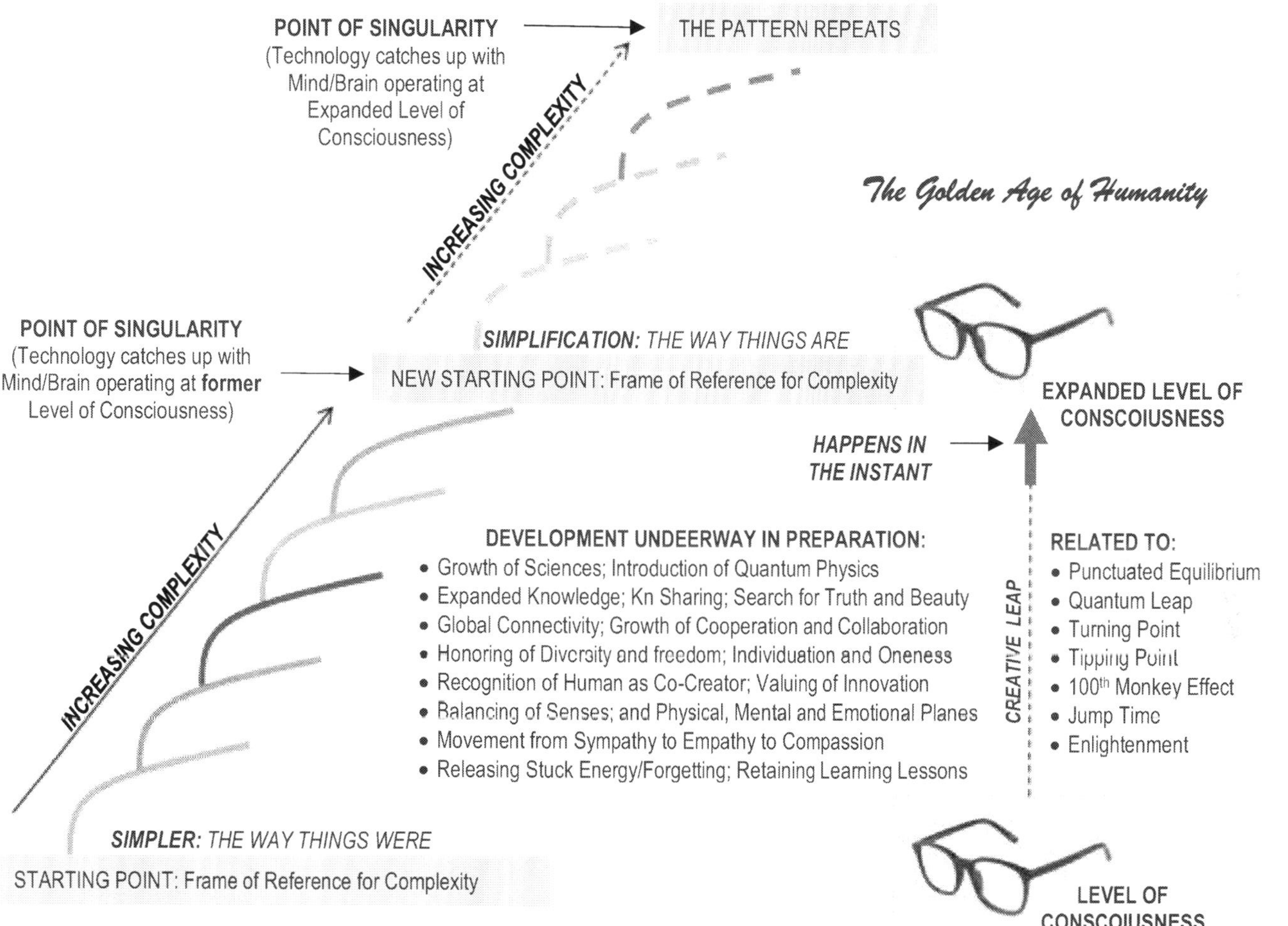

Figure 4. Complexity begets complexity, from which emerges simplicity.

The Future of Work

There is no question that technological advancements—particularly AI and human augmentation, and the potential of developing artificial general intelligence (AGI), which would more fully exhibit human-like intelligence—are transformational at all levels of the human experience. We agree with Kissinger, Schmidt, and Huttenlocher when they advocate that at the core this "transformation will ultimately occur at the philosophical level, transforming how humans understand reality and our role within it",[150] all driven by the human mind.

In 2018, the MIT Task Force on the Work of the Future was commissioned to understand the relationship between work and emerging technologies, specifically to explore realistic expectations and future strategies to enable a "shared prosperity".[151] It is important to recognize that while technology continues to have momentous impacts on our lives, the unfolding of robotics and AI is an evolving, gradual process. Even spectacular advances reflect substantial time lags from discovery to widespread adoption.[152] In recent research, Autor and his colleagues found no "compelling historical or contemporary evidence" that suggested technological advancements would make humans jobless.

> *On the contrary, we anticipate that in the next two decades, industrialized countries will have more job openings than workers to fill them, and that robotics and automation will play an increasingly crucial role in closing these gaps.*[153]

This does not mean that AI and automation won't continue to impact the *types* of jobs and skills needed by humans to move into the future, which are very much dependent on economic incentives as well as policies and decisions put forward by public and private organizations. In the development of various technologies—both initially and as they mature—human choices are necessary to direct the outcomes, with engineers encoding "social relationships and preferred futures into the machines they build."[154] Further, a reliance on human-generated data also produces biases, which adds "noise" to the system that can only be recognized and mitigated by human intervention.

Acknowledging that humans are complex adaptive systems, with the mind/brain designed to adapt to new emerging situations—a depth of ability and flexibility which to date has not been replicated through AI—humans will remain necessary in decision processes, with the use of judgment very much involved with emotions (see Chapter 8). Humans also excel at a variety of other tasks. As Autor and his fellow researchers found, "Humans still excel at social interaction, unpredictable physical skills, common sense, and, of course, general intelligence."[155]

The self is core in the ICALS research, and managing self is the underlying theme of this book, with the success factors of critical thinking, creative thinking, innovative thinking, social engagement, attention/intention, morality, and humility, all connected with a deeper understanding of the mind/brain and the human potential unleashed through that understanding. Similarly, in a recent offering developing a theory of hyper learning—the idea of reinventing ourselves again and again—Hess writes that, consistent with our success factors, the human tasks are,

> *... exploring the unknown and novelty by being creative, imaginative, and innovative; engaging in higher-level critical thinking; making decisions in environments with lots of uncertainty and little data; and connecting with other human beings through high emotional engagement and effective collaboration.*[156]

In other words, at least for the present and near future, thinking in ways that computers can't think such as developing models of the world through reference frames, applying what is learned from one task to another, knowledge representation (representing facts in useful ways), and the shifts engaged in Knowledge Capacities (see Chapter 5). As you will see, this thought is highly complementary to the findings in the ICALS study and subsequent development of the ICALS model.

In *The Profundity and Bifurcation of Change (Part V): Living the Future,* the "bifurcation" is represented as a split in the road, a point of choice about who we—as a humanity and as individuals—are becoming, not who we are.[157] Acknowledging the duality of human existence, the question is posed: *Are such negative experiences as death, survival, and fear necessary to sustain the depth of experiential learning happening today?* Considering the political and economic upheavals, the threat of climate change, all interwoven with a global pandemic accompanied by a rising death toll, never was that question more meaningful. The *Profundity and Bifurcation* text suggests that this is a human choice, that conflict is no longer required to choose change, but rather *the excitement of exploring the unknown* can drive us forward with the attractors of exploring, discovery, anticipation, testing ourselves, creating, love of adventure, curiosity, innovating, and challenge. THESE are the qualities that humans bring to the world, which, in a CUCA environment, cannot—at least for the foreseeable future—be replicated by machines.

In their book, *Humility is the New Smart: Rethinking Human Excellence in the Smart Machine Age*—which we recommend strongly—Hess and Ludwig see humility as a mindset that "results in not being so self-centered, ego defensive, self-enhancing, self-promotional, and closed-minded—all of which the science of learning and cognition shows inhibit excellence at higher-order thinking and emotionally engaging with others."[158] We agree. As working definitions for this discussion, which is extended in Chapter 11 on humility, we refer to the work of Joshua Hook.[159] At the individual level, humility is having an accurate view of yourself—not too high and not too low, which would include knowing your strengths and abilities as well as your weaknesses and limitations and being honest about them to yourself and others. At the interpersonal level, humility is being other-oriented rather than self-focused, which would include development of empathy. Table 2 in Chapter 11 shows characteristics compatible with and counter to humility.

In this important work, Hess and Ludwig point out that today's understanding of "smart" is quantity-based, and, of course, it would be impossible for any human to outdo a smart machine if the knowledge standard is in terms of quantity. Thus, the value of humans is not in *how* much you know but in the *quality* of what you know, and that quality has to do with thinking, learning and how you emotionally engage with others. Hess and Ludwig refer to this quality as "NewSmart", which leads them to what they call the SMA (Smart Machine Age) skills of **critical thinking, innovative thinking, creativity, and high emotional engagement with others**. These are consistent with MQI research and ICALS findings.

Hess and Ludwig further argue that in today's environment most adults have not had formal training in *how to think, how to listen, how to learn and experiment through inquiry, how to emotionally engage, how to manage emotions, how to collaborate, or how to embrace mistakes as learning opportunities.*[160] We agree. While certainly all of these topics have been the focus of research and best-selling books—for example, Goleman's extensive work on emotional intelligence—the educational standard based on grades over mastery and a business culture of aggression and competitiveness with avoidance of failure at all costs have prevented this continuous focus on quality.

The authors identify four key behaviors to excel at these SMA skills. These are: Quieting Ego; Managing Self (one's thinking and emotions); Reflective Listening; and Otherness (emotionally connecting and relating to others).[161] These behaviors require shifting our current mental models focused on ego and self to a model which emerges from *the choice of humility*, acknowledging our limitations and becoming outwardly focused, honoring the experiences and learning of others. Thus, **humility becomes, indeed, the new smart in the age of smart machines**.

In Chapter 5, as we explore SELF as the locus of learning—as foundational to the Intelligent Complex Adaptive Learning System—we look at the skill sets needed from that viewpoint: Managing Self, Knowledge Capacities, and Presencing. When exploring social learning in Chapter 6, we look at the skill sets needed from that viewpoint: Managing Relationships, Otherness, and Knowledge Sharing. In Chapter 8 as we explore the merging of experiential learning and existential learning, we recognize the very human skills of Synthesizing and Planning, which have long been acknowledged as critical to human progress. These are human skills that continue to provide both capacity and capability as we move into the future. In Chapter 9 we explore the human gift of Humility.

UNLEASHING Litmus Test #1*

*After reading **Chapters 1-3**, reflect on each question for one minute (Reflective Observation) prior to answering silently, verbally or in writing.*

1. **AWARENESS** means something has come to your attention, it is perceived, it has been mentally engaged.
   Ask: What ideas presented in these introductory chapters have caught my attention and raised my awareness?

2. **UNDERSTANDING** includes your perception of the situation—the who, what, where, when, and why, and the anticipated results. As the situation becomes more complex, you need to re-create your understanding.
   Ask: What information expanded my understanding of my SELF and the environment in which I live and work?

3. **BELIEF** means you accept what you are aware of as true and understand it exists. Beliefs which dominate other patterns are prominent. Strong patterns are created by experiences and are closely related to emotions.
   Ask: Do I believe that it is necessary to expand my capacity and learn new skills to survive in a changing and complex world?

4. **FEELINGS** are the foundation of learning—positive feelings make actions important to you and worthy of your efforts. Reason cannot operate without emotions.
   Ask: How do I feel about what I am reading? Is change necessary to operate successfully in the current environment?

5. **OWNERSHIP** implies a personal commitment for you to take responsibility and act.
   Ask: Do I recognize that my learning is my responsibility and, as a complex adaptive system, that the choices I make today will change my life tomorrow?

6. **EMPOWERMENT** refers to self-empowerment, that is, having the *knowledge* to make the necessary change, and the *courage to act* on what you have learned?
   Ask: Am I open to learning how to use the full capacity of my mind and heart? Am I willing to make the life changes necessary to do this?

*These questions are based on the Bennet Individual Change Model.[162] CHANGE comes from within, that is, unleashing your mind is YOUR choice.

Chapter 4
A New Theory of Learning

Redefining Experiential Learning ... The Intelligent Complex Adaptive Learning System ... Self-Organized Learning ... Exploring Experiential Learning as a CAS ... The ICALS Model ... Characteristics of the ICALS Theory ... In Summary

FIGURES: (5) A concept diagram highlighting some interactions among the 13 neuroscience areas and the expanded experiential learning model; (6) The Intelligent Complex Adaptive Learning System experiential learning model.

While our first inclination when talking about learning is to pull up a vision of the brain, which is a molecular structure and the fluids that flow within and through that structure, it is the mind, created by neurons and their firings and connections, that is the totality of the *patterns in the brain and throughout the body*, which encompasses all our thoughts, ideas and perceptions. Thus, as introduced in Chapter 1, we use the term mind/brain to refer to both the structure and the patterns emerging within the structure *and* throughout the nervous system.[163] Note that while the brain has historically been recognized as the seat of control, the body-mind in its entirety acts as an information network with no fixed hierarchy.[164]

As we interact with life, our neuronal circuitry rewires itself in response to stimulation. Neurons are not bound to each other physically and have the flexibility to repeatedly create, break and recreate relationships with other neurons, the process of plasticity. Axons and dendrites—which are appendages treelike in nature—enable the exchange of information among neurons. As a matter of reference, it is estimated that the average brain contains 10 billion neuron cells, with each connected through synaptic connections or small gaps through which neurotransmitters can flow to about 10,000 other neurons.

The total possible synaptic connections in the brain is calculated in the hundreds of trillions. The patterns of neuronal connections—the flow of small electrical impulses through the branch-like axons and dendrites, together with the flow of molecules through the synaptic junctions—create the patterns within the mind/brain. The dendrites (collecting inputs) and axons (providing outputs) are unique properties of neurons. Further, neurons create spikes, what are called "action potentials", which are patterns of changes in electrical signals starting near the cell body and traveling along the axon until reaching the end of every branch, at which point it can make connections to other neuron dendrites (synaptic connections). This is consistent with the idea of Hebbian learning which forwards that when we learn something connections are strengthened and when we forget something connections are weakened (*neurons that fire together, wire together*). Forgetting occurs when unused connections disappear.

Redefining Experiential Learning

Humans are experiential learners. Building on the intellectual thought of John Dewey, Kurt Lewin and Jean Piaget—with somewhat different terminology and diagrams but with closely related functional processes— in 1984 David Kolb detailed "the process whereby knowledge is created through the transformation of experience."[165] He considered this work a *theory*, not just a model. Kolb's intention was to address learning from experience in all of its forms and situations. He believed that, "Knowledge is continuously derived from and tested out in the experiences of the learner",[166] and felt that experiential learning complemented both existing cognitive and behavioral theories of learning. Cognitive theories heavily emphasize abstract symbols and their manipulations, somewhat a semiotics approach, and behavioral theories ignore consciousness and subjective experiences.

Kolb's model is characterized as a cycle involving four adaptive models, namely concrete experience, reflective observation, abstract conceptualization, and active experimentation. Note that "the very act of thinking is a form of movement",[167] or experience. The model also contains two distinctive dimensions made up of concrete experience/abstract conceptualization along the vertical axis and active experimentation/reflective observation along the horizontal axis. The learning process derives from the transactions among these four modes and the resolution of the adaptive dialectics between them. Dialectics plays an important role in the model, because each pair of opposite modes is considered to be dialectically related.

While Kolb's use of "concrete experience" as one of his four modes of learning was meant to provide an integrating perspective on learning which combined experience, perception, cognition and behavior—taking a cognitive psychology approach—his model was based primarily on observation. This is understandable since the technological advancements that enabled the creation and sophistication of brain measurement instrumentation did not occur until the turn of the century. With the advent of instrumentation such as functional magnetic resonance imaging (fMRI), which is used to produce precise measurements of brain structures,[168] the electroencephalograph (EEG), which is another noninvasive technique that measures the average electrical activity of large populations of neurons,[169] and transcranial magnetic stimulation (TMS), which uses head-mounted wire coils that send very short but strong magnetic pulses directly into specific brain regions that induce lower-level electrical currents into the brain's neural circuits, turning on and off specific parts of the human brain,[170] researchers were finally able to explore the unconscious mental life, expanding our knowledge of adult learning from the inside-out.

In 2002 biologist J. E. Zull expanded Kolb's model, suggesting that there was a related view of the human brain and that Kolb's learning cycle arose naturally from the structure of the brain. The superposition of Zull's model over the Kolb model demonstrates the relationships between the modes of thinking from Kolb and the physiological sections of the human brain from Zull. The sensory and post-sensory part of the brain related directly to the incoming information from concrete experience. This would include all methods of sensing the external world. Reflective observation, or reflection, occurs in the temporal integrating cortex or what is often called the back cortex. Although it does not lie physically in the back of the brain, it is the back of the cortex. This is because the cortex folds in on itself as the adolescent develops. The various incoming signals are combined to form an integrated representation of external reality. Reflection on this re-presentation begins the process of understanding and meaning.

The frontal integrative cortex relates to what Zull calls abstract hypothesis and Kolb called abstract conceptualization or comprehension. The frontal integrative cortex is also referred to as the prefrontal cortex. It is here that the mind creates a deeper knowledge or understanding and additional insights related to the incoming concrete experience. This prefrontal cortex also looks for ways of influencing the external situation as a guide to what action will have the most effective outcomes (which ties directly to the definition of knowledge, the capacity to take effective action). This area is what Goldberg calls the executive part of the brain.[171] Zull also points out the relationship between active experimentation and the premotor and motor physiology. For many years it was thought that here is where the body initiates the execution of actions determined by the prefrontal cortex through comprehension and higher-level thinking. As Zull summarized, "Concrete experience comes through the sensory cortex, reflective observation involves the integrative cortex at the back, creating new abstract concepts occurs in the frontal integrative cortex, and active testing involves the motor brain."[172]

However, more recent research finds this description of movement from the sensory regions to the motor region misleading. Scientists have found cells throughout the "newer" brain (the neocortex, which recall makes up 70 percent of brain volume and is recognized as the organ of intelligence) all appear to project to the "older" brain related to movement, with evidence that "the complex circuitry seen everywhere in the neocortex performs a sensory-motor task."[173] Thus, it appears that there are no specific areas that could be described as motor or sensory regions.

The Intelligent Complex Adaptive Learning System

In 2010, nuclear physicist David Bennet completed a ten-year study of emerging research in neuroscience with a focus on adult experiential learning. Pondering on the human as an integrated, biological and complex system entangled across the physical, mental, emotional and spiritual dimensions, Bennet undertook a scientific analysis of thinking and learning in the human brain through the lens of consilience. Kolb's model and Zull's overlay provided a baseline for development of an expanded experiential learning model, which supports the Intelligent Complex Adaptive Learning System (ICALS).[174] The nine assumptions this work is built upon are in Appendix C.

A first step in understanding the expanded model is to have a common understanding of the terms used to describe that model. A system is *a group of elements or objects, their attributes, the relationships among them, and some boundary that allows one to distinguish whether an element is inside or outside the system.* Elements of a system may be almost anything: parts of a television set, computers connected to a network, people within an organization, neurons within a brain, patterns of the mind, ideas within a system of thought, etc. The nature and number of elements and their relationships to each other are very important in determining a given system's behavior. Almost everything can be viewed as a system. For example, the following are all systems because they have many parts and many relationships: automobiles, ER teams in a hospital, cities, organizations, ant colonies, animals, and individuals.

While complexity was introduced as part of the CUCA environment explicated in Chapter 2, the term itself was not clearly defined. The Santa Fe Group says complexity "refers to the condition of the Universe which is integrated and yet too rich and varied for us to understand in simple, mechanistic or linear ways ... complexity deals with the nature of emergence, innovation, learning and adaptation".[175] Considering the immense number of interconnections that exist among the 100 billion neurons in the brain with roughly 10,000 connections per neuron, it seems clear that the mind/brain/body is a highly complex system.

Adaptation is *the process by which a system* (in our context, an individual or organization) *improves its ability to survive and grow through internal adjustments*. Adaptation may be responsive, internally adjusting to external forces, or it may be proactive, internally changing so that it can influence the external environment. Humans do both.

Intelligence was introduced in Chapter 1 as *the capacity for reasoning and understanding or an aptitude for grasping truths* with Wiig[176] broadening this to include an individual's ability to think, reason, understand and act. Learning is the creation of knowledge, with knowledge defined as *the capacity (potential or actual) to take effective action* or, as defined by early Western philosophers, *justified true belief*. Stonier[177] and McMaster[178] consider intelligence as the ability to set and achieve goals. This is consistent with the concept of taking effective action.

Hawkins connects intelligence to prediction when he posits, "It is the ability to make predictions about the future that is the crux of intelligence."[179] Thus, we are focusing on *how the individual interacts within and with its environment* in anticipation of the outcomes of that interaction. In this context, intelligent activity represents a perfect state of interaction where intent, purpose, direction, values and expected outcomes are clearly understood and communicated among all parties, reflecting wisdom and achieving a higher truth.[180] Knowledge is in service to intelligent activity. Because the effectiveness of all knowledge is context sensitive and situation dependent, knowledge is shifting and changing in concert with the environment and the demands placed upon us in different situations. The incompleteness of knowledge that is never perfect serves as an incentive for the continuous human journey of learning and the exploration of ideas.[181]

Self-Organized Learning

Enhancing the learning process raises the question of learning from what perspective? Taking a systems perspective brings home the point that it is a *self-organizing* complex adaptive system. Bennet and Bennet[182] have proposed that "a self-organizing complex system is one in which the agents (individuals) have a high degree of freedom to organize themselves to better achieve their local objectives." Dropping down one level to the individual learner, the agents become various subsystems of the mind/brain/body. As Battram forwards,

> In 'self-organized learning' the self-organizing is taking place inside the learner's brain (which is a complex adaptive system in its own right) rather than in a networked group of individuals.[183]

We expand Battram's statement to include the body because the brain and the body are tightly connected through the flow of chemicals such as hormones and neuropeptides.[184] In addition, the neuronal signals flow throughout the central nervous system and the flow of blood continuously provides energy to the brain and the body. Simultaneously, the mind and brain are connected via the physiology of the brain and the neuronal patterns created through incoming signals and the thinking processes.

The environment is closely connected to the mind/brain/body via incoming information and exchanges of energy with the body and from physical interactions between the body and its environment. Thus, a holistic view indicates that the environment, mind, brain, and body all represent entangled complex adaptive systems which co-evolve through various levels of sensing, information and energy transfer, and some level of mutual structural adaptation. In considering the brain as a self-organizing system, Buzsaki explains that,

> The brain is perpetually active even in the absence of environmental and body-derived stimuli. In fact, a main argument put forward in this book is that most of the brain's activity is generated from within, and perturbations of this default pattern by external inputs at any given time often causes only a minor departure from its robust, internally controlled program. Yet these perturbations are absolutely essential

for adapting the brain's internal operations to perform useful computations.[185]

It is clear that individual learning takes place via significant interplay among the various subsystems of the body and the environment. Many systems and subsystems of the individual work together to create an effective learning system. While the conscious self may represent one unifying perspective of the individual through consciousness, it does not control all subsystems of the mind/brain/body. Feelings, memories, belief systems, past knowledge, motivation, and immediate goals influence learning effectiveness.

Understanding the learner as a self-organizing Intelligent Complex Adaptive Learning System gives caution to teachers and organizations that would control, direct, or mandate learning. Such a system can rarely be controlled by force—rather, it must be guided and nurtured through supportive influence. Thus, a theory of adult learning has to take into account *the learner's capacity to self-organize and internally guide his or her learning* as well as the major factors of the mind/brain/body that impact the learning process, and the individual's autobiography, motivation, and beliefs about his or her learning efficacy.

Since learning is context sensitive and content dependent, and unpredictability is a characteristic of complex systems, there is no assurance of *how much* learning will occur. However, the results of this research indicate there are specific aspects of the learning process that can *reduce that unpredictability*. For example, learning will be enhanced when it occurs near the top of the inverted "U" of the arousal/stress curve (see Figure 10 in Chapter 6), indicating that the learner has good motivation but not too much stress. Another example is that *the beliefs and attitudes of the learner will affect learning and can be under the control of the learner if he or she so chooses* (see Chapter 6).

Considering the roles of emotion, memory, the unconscious, and stress in experiential learning (see Chapters 6-8), one can understand why learning is not always a rational process. From a creativity viewpoint, playing with ideas and using intuition, feelings, and anticipating the future outcome of actions to create potential solutions to problems provides a significant advantage to the learner (see Chapter 3).

Exploring Experiential Learning as a CAS

The research emerged 13 areas of neuroscience findings critical to adult experiential learning. These can be divided into three groups. The first is made up of the unconscious, memory and emotion, and represents a foundation that is always involved in learning. The second group represents findings that influence learning in specific situations. It includes stress, creativity, mirror neurons, anticipating the future and social support. The third group is related to the capacity for, and enhancement of, lifelong learning. This group is made up of social interaction, epigenetics, plasticity, physical and mental exercise, and aging.

Figure 5, shown on the following page, is a concept diagram providing a simplified picture of some of the interconnections among the 13 areas of neuroscience findings and an expanded experiential learning model. The arrowheads of the various connections indicate the direction of influence. Starting at the lower left with the box labeled "active experimentation," actions influence and are influenced by the social support and social interaction areas of neuroscience by tracing the paths from environment to concrete experience to reflective observation to abstract conceptualization back to active experimentation. Neuroscience findings 1 through 9 support the operational complex adaptive learning system (CALS) and areas 10 through 13 (epigenetics, plasticity, exercise, and aging) support sustainability of the CALS.

Two other small boxes at the top and to the right labeled "observing" and "reading" remind us that an action may not be a social interaction. Each of the boxes labeled "social support" and "social interaction" and the active experimentation box at the lower left interact with the environment and the environment interacts with them.

The neocortex—a thin layer of neural tissue appearing to be arranged in layers with dozens of regions—is where, as previously noted, 70 percent of brain activity occurs. While the brain grew to its current size through evolution by adding new brain parts over old brain parts, the neocortex (or "new" brain) advanced quickly to its size by making copies of a basic circuit, with every part of the neocortex working on the same principle.[186] Thus, despite the functional differences of various regions, the circuitry in the neocortex created by neurons and their connections looks remarkably similar. Mountcastle proposed that what makes the regions different is not their function but their connections.[187] While these connections are primarily vertical (forming cortical columns), appearing to have an orderly hierarchical construct, there are also distant horizontal connections, which suggests "information goes all over at once."[188] In research coming out of Numenta, a neuroscience research company, it was discovered that in the neocortex "the majority of connections between regions do not fit into a hierarchical scheme at all",[189] and scientists have been unable to determine exactly what the majority of cells are doing. Further, various regions exist at different levels and connect to multiple hierarchical levels.

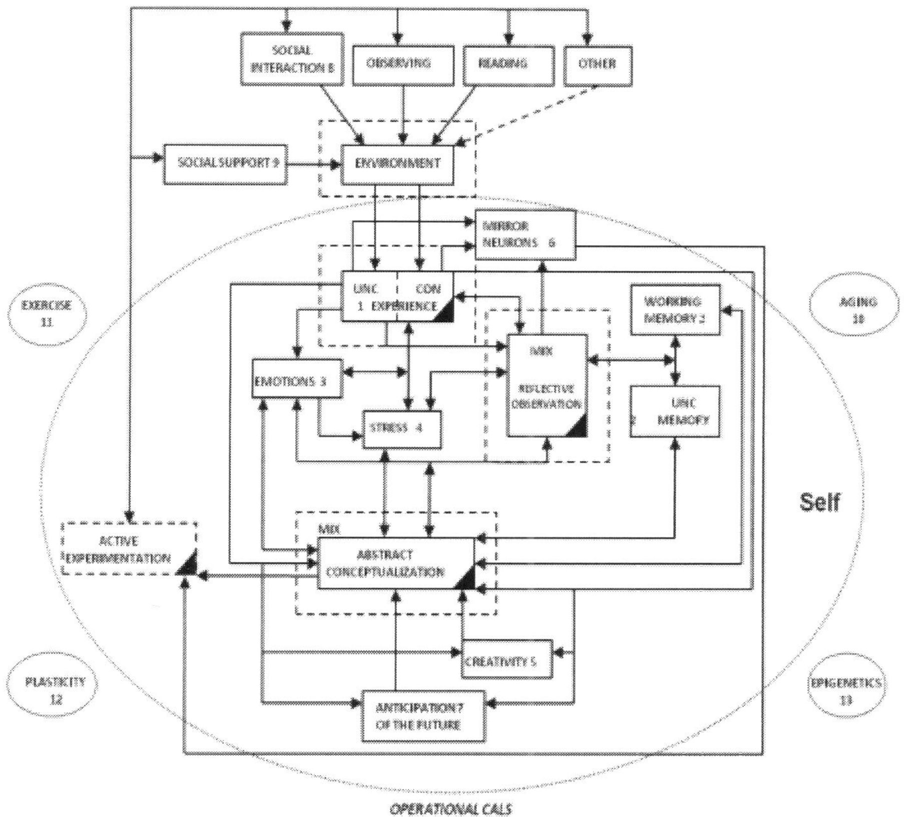

Figure 5. A concept diagram highlighting some interactions among the 13 neuroscience areas and the expanded experiential learning model. This highly simplified model illustrates the self-organization and complexity of the human mind/brain/body.

The famous mapping of the neocortex of a macaque monkey in 1991 created a layered image much like Figure 5, although more complex in its layering. In the Figure 5 mapping, the rectangle boxes represent different regions of the neocortex, with the lines representing information flows from one region to another. The point here is that in this complexity there is certainly a high element of self-organization that would no doubt be specific to the individual. Even though this figure is highly simplified, we can follow the arrows around and create many potential feedback (or feed-forward) loops. Most likely, these feedback and feed-forward loops provide the associations that promote the continuous self-organization of the system. In any case, the mind/brain/body represents a self-organizing, complex adaptive system.

Going back to the environment box at the top of Figure 5, the two major areas coming down into the box labeled "experience" can be seen, with the left side unconscious experience and the right-side conscious experience. This reminds us that the individual learner has both a conscious and an unconscious experience underway. Note that from the box labeled "experience" there are a large number of arrows going out of and coming into such neuroscience findings areas as emotions, stress, mirror neurons, and the reflection process.

In turn, the reflective observation process interacts with working memory, unconscious memory, stress, emotions, and abstract conceptualization. The abstract conceptualization has a large number of inputs and outputs, interacting with stress, emotions, unconscious memory, working memory, mirror neurons, creativity, anticipation, and active experimentation as well as the experience mode. Although not shown in the diagram, the three major areas of unconscious, memory, and emotions are connected to almost everything in the ellipse.

The ICALS Model

As can be seen, the term Intelligent Complex Adaptive Learning System (ICALS)—which is what each person must be to survive—describes the multiple and highly complex influences, connections and relationships within the mind/brain/body. See Figure 6. This model is the result of a sequence of model upgrades, beginning with the original Kolb model of experiential learning published in 1984 (built on the earlier work of Dewey, Lewin and Piaget). This model was then overlayed with Zull's 2002 model, which brought in the brain and various parts of the brain associated with the four modes of Kolb's model. This approach was further expanded to include the 13 neuroscience findings areas (the unconscious, memory, stress, creativity, mirror neurons, anticipating the future, social support, social interaction, epigenetics, plasticity, physical and mental exercise, and aging). The new experiential learning model shown in Figure 6 includes the addition of self and a fifth mode of learning, social engagement (social interaction and social support).

The four boxes circled around Self represent Kolb's original four modes of learning. Although the sequence appears clean, it was recognized by Kolb that as learning occurs a learner experiences many direct connections between and among all of these modes. Clearly, the human being is far beyond a simple system, ALWAYS multiprocessing, that is, with the unconscious working 24/7, largely in support of the conscious self (see Chapter 6) except during the "at rest" state (see Chapter 10). The larger circle moves from Active Experimentation to Social Engagement—a fifth mode added to Kolb's original model—to Reflective Observation, with a direct interactive link with Concrete Experience. Social Engagement opens up a dialogue with the environment, bringing social support and social interaction into the sphere of the self as learner. This focus is critical as we have learned more about the workings of the mind/brain and have recognized the involvement of self in social learning. There is no greater example of this than the current environment, where social media is largely determining attitudes, norms, behaviors, beliefs, and truth.

Concept mapping—a process to explore relationships between concepts—is used to organize and structure knowledge. Developed by Cornell University in the 1970's,[190] concept mapping is based on the cognitive theories of Ausubel[191] which stress that new knowledge is built on prior knowledge. This theory is, of course, consistent with the way the mind/brain works in terms of associative patterning, although it was promulgated long before the turn of the century when we were able to map the workings of the mind/brain.

A set of Conscious Look Books authored by David Bennet focuses individually on the thirteen areas of neuroscience findings based on the ICALS research and is available on Amazon. These books—written for the graduate of life experience—are more conversational in nature and provide another resource for discovering and unleashing the potential of the human mind/brain. The 13 concept maps are included in this series.

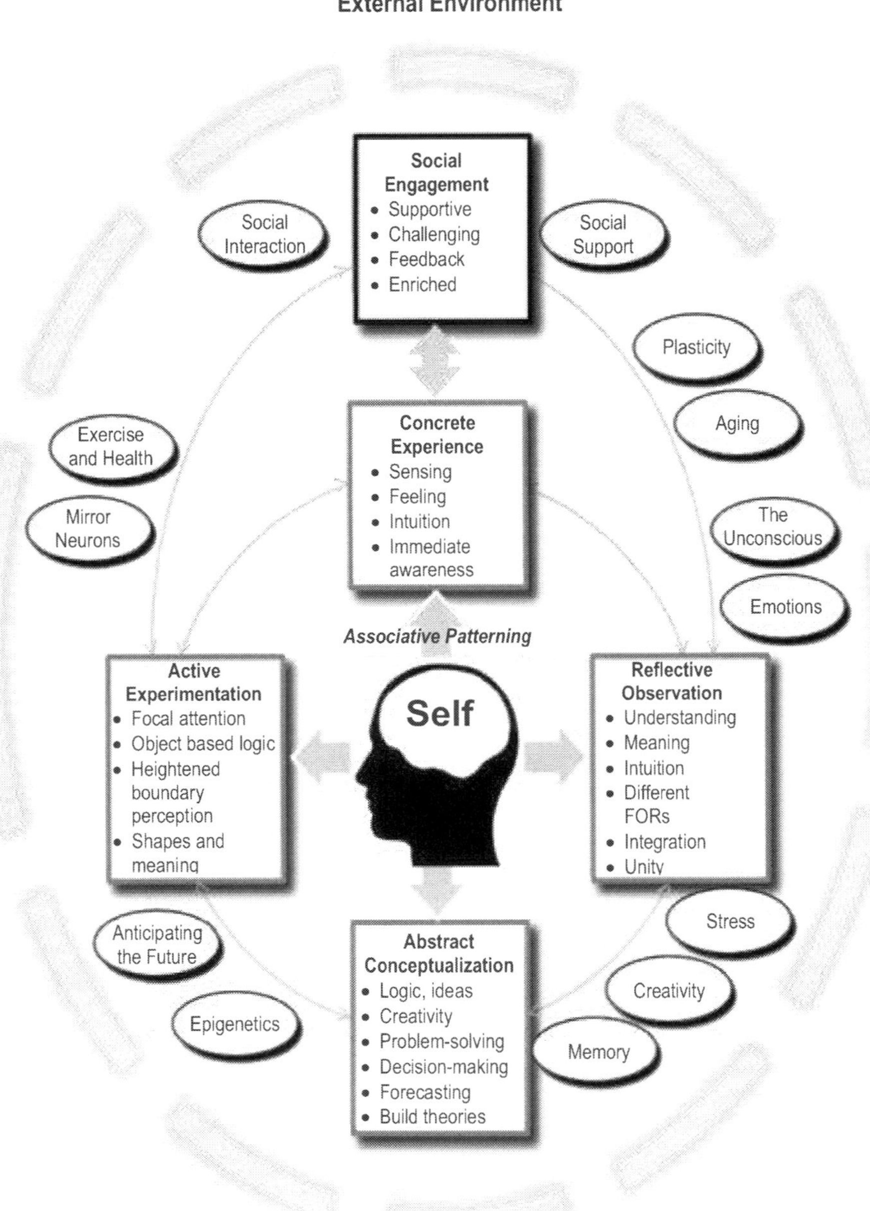

Figure 6. The Intelligent Complex Adaptive Learning System experiential learning model.

To increase the effectiveness of the concept mapping process, it was necessary to develop sub-elements of the four modes, that is, **characteristics that described what is occurring when this mode is engaged**. To surface these characteristics, referred to as sub-elements, we look to the work of Kolb[192] and Zull[193] as well as other educators and theorists who have explored the experiential learning model. Building on the descriptions of the four modes in this body of work, the sub-elements of Kolb's initial four modes are identified as follows:

> **Concrete Experience** (grasping through apprehension). For this mode the identified characteristics or sub-elements are: *sensing; feeling; awareness; attention; and intuition.*

> **Reflective Observation** (transformation via intention). For this mode the identified characteristics or sub-elements are: *understanding; meaning; truth and how things happen; intuition; integrate; and look for unity.*

> **Abstract Conceptualization** (grasping via comprehension). For this mode the identified characteristics or sub-elements are: *concepts, ideas, logic; problem solving; creativity; build models and theories; anticipation of outcome of action; control, rigor, and discipline.*

> **Active Experimentation** (transformation via extension). For this mode the identified characteristics or sub-elements are: *act on the environment; focus attention; object-based logic; heightened boundary perception; and sensory feedback to brain.*

Through this process, the fifth mode of social engagement became clear. While Kolb noted the influence of the environment on internal learning, today we recognize that the environment is actively (and continuously) engaged in the learning process. For example, we now understand that an enriched environment can produce a personal internal reflective world of imagination and creativity, affecting both active experimentation and concrete experience. Since the mind is an associative patterner, the things and people in our environment are very much triggering our thoughts. Further, we also now know that we are social creatures, just beginning to wake up to "the complexity of our own brains, to say nothing of how brains are linked together … all of our biologies are interwoven".[194] Given this understanding—and that which is explored in Chapter 9—sub-elements can be grouped into social support and social interaction.

Social Engagement (grasping through direct comprehension; transformation via association). Connected to social support (which includes enriched environment, social bonding and affective attunement), characteristics or sub-elements include: *open mind; risk-taking; willingness to listen and learn; reducing stress and fear; creating resonance with people and ideas; and contributing to evolution and sculpting of the brain*. Connected to social interaction (which includes physical learning mechanisms naturally evolved in the brain and energetic unconscious exchanges such as occur through mirror neurons), characteristics or sub-elements include: *accelerating learning and creativity; enhancing understanding, meaning, truth and how things work; developing a shared language; supporting use and understanding of concepts, metaphors, anecdotes and stories.*

The two descriptors in parentheses following the name of the mode deserve an explanation. Unlike the original four modes, the mode of Social Engagement has two aspects, that is, it includes both social support and social interaction (see the characteristics connected to each in the paragraph above). In this context, "grasping through direct comprehension" encompasses unconscious communication, enabling direct (and faster) learning through bypassing the conscious mind (see Chapter 9). As we now understand through the mathematics of transreal numbers (a concept expanded in Chapter 8), this learning includes the transfer of non-propositional objects such as sensations, feelings, perceptions and intuition. Note that "unconscious communication" is inclusive of energetic exchanges connected to specific individuals as well as the larger consciousness (quantum) field of which we are a part. The second descriptor, "transformation via association" refers to both conscious and unconscious exchanges in the environment as well as the associative patterning within the brain occurring in response to input from the external environment (introduced in Chapter 3).

At the top level of the ICALS model, the addition of *Self* as an underlying foundation along with acknowledgement of the role of the environment in the learning process—and adding *Social Engagement* (including both social interaction and social support) as a fifth mode of learning—fully reflect the Bennet findings. As referenced above, this fifth mode brings in the importance of associative patterning, which supports all five modes from the viewpoint of self.

Characteristics of the ICALS Theory

Kolb forwarded five major characteristics that were essential to the Kolb experiential learning model: (1) Learning is best conceived as a process, not in terms of outcomes; (2) Learning is a continuous process grounded in experience; (3) The process of learning requires the resolution of conflicts between dialectically opposed modes of adaptation to the world; (4) Learning is a holistic process of adaptation to the world; and (5) Learning invokes transactions between the person and the environment.[195]

The ICALS theory is consistent with Kolb's characteristics (1) and (4). Characteristics (2) and (5) have been expanded to include the human capacity to dialogue with one's self (see the descriptions below). Kolb's characteristic (3), asserting that the process of learning requires the resolution of conflicts between dialectically opposed modes of adaptation to the world, is neither supported nor unsupported by the ICALS approach. Although there are different ways of viewing the world and different philosophies relative to modes of adaptation to the world, looking from a neuroscience perspective, we find nothing that would support Kolb's belief that the modes of thought were dialectically opposed, although they certainly *may* be different ways of thinking. For example, considering the past-present timeframe of the reflection mode and the present-future perspective of the comprehension mode, these two modes would appear to be more complementary than dialectically opposed.

Further, as introduced earlier in this chapter, the ICALS model includes the fifth mode of *social engagement* and brings in the process of associative patterning, which supports all five modes from the viewpoint of self. These processes certainly work in tandem with the original four modes and, since all knowledge is context sensitive and situation dependent,[196] may or may not affirm Kolb's belief that learning requires the resolution of conflicts between dialectically opposed modes of adaptation to the world. Accordingly, Kolb's characteristic (3) is not considered a characteristic of the ICALS theory.

An additional characteristic essential to the ICALS theory, as previously forwarded, is: Learning is a self-organizing system. This is included as characteristic (5) below. Considering these changes, here are five major characteristics essential to the ICALS theory:

1. Learning is best conceived as a process, not in terms of outcomes.

That learning is a *process* is emphasized in Kolb's model to differentiate it from the behavioral theory of learning and the idealist approach to education, both of which consider learning to be a product or outcome, not a process. Their epistemology derives from the concept that there are "mental atoms" or simple ideas that are constant and their combinations create patterns of thought.[197] Kolb's model takes learning to be the result of a potentially complex process made up of the interactions of four modes, two that represent knowledge and two that create knowledge through transformation.

This is consistent with the ICALS theory as seen from a neuroscientific view of the mind/brain and the associative patterning process, that is, recognition that knowledge, context sensitive and situation dependent, is being recreated for the instant at hand, thus continuously shifting and changing in response to new associations and influences of the environment. Recall that learning is the process of creating knowledge, that is, the process of creating the capacity (potential or actual) to take effective action, part of a continuing journey toward intelligent activity.[198]

2. Learning is a continuous process that is either grounded in experience or an emergent property of consciousness.

Builds on characteristic (1) above. Knowledge is continuously developed from and tested in the experiences of the learner (the associative patterning process). New ideas emerge from this combining/associating of incoming information from the environment with internally stored information.[199]

Consciousness is a process, a sequential set of ideas, thoughts, images, feelings and perceptions and an understanding of the connections and relationships among them.[200] It is the sum total of who we are, what we believe, how we act and the things we do, so it's all of our actions, thoughts and words.[201] William James was amazed at the continuity of human consciousness.[202] Dewey noted that, "the principle of continuity of experience means that every experience both takes up something from those which have gone before and modifies in some way the quality of those which come after." [203]

We agree that a significant contribution to learning occurs through concrete experience. However, another way of learning is through internal dialogue with oneself. Important parts of the mind/brain such as the unconscious, the emotions and the memory systems play a strong role in creating knowledge. There are useful techniques for the conscious mind to interact and communicate with the unconscious, the emotions, and memory without immediate input from concrete experience. For example, artists, writers, and theoretical scientists do not necessarily require continuous concrete experience to enhance their learning. Gedanken experiments—Einstein's term for conceptual thought experiments—can create new ideas and knowledge.

While having a dialogue with one's self might be characterized as an "experience" (and later we will explore existential experiences), this is not the intent of the descriptor of concrete experience. Therefore, we have expanded Kolb's original characteristic to include the learning occurring through self-dialogue, an emergent property of consciousness.

3. *Learning is a holistic process of adaptation to the world* (formerly Kolb's fourth characteristic).

Experiential learning is a holistic concept that represents the major process of human adaptation to the social and physical world. The Kolb learning model is holistic in the sense that learning and the knowledge it creates emerge from the interplay and interaction of the four modes. Experiential learning also includes such adaptive concepts as creativity, problem solving, decision-making, and attitude change.[204] Learning is also holistic in that it exists in all individual and social environments and encompasses all ages of life.

By holistic Kolb meant learning "seeks to describe the emergence of basic life orientations as a function of dialectic tensions between modes of relating to the world."[205] While certainly partially correct, it neglects actions the learner can implement that can influence the world. Modern complexity theory considers co-evolution and autopoietic interactions as normal evolutionary processes of interacting complex systems.

While the underlying neuroscience findings on which the ICALS model is built reinforces Kolb's acceptance of learning as a holistic process relative to basic life orientation, we take the position that this is due to the *complementarity* of the reflective observation and abstract conceptualization modes instead of their dialectic tension. Further, this work broadens the system to include the learner (self) and the external environment along with the fifth learning mode of social engagement, which can be collectively viewed as a complex adaptive learning system.

Further, from a neuroscience perspective, the human mind/brain has tremendous flexibility–plasticity–the ability to learn just about anything. Mountcastle proposed that there was a fundamental algorithm responsible for perception and intelligence, and that the unit of intelligence is the cortical column in the new brain (the neocortex).[206] Further, if we are able to understand how one of these columns works, then we will understand how intelligence works–"as below, so above", what we find throughout nature. As Hawkins describes, "Being able to learn practically anything requires the brain to work on a universal principle."[207]

4. Learning—in the conscious and unconscious—invokes transactions between the person and internal and external environments.

This is an expansion of the former characteristic (5) forwarded by Kolb. As Kolb says, "The casual observer of the traditional educational process would undoubtedly conclude that learning was primarily a personal, internal process requiring only the limited environment of books, teacher, and classroom."[208] In his experiential learning model, Kolb emphasizes that experience is both personal and internal, and social and externally driven. The use of the word *transaction* has a special denotation; meaning a fluid, interpenetrating relationship between the person and the environment such that when they become related, both are changed. This is very similar to the autopoiesis concept proposed by Maturana and Varela in their theory of coevolving complex adaptive systems.[209] The grasping or creating of figurative representations of that experience through apprehension and comprehension, coupled with the transformation of those representations through reflection and active experimentation, create understanding, meaning, insight, creativity, intuition, and the ability to anticipate the outcome of one's actions, that is, knowledge (the capacity to take effective action).

As in characteristic (2) above, we agree with Kolb as far as concrete experience and action are concerned. However, this is not so clear when one considers learning being generated between the conscious and unconscious mind. For example, consider Poincare's statement relative to the volleyball effect during the creative process in which thoughts and concepts are continually passed between the conscious and the unconscious mind resulting in new ideas, that is, learning.[210] Further, there is a sense of knowing that emerges from the unconscious that may or may not be linked to the external environment, although it certainly emerges from the internal environment of self.[211]

Note the addition of "in both the conscious and unconscious". Throughout this text you will discover the powerful continuous learning that occurs in the unconscious. The process of associative patterning itself occurs at the conscious and unconscious levels.

5. *Learning is a self-organizing system.*

Envision a situation where the individual learner and the environment are mutually interacting and adapting to each other. Adaptive implies that the system is capable of studying and analyzing the environment and taking actions that internally adjust the system and externally influence the environment in a manner that allows the system to fulfill local and higher-level goals. This is the process of co-evolving with our environment such that an individual moves in a desired direction or towards a desired goal. A complex adaptive system that is self-producing (or self-preserving) is referred to as an autopoietic system.[212]

In the context of a learning system, self-producing means that as learners create and adjust their internal patterns (thoughts, concepts, beliefs) in response to incoming information, they will maintain, modify, or enhance their concepts of self and individuality. In the process of reflecting and comprehending in response to the external world, the learner undergoes changes in both brain patterns and neural architecture, what is called neuroplasticity. The creation of these new patterns (learning) represents an emergent phenomenon (information and knowledge) that results from numerous internal pattern associations. There are many interactions and a high level of communication among the various patterns, neurons, and chemicals of the mind/brain/body of the learner. For example, mind patterns are continuously associated with one another, the emotional system tags all incoming information at some level, and thoughts and ideas bounce back and forth among the cortical columns of the neocortex between the prefrontal cortex and the integrated cortex.[213]

In Summary

The Intelligent Complex Adaptive Learning System takes into account the following aspects of the mind/brain/body: the self-organizing complex adaptive behavior; the fundamental neuroscientific subsystems as identified in this study; the key cognitive parameters such as those identified in the work of Kolb, Dewey, Lewin and Piaget; the environmental influences (both human and physical); and the power of the self-conscious mind to achieve a high capacity to learn through developing a knowledge of meta-learning. The neuroscience findings in the 13 areas of focus emerging from this research are, where appropriate, *embedded throughout this book.*

Along with considerable research, substantial modeling, and a high volume of writing, nine guidelines for learners also emerged, offering a high-level perspective for adult learners as they acquire and apply knowledge and consider implementing their own self-development. These guidelines are concerned with (1) the infinite potential of the human mind, (2) the powerful role of beliefs, (3) the influence of the environment, (4) the responsibility of knowledge, (5) the power of the unconscious, (6) the wisdom of age [we need to reread that one!], (7) the drive for certainty, (8) the sacredness of values, and (9) the paradox of life. See Appendix D.

An emergent finding that is critical to your learning from this text is: **Voluntary learning promotes theta waves that correlate with little or no stress and positive feedback**. Theta waves are low frequency (4-8 hertz) electrical waves generated within the brain and detected by an electroencephalograph (EEG) that measures electrical activity near the surface of the brain via electrodes on the scalp. There are a number of suggested correlates to theta waves; for example, intentional or voluntary movements, high-level processing of incoming signals, selective attention, arousal, and decision-making.

As Buzsaki posits, "an inescapable conclusion is that 'will' plays a crucial role in theta rhythms."[214] *Voluntary learning, as distinct from forced learning, has little or no stress attached to it and therefore fosters a much higher level of learning.* Specifically, positive learning would impact sensing, feeling, awareness, attention, understanding, and meaning as they relate to the adult learning model. This is shared here as we move forward grounded in the ICALS theory and model and connecting neuroscience findings in the ICALS study to the success factors and skill sets needed for an unprecedented future.

As this work finally comes into being—transforming it from the inner depths of our souls, the findings of scientists who are charging ahead of the crowd, and the screaming needs of a CUCA world—we genuinely enjoyed a sigh of relief before realizing that there is still so much more to do, so much more to learn. We begin with thought and thinking.

Chapter 5
Thoughts and Thinking

Thought … Conceptual Thinking … Critical Thinking … Knowledge Capacities … Creative Thinking … Innovative Thinking … In Summary

SUCCESS FACTORS: Critical Thinking, Creative Thinking, Innovative Thinking

SKILL SET: Knowledge Capacities

FIGURES: (7) Ways humans operate in the world; (8) Exploring the relationships among information, knowledge, creativity and innovation.

TOOLS: (1) Situational Backtalk; (2) Activating New Resources.

The human ability to control our attention, remember, abstract, and reason are largely what sets us apart from other animals. With these abilities humans have built cultural systems such as language, religion, art and science. Whether at the individual, organizational, country or global level, we require ordered systems with rules to engage in intelligent activity, which at the highest level is called civilization. Csikszentmihalyi describes this quite well. As he says,

> *It is good to have rational, logical structures by which to order thoughts and actions. Much of what we call civilization consists of attempts at rationalizing life, so that actions can be predictable and reasonable. But civilization is a fragile construction that needs constant protection and care. Without it, the mind will not behave logically.*[215]

However, civilization requires more than being logical. Is war logical? Is the economy of today logical? Are political stands logical? Csikszentmihalyi argues that rational behavior cannot happen by itself, that the creating and preserving of ordered systems of rules requires an investment of *psychic* energy, to which all people have access. This psychic energy takes the form of intuition, empathy, wisdom and creativity, a changing part of the human evolutionary process. As he continues,

> *We must foster intuition to anticipate changes before they occur; empathy to understand that which cannot be clearly expressed;*

wisdom to see the connection between apparently unrelated events; and creativity to discover new ways of defining problems, new rules that will make it possible to adapt to the unexpected.[216]

While we agree, it is critical to recognize that it takes the mind and the development of our mental faculties to create knowledge, that is, to use our learning to *effectively act on and in the world.* We are mentally focused in the physical. Neurons fire not only in our minds, but throughout our body, with connective tissue to all parts of our body. Everything about mental work can be applied to the physical world. Life is about developing our mental faculties and improving our mental field as we experience, with our thought directly tied to our actions. It is only then that we are ready to tap into the higher energy of the intuitional field and, as a co-creator, have the ability to bring those ideas into our current reality.[217]

Thought

Thought is a combination of choices, connecting a choice to other choices previously made, and affecting further choices by limiting or reducing those choices, in all of which the unconscious plays a primary role. As Tallis forwards:

> *Virtually every aspect of mental life is connected in some way with mental events and processes that occur below the threshold of awareness ... the profound importance of unconscious procedures, memories, beliefs, perceptions, knowledge, and emotions is recognized universally.*[218]

One of the neuroscience findings in the ICAS study is that **thinking is mostly unconscious**. Unconscious thinking may lead to abstract conceptualization, the role that the unconscious plays when it comes to creating concepts dealing with ideas and logic, or building models and theories. As Lakoff and Johnson have noted, "All of our knowledge and beliefs are framed in terms of a conceptual system that resides mostly in the cognitive unconscious."[219]

Energy and thought are in relationship. Einstein proved that everything in the material world—animate and inanimate, organic and inorganic—radiates energy. We literally swim around in an enormous energy field full of continuously flowing entangled subfields. The biochemical molecules that make up physical human bodies are a form of vibrating energy, with the human biofield consisting of electric magnetic fields of photon emissions, extending out into the world around us. From one viewpoint, human thought and behavior are part of what appears to be a large simulation. Information from the environment is transferred through receptor and effector proteins, much like a set of antennas attached to the outer membrane of every cell. Forces directly affect the physical, mental and emotional levels of life, whether we are aware of it or not.

Energy follows thought. The material world is an effect, not a cause. Everything in our material reality was a thought first. Change occurs from the inside out. As the Dalai Lama says so beautifully, "In order to change conditions outside ourselves, whether they concern the environment or relations with others, we must first change within ourselves."[220] Thus, thoughts and images generated within ourselves have a profound creative and motivating power in human consciousness, with the heart-mind (thought and feelings related to that thought AND/OR feelings and the thought related to those feelings) *controlling energy and building form*. These are not physical forms, but rather forms made of emotional-mental matter.

Thought forms are sent out into the environment where they attract sympathetic vibrations, those vibrations that resonate with the thought and feelings being produced. Just as we have recognized that our mind/brain is an associative patterner, we now understand that we are also an associative attractor, that the thoughts our mind/brain is producing attract other thoughts vibrating at the same frequencies. This can be quite difficult to accept for those who suffer from the Cartesian dichotomy between matter and mind and desire to bring everything to the physical level. However, even using that frame of reference, we now understand from neuroscience the power of the mind/brain—that our thoughts actually change the structure of the physical brain (as well as impacting all of our other human systems), and that the structure of the physical brain very much affects our thoughts.

Each thought that is definite, differing in both density and quality, has a double effect, causing both a floating form and a radiating vibration. The qualities have significance for the meaning of the thought form, with different colors, shapes and distinctness of outlines. Color relates to the emotional quality, form relates to the intent, and distinctness of outline relates to the degree of concentration of the thought. The radiating vibration is quite complex, with every feeling shift associated with thought producing a permanent effect. There is a significant body of work available on thought forms.[221]

In the early days of evolution, humans didn't think as well as they do today. As suggested by the structure of the brain, there would have been thought using narrow basic concepts and a small number of thought forms, although no doubt there were a few universal concepts about basic things like survival, food, and the forms in the surrounding environment. An emergent neuroscience finding in the area of plasticity in the ICALS research acknowledges the direct relationships between thought and the structure of the brain.

Thoughts change the structure of the brain, and brain structure influences the creation of new thoughts. This feedback loop highlights the recognition of the interdependence and self-organization of the mind and brain in the sense that each influences the other. This also reminds one of the importance of both physical and mental health. This duality can help awareness, understanding, meaning, creativity, and anticipation, all sub-elements of the adult experiential learning model.

Thus, today accelerated mental development is producing a higher level of mental thought. As introduced in Chapter 1, there are two levels of mental thought: logic, built on cause-and-effect (lower mental thought), and concepts that emerge from patterns (higher mental thought). Logic uses the past-present-future, in that order, with the primary focus on the past, and if you put everything together it supports itself, consistent with experience and with no conflicting information coming in through the senses. Lower mental thought begins with examples from experience, bringing examples together that result in new information. When those patterns are discovered in life, an individual has moved into the higher mental thought of conceptual thinking. From this viewpoint, we are looking for logical examples that demonstrate the concepts and the relationships among concepts that are emerging in the mind. As we discover examples that lie outside of a concept, the concept is no longer true and must be shifted to include these new examples, ever asking *Why?* and expanding our level of truth as well as our consciousness. Thus, it can be recognized that thought at the conceptual level vibrates at a higher frequency than thought at the cause-and-effect related logical level.

The development of a predisposition to lower mental thought forms actually diminishes the capacity for higher thought forms to function, and increasingly limits an individual's ability to generate higher thought forms, lowering the consciousness level of the individual generating thought. This happens because thought forms linked through logic limit the ability to perceive the larger truth of conceptual thinking. Recall a time when you've gotten stuck in the detail of an activity or event and how difficult it is to look at the larger picture. An example in today's environment is "fake news", the bombardment of misinformation and disinformation, which limits an individual's ability to discern a larger truth. Nonetheless, once we are aware of this danger, we have an element of choice and agency associated with both our personal thought and feelings.

Conceptual Thinking

Let's explore conceptual thinking a bit deeper. In *The Tao of Personal Leadership*, Dreher writes:

> *To succeed in any field, we must look to those skills that make us fully human: the ability to learn continuously throughout life, to communicate with others, to come up with creative new solutions, and to deepen our understanding, looking to the larger patterns within and around us.*[222]

As discussed above, when we are able to discover patterns in life, we have moved into the higher mental thought of conceptual thinking. From this viewpoint, we are looking for logical examples that demonstrate the concepts and the relationships among concepts that are emerging in the mind. As we discover examples that lie outside of a concept, the concept is no longer true and must be shifted to include these new examples, ever asking *Why?* and expanding our level of truth. The *Five Whys*—a simple problem-solving approach to get to the root cause of a problem, drilling down five levels by asking *Why?* at each level—is a truth-seeking tool.

As can be seen, higher mental thought (conceptual) builds on lower mental thought (logic). This is a shared process; the more examples that can be brought to bear, the greater the potential to find a higher level of truth. As these examples are shared in a cooperative and collaborative fashion, ever open to a diversity of insight and reflection, there is a balancing of thought and perceptions that occurs and an expansion of consciousness.

It is more difficult to recognize our mental senses than our physical senses. For example, when you look at another individual what do you see? A first response might be descriptive characteristics of the individual and the backdrop against which you are seeing them. But we now know that we sense through three planes simultaneously, and that any emotional tags connected with that person or situation will be connected to that visual and "felt". Because mental thoughts are not only past memories and learning but very much connected to the NOW, and of an emergent quality, and because the conscious mind is of a linear nature, information being processed by our mental plane is harder to separate out. Yet, it is very much there.

Concepts as formulas facilitate rapid processing (largely in the unconscious) of large amounts of information coming in through all our senses on all three planes (physical, emotional and mental). Rather than addressing each sense separately, these formulas help build a predictive approach to accurately determine the outcome of various actions and interactions. With recent research findings, we now know this is continuously occurring in the cortical columns of the neocortex. Discernment and discrimination—representing the ability to identify and choose what is of value, and the equally difficult ability to toss aside that which is not of value—enable our thinking to challenge a concept and try to apply it to different situations. As an example:

> *Within a specific domain of knowledge, when this action is taken within this situation and context, this is what is going to happen.*

OR, as a higher-order pattern,

> *Within a specific domain of knowledge, when this **type** of action is taken within this **type** of situation and context, this is the **type** of response that will occur.*

In the latter case, *type* refers to sameness in identity, although this may be a loose and popular identity,[223] that is, similar. Each example would represent a token of this type. This sameness in identity has multiple possibilities. It may mean that two things are different parts of some wider unity that includes both, or they are both different members of the same class of things, or they are different parts of a resemblance structure, or they fall under the same predicate or concept.[224] While we do not hold with the *Realist* belief that two things of the same type are strictly identical; simultaneously, we do not agree with the *Nominalist* belief that there are no strict identities reaching across different tokens of type. Further, we disagree with John Locke's description that all existing things are particulars.[225] In other words, **there *are* universals, albeit these universals may take the form of heuristics.**

Continuing with our example, the theory of universals looks at resemblances of identity in terms of properties and relations. From these properties and their relationships, natural classes of things emerge. We add the term conditions to this description of universals, and defer to Armstrong's description that universals are the substance of the world,[226] with substance something that is capable of independent existence. As a path to understanding the role of universals in the laws of nature, we look to *natural class* explanation in pronology, which includes individuals, first-order tokens and classes of higher order (with definite truth conditions). The well-established discipline of *set* theory supports the concepts of classes that have definite truth conditions. Further, natural class theory requires *relations among* individuals and classes of individuals, with the naturalness of the class a property of the class.

While there are certainly solid arguments against the acceptance of universals, we are in agreement with Armstrong that (1) the less than exact resemblance of universals is analyzable in terms of an incomplete identity of universals, which themselves are universals, and (2) that there are irreducibly higher-order relationships holding *between* universals, both of which help substantiate the existence of universals.

While all this sounds a bit complex, as demonstrated in the paragraphs above, and by way of a quick summary, we are using our conceptual thought to create changes in our interactions in a continuous loop of discovery: creating formulas from the examples we perceive, applying those formulas, discovering more examples through searching out similarities and differences in those formulas, and continuously shifting and changing the formulas to find a higher truth, that is, discovering the nature of nature in terms of heuristics. **In this journey of discovery, we are using our mental faculties to co-evolve with—and co-create—the world within which we interact**, and we are inviting you as the reader or listener along on the journey. The number of concepts that we create is directly related to the number and level of interactions that we have with our environment. *The more we think, the more we can think.*

Critical Thinking

Critical thinking, which includes the integrative competencies of systems thinking, complexity thinking and knowing, as well as Knowledge Capacities introduced below, provides ways for individuals to make sense of and evaluate the value of data and information, whether informational or opinion-based. In critical thinking, we are always searching for larger concepts, larger truths. Reconstruction, assessment, and evaluation are used to discern the information; to determine its status as deductive, inductive, or irrational; and to judge its accuracy and reliability. Deductive reasoning is a logical progression, with the conclusion based on multiple premises assumed to be true. Inductive reasoning starts with an observation or question, and then, by examining related issues, tries to figure out a theory. Both are involved with the search for truth.

What has more recently been discovered through the use of mathematics—specifically, transreal numbers—is that learning includes forms of apprehension of reality not occurring through analytical means. This means that it is possible to evaluate non-propositional objects (such as sensations or feelings) as having a truth value, bringing heretofore discarded "truths" into the realms of deductive and inductive reasoning.[227] This concept is expanded in Chapter 8.

Critical thinking includes, but is not limited to:

- Open-mindedness.

- Engagement in the constructive challenging of ideas and concepts to make balanced decisions and extract insights.

- An awareness that groups and other individuals can shape issues and may have their own agendas and interests in mind, and the ability to identify those agendas and interests.

- The ability to make detailed observations, question, analyze, make connections, and try to make sense of a situation, a set of behaviors, or a single piece of information out of context.

- Recognizing that there is no single right or wrong way to interpret information, but many different ways, and the ability to explore each interpretation in terms of its strengths and weaknesses.

The intent of critical thinking is to allow individuals to form their own opinions of information resources based on an open-minded evaluation of the features and components of those resources by using, for example, problem-solving skills. These skills would include listening, considering other points of view, negotiating, and evaluating one's own mental models, including beliefs, values, and perceptions. Critical thinkers consider not only the data and information, but the *context* of the resource, questioning the clarity and strength of reasoning behind the resource, identifying assumptions and values, recognizing points of view and attitudes, and evaluating conclusions and actions. See TOOL 1. Critical thinking is sorely needed in today's environment of propaganda, disinformation and misinformation.

****

TOOL 1: Situational Backtalk

The knowing of a situation is a unique mental skill. Situational backtalk is a tool for perceiving or sensing surroundings to gain information. Recall that all knowledge is situation dependent and context sensitive. The back talk of the situation refers to both the situation *and* the context, shifting your framework to look from the point of view of the situation, and taking into account people, relationships, networks, events, culture and structure.

In systems theory language, you would identify the system in trouble and its subsystems, exploring the relationships among these systems and the state of boundary conditions, etc. In complexity language, you might look for and explore feedback loops, emergent properties, nonlinearities, time delays, trends and patterns, events and processes, sinks and sources, and so forth.[228]

No matter what language best makes sense to you as the idea generator, the decision-maker, the bottom line is to look at (and listen to) the current situation at hand, not to rely on the way things have been done in the past. This is a rethinking of the way to look at the world emerging from the ability to *listen with humility*. (See Chapter 11.)

If you can't see the world from the perspective of others, you cannot understand others; if you can't see the world, it's very hard to know how to cope with it.[229]

STEP (1): Be sure you have read all the reports and/or support papers on the situation or incident on which you choose to focus. Look at and understand the importance of any available pictures, memos, emails, etc.

STEP (2): Disconnect yourself from all distractions, surrounding yourself with any artifacts of the situation at hand. This may include situating yourself physically in the place where an event occurred, or having those pictures, reports and/or support papers, etc. spread out in front of you.

STEP (3): Physically close your eyes and take several deep breaths, relaxing your body.

STEP (4): Starting at the very beginning, that is, *before* the situation or incident occurred, building the context, then slowly re-create the situation or incident in your head. If you need to check the materials around you for details, open your eyes and do so, then start the process again from STEP (3), adding additional details as perceived.

STEP (5): When you are comfortable with your re-creation of the process, begin at STEP (3) again, only this time with your eyes open and taking notes as you re-create the process in your head. As you move through the process, every time there is a change of some nature briefly close your eyes and look at the details surrounding that change. Then open your eyes and record those details in your notes.

STEP (6): Review your notes, paying special attention to the details surrounding points of change. If clarification is needed, repeat STEP (5) with a focus on those details. *Ask:* What can be learned from the surroundings? Are there any items that may have had a strong impact on this situation?

STEP (7): Record the names and/or descriptions of other people involved in the situation or incident.

STEP (8): Focusing on one individual at a time, take a few minutes to think about—and write down in your notes—a general description of how each individual may see the world differently than you. Then move through your re-creation and try to see how that individual, *through the lens of their unique world view*, would perceive the situation. Make notes as insights emerge. Do this step for each individual involved in the situation or incident. *Then ask:* Which ones had a strong impact on this situation? What was that impact? What are the relationships among these individuals? Which ones are affected by this situation (noting short-term and long-term effects)?

STEP (9): Step back and connect the dots, that is, bring all of this together. *Ask:* What new has been learned by this exercise? How can this benefit the situation? What are the necessary next steps that should be taken? And when you are ready—mentally, emotionally, and spiritually comfortable with those next steps—ACT.

****

With the advent of the internet, and the increase in the amount of data and information available, critical thinking becomes essential to functioning in the world. Halpern states:

> [Critical thinking] is used to describe thinking that is purposeful, reasoned and goal directed—the kind of thinking involved in solving problems, formulating inferences, calculating likelihoods, and making decisions when the thinker is using skills that are thoughtful and effective for the particular context and type of thinking task.[230]

This includes meta-thinking, or *thinking about our own thinking and decision processes* in order to continually improve them.

The successful self, a critical thinker, has a colossal collection of different approaches to deal with different situations and contexts. As Minsky describes:

> *If you "understand" something in only one way, then you scarcely understand it at all—because when you get stuck, you'll have nowhere to go. But if you represent something in several ways, then when you get frustrated enough, you can switch among different points of view, until you find one that works for you!*[231]

This "switching" can be done consciously (as is the case in the Situational Backtalk Tool and in terms of the Knowledge Capacities discussed below and in Appendix E) or it might originate in the unconscious. For example, Middlebrook says an individual is not a single self, but rather many selves, which shift and change as the individual moves from situation to situation.[232] *We become what the situation demands.* We use the terminology "sub-personalities" to describe these sets of chunked knowledge that come to the fore when they are needed.[233] Brown offers that sub-personalities are "patterns of feelings, thoughts, behaviors, perceptions, postures and ways of moving which tend to coalesce in response to various recurring situations in life."[234] The more an individual experiences similar situations, the stronger these sub-personalities become, eventually quite capable of driving actions, which may or may not be recognized and which may or may not in the instant be changeable by a conscious decision or act of will. As Rowen describes, this experience of being *taken over* by a part of ourselves "lasts as long as the situation lasts—perhaps a few minutes, perhaps an hour, perhaps a few hours—and then changes by itself when we leave this situation and go into a different one."[235]

With a larger awareness, our conscious mind has choices in terms of ways to think and act. As Minsky says, "Each of our various Ways to Think results from turning certain resources on while turning certain others off—thus changing the way one's brain behaves."[236] AND the way one acts in response to that thinking! Resources refers to structures and processes ranging from perception to action (see TOOL 2 below). Ways to think include self-conscious emotions and feelings, self-reflective thinking, reflective thinking, deliberative thinking, learned reactions and instinctive reactions. Note that self-conscious emotions and feelings are included as a way of thinking. This has been affirmed in sentient beings through the mathematics of transreal numbers.[237]

****

TOOL 2: Activating New Resources

Next time you have a problem facing you, try one of these approaches to change your resources. The **Ways to Think** below are some of those suggested by Minsky[238] accompanied by our explications.

Reasoning by Analogy. Connecting to a similar problem in the past, note the patterns of activity and relationships among activities, and apply those patterns to the new situation. Remember, knowledge is situation dependent and context sensitive. The learning is in understanding the patterns and relationships in one situation and being able to extrapolate those to another situation. When this pattern thinking occurs across different domains of knowledge, it enters into the realm of wisdom.

Dividing and Conquering. A larger problem may be made up of a number of smaller problems, which are easier to resolve. A corollary to this in complexity thinking is to simplify. While simplification can lead to only a partial solution, a solution to part of the problem can serve as a stepping-stone for resolving the larger problem.

Reformulating. Look at the issue through different lenses. Minsky suggests looking for more relevant information. As he examples, "We often do this by making a verbal description—and then 'understanding' it in some different way!"[239] Reformulating also plays a role in applying different Knowledge Capacities.

Planning. This is a goal setting and/or scenario planning approach and exploring how they affect each other. Working toward sub-goals and exploring scenarios helps build an understanding of what actions to take. See Chapter 10 for a detailed discussion of scenario planning.

Elevating and Demoting. When there are too many details bogging you down, try thinking about the situation in more general terms. Move from specifics to generalities (a higher concept) and look at the problem from that viewpoint. Conversely, if your problem is too vague, then try making it more concrete by adding detail.

Self-reflection. Address questions to your self. *Ask:* Why is this problem so hard? What might we be doing wrong? How might we make this easier? What can we do differently? Who might we ask to partner in solving this problem? So many of our answers reside within if we just take the time to dialogue with our selves.

Logical contradiction. Take the opposite stance (this is a favorite, and is similar to the Knowledge Capacity of Reversal in Appendix E). As Minsky says, "Try to prove that your problem cannot be solved, and then look for a flaw in that argument."[240] What fun! A related approach is to set up a debate. This is the inducing resonance approach (see Chapter 8).[241]

External Representations. Writing ideas down—making notes and/or creating records, or building a PowerPoint presentation—helps organize and keep track of your thoughts. The very act of creating diagrams and graphics helps in understanding the relationships among elements of a problem. Even if you don't find a solution, you will better understand the problem and be able to better articulate it to others.

Imagining. Use your imagination to engage the problem. When we imagine we aren't taking any risks. As you are imagining, *Ask:* How does it feel? Following are three specific ways to use your imagination (and feel free to *imagine* other ways that would be fun!)

> ***(1) Wishful Thinking***. Imagine all the resources (including time) that you want. What would you do? If you can't imagine it, reformulate the problem. Try simulating actions inside your mental models.

> ***(2) Impersonation***. Imagine you are someone else—a teacher, a scientist, a factory worker. *Ask:* What does this problem look like from this viewpoint? What actions would make sense?

> ***(3) Storytelling***. Imagine you are a storyteller in the future, telling the story about this problem and how it was solved. Take your time and enjoy the process. If you get stuck, imagine an adult or child asking a

simple question, and then continue on until you get to the end, whatever you imagine that end to be. Above all, have fun!

****

The need for Critical Thinking is not new—howbeit a critical necessity in today's environment—such that there are many excellent texts available. Thus, it is not the intent here to cover the sphere, but rather to remind us that Critical Thinking continues to be "critical" for individual (and organizational) success—and indeed, often for survival—in our current and future world.

Knowledge Capacities

A less known body of research is that on Knowledge Capacities, which provides focused thinking approaches in support of Critical Thinking. In a CUCA environment where surprises emerge and must be quickly handled, capacity is more important than capability for sustainability over time, and this includes in the planning stage. For example, when you plan an outdoor event, you create an alternate plan for inclement weather. When you're raising a family, you try to put aside a bit for kid's learning experiences such as a school visit to the museum or a trade school. When you study a domain of knowledge in college, you learn the theory right alongside pragmatic applications.

One of the authors, who taught nuclear physics for the U.S. Department of the Navy, recalls a story about a time when, in response to needed budget cuts, classes dealing with theoretical physics were on the chopping block. However, a short scenario quickly demonstrated the need for theory. Imagine a nuclear submarine with a full crew on board, deep under the North Pole when the reactor fails. While there are a large number of possibilities for failure, this particular situation did not appear in the textbooks. Ask yourself, do you want a nuclear officer well-trained in pragmatic responses, or do you need someone who also has the capacity to explore, who understands the theory and complicated, even complex, relationships within the core which could have caused this failure? Both theory (capacity) and practice (pragmatic actions) are important and necessary in a CUCA environment.

Knowledge Capacities are sets of ideas and ways of acting that are more general in nature than competencies, more core to a way of thinking and being, that change our reference points. *They provide different ways for us to perceive and operate in the world around us.* The analogy here would be building an infrastructure of sorts relating to the mind/brain (information, knowledge and the structure and connection strengths of neurons within the brain).

Knowledge Capacities, developed by combining senses, complement six different ways that humans operate in the world. These are: looking and seeing, feeling and touching, perceiving and representing, knowing and sensing, hearing and listening, and acting and being. Each of these sets has two concepts introduced because, while they are related, there is clarity added by coupling the concepts. Each area is briefly addressed below. See Figure 7.

With our compliments (and apologies) to *Encarta World English Dictionary*, we attach specific meanings to these six ways humans operate in the world, meanings that suggest ways of observing and processing the events that occur in our lives. Knowledge Capacities are all about expanding the way we see those events in order to raise our awareness and, in the case of problem solving and decision-making, offering new ideas and an expanded set of potential solutions.

Looking and Seeing: To direct attention toward something in order to consider it; to have a clear understanding of something. (*Examples*: Shifting Frames of Reference, Reversal)

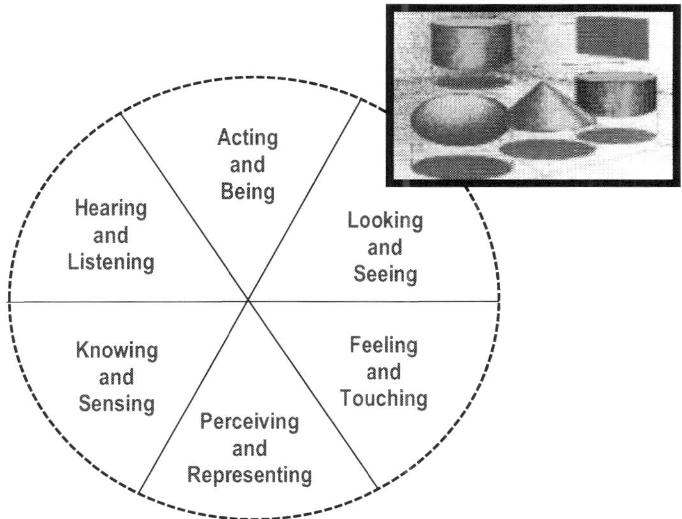

Figure 7. Ways humans operate in the world.

Feeling and Touching: The sensation felt when touching something; to have an effect or influence on somebody or something; to consider the response of others being touched. (*Example*: Emotional Intelligence)

Perceiving and Representing: To acquire information about the surrounding environment or situation; mentally interpreting information; an impression or attitude; ability to notice or discern. (*Examples*: Learning How to Learn, Comprehending Diversity, Symbolic Representation)

Knowing and Sensing: Showing intelligence; understanding something intuitively; detecting and identifying a change in something. (*Example*: Engaging Tacit Knowledge)

Hearing and Listening: To be informed of something, especially being told about it; making a conscious effort to hear, to concentrate on somebody or something; to pay attention and take it into account. (*Example*: Active Listening, Humility)

Acting and Being: To do something to change a situation; to serve a particular purpose; to provide information (identity, nature, attributes, position or value); to have presence, to live; to happen or take place; to have a particular quality or attribute. (*Examples*: Orchestrating Drive, Instinctual Harnessing, Organizational Zoo)

To build a further understanding, some short examples of Knowledge Capacities are provided as Appendix E. Specifically, these examples include: Learning How to Learn (perceiving and representing); Shifting Frames of Reference (seeing and looking); Reversal (seeing and looking); Comprehending Diversity (perceiving and representing); Orchestrating Drive (acting and being); and Symbolic Representation (perceiving and representing). Humility (hearing and listening) is the focus of Chapter 11.

As we mature and age—and as our senses lose some of their effectiveness—the ways we perceive and operate in the world change, and with them a preference for higher engagement of the intuitive emerges. Thus, for example, the Knowledge Capacity of "Knowing and Sensing" would dominate decision-making. This is consistent with the following neuroscience finding in the area of aging: **Despite certain cognitive losses, the engaged, mature brain can make effective decisions at more intuitive levels**.

Historically there has been a widespread belief that as individuals grow older, their mental powers decrease due to the continuous loss of neurons. Recent research contradicts this myth. It is now clear that, barring diseases, accidents, or genetic issues, so long as individuals use their mind/brain and continue to actively exercise and think, their brain capacity will maintain a good capability in healthy bodies. Although neurons do continue to die as one ages, they also can be created in some parts of the brain as long as it stays active. However, with the diminishing of senses, the intuitive—which has been developed through years of living—plays a greater role. Goldberg points out that the aging brain can accomplish mental feats that are different than younger brains.[242] Further, he makes the point that although older people forget names, facts, and words, they have the capability of remembering high-level patterns and meaningful insights that we often consider as wisdom. This capability would impact many of the sub-elements of the adult experiential learning model.

Creative Thinking

Creativity is the emergence of new or original patterns (ideas, concepts, or actions). Mihaly Csikszentmihalyi, Professor of Psychology at the University of Chicago, notes that the term creativity originally meant to bring into existence something genuinely new that is valued enough to be added to the culture.[243] Quantum physicist Amit Goswami offers his definition of creativity as the creation of something new in an entirely new context; with *newness of the context* the key.[244]

Theoretical biologist Rupert Sheldrake says that creativity is a profound mystery precisely *because* it involves the appearance of patterns that never existed before.[245] And Plato viewed creativity as both mysterious and divine:

> *For the poet is an airy thing, a winged and a holy thing; and he cannot make poetry until he becomes inspired and goes out of his senses and no mind is left in him ... not by art, then, they make their poetry ... but by divine dispensation.*[246]

The romantic could substitute the word "exceptional" for "divine", glorifying creative people as gifted with talent (insight or intuition) that others lack.

As a working definition, we see creativity as emerging new or original ideas or seeing new patterns in some domain of knowledge.[247] In other words, creativity can be considered as *the ability to perceive new relationships and new possibilities*, seeing things from a different frame of reference, realizing new ways of understanding, or having insight.

Everyone is creative. As a simple way of understanding this, Boden breaks creative thought (or creative people) into two types: P-creative (psychological or personal) and H-creative (historical).[248] P-creative ideas are fundamentally novel with respect to the individual mind, the person who has them, and H-creative ideas are historically grounded, fundamentally novel with respect to the whole of recorded human history. The point is that the H-creative ideas, which by definition are also P-creative, are the ones that are socially recognized as creative, but P-creative ideas are *possible in, and do occur in, every human being*.

Neuroscience findings support that everyone is creative, and thinking more creatively is a choice, with creativity an inherent capability of every mind/brain.[249] Learning consists of building new patterns within the mind that represent ideas, concepts, and capabilities, and processes that are internally imagined (creative imagination), or represent some interpretation of external reality. According to Stonier, thinking involves *the association of different patterns* within the mind/brain.[250] Since memory is distributed throughout the brain, these associative processes may contain random associations that can result in new ideas, concepts, or approaches. This is considered as ordinary creativity and can be very valuable as a form of self-learning. Two related findings from the ICALS research in the area of Creativity are:

Free-flow and randomly mixing patterns create new patterns. This relates directly to creativity and is one way to describe the interaction among neuronal networks and patterns to create new ideas. As described, this is a random emergent process stimulated by the continuous activity of the mind. While this activity is outside of conscious internal focusing or environmental external triggering, it is based on past exposure, experience and learning; thus, future activity may be nurtured.

Accidental associations can create new patterns. As an extension to the above finding, this makes the point that creativity can happen by accident *within an active mind that plays with ideas, connections, and their relationships*.

These new patterns may represent insights or increased understanding of a situation, or perhaps a clarification of the meaning of a situation. They may come from conscious consideration or the unconscious mind "playing with" ideas that are experienced during the night. Since much thinking is done by the unconscious during sleep, learning can be enhanced by making use of the mind's capability for creativity. Such practices as meditation, lucid dreaming and hemispheric synchronization can serve to improve creativity and problem solving. For example, a third finding in the area of Creativity is: **Meditation quiets the mind**. Again, meditation requires some level of control and discipline, and it can significantly enhance the ability to focus attention. In this context, quieting the mind means to reduce the noise that "bedevils the untrained mind, in which an individual's focus darts from one sight or sound or thought to another like a hyperactive dragonfly, and replace it with attentional stability and clarity."[251] Quieting the mind allows the mind to focus attention on potential relationships that may create new ideas. See TOOL 10, Quieting the Mind, in Chapter 10.

Closely related, a finding in the area of exercise and health is that **meditation and other mental exercises can change feelings, attitudes and mindsets**. Goldberg suggests that mental exercise can stimulate neuronal growth, and this might be a pattern for growth in other parts of the brain as well as supporting creativity.[252] Mental exercises would likely influence many aspects of the adult experiential learning model.

It would be difficult to cover the many approaches that have been taken to explore the subject of human creativity. Just to provide an idea of the massive amount of research, these include mystical approaches such as Rudyard Kipling's Daemon;[253] pragmatic approaches such as the lateral thinking of Edward deBono;[254] the psychodynamic approach introduced by Freud;[255] psychometric approaches, which involved laboratory testing as exampled by Guilford[256] and Torrance;[257] cognitive approaches such as those forwarded by Finke, Ward and Smith,[258] which described a generative phase and exploratory phase to creativity; social-personality approaches focused on personality and motivational variables in combination with the sociocultural environment as exampled by Amabile[259] and Eysenck;[260] evolutionary approaches instigated by Campbell[261] and picked up by Perkins[262] and Simonton,[263] saying there were two steps to creativity, blind variation and selective retention; and confluence approaches such as the work forwarded by Csikszentmihalyi,[264] Amabile,[265] and Gruber.[266] So, there is much literature out there to explore.

Andreasen suggests there are five circumstances that create *cradles of creativity*.[267] These are (1) an atmosphere of intellectual freedom and excitement; (2) a critical mass of creative minds; (3) free and fair competition; (4) mentors and patrons; and (5) at least some economic prosperity. And we add an enriched environment, which is one of the findings in the area of creativity in the ICALS research.

An enriched environment can produce a personal internal reflective world of imagination and creativity. As part of the environment, this item may affect a number of the aspects of concrete experience and active experimentation. A rich environment entering the mind/brain/body through experiences increases the formation and survival of new neurons, stimulating the mind to associate patterns and create new possibilities. The literature suggests that an enriched environment contains many interesting and thought-provoking ideas, pictures, books, statues, etc. Byrnes says that in such an enriched environment there are physiological changes in the brain, specifically, thicker cortices are created, cell bodies are larger, and dendritic branching in the brain is more extensive.[268] These changes have been directly connected to higher levels of intelligence.[269] While a natural/nature setting is often associated with an enriched environment, in today's world, the context of an enriched environment can include the augmentation of the mind/brain with technologies.[270]

Gell-Mann demonstrates a cradle of creativity by describing a shared experience in conceiving creative ideas, which, as introduced above, is one of Andreasen's five circumstances. Gell-Mann was one of a small group of physicists, biologists, painters and poets that joined together in 1970 to talk about their experiences in getting creative ideas. As he describes,

> *The accounts all agreed to a remarkable extent. We had each found a contradiction between the established way of doing things and something we needed to accomplish: in art, the expression of a feeling, a thought, an insight; in theoretical science, the explanation of some experimental facts in the face of an accepted 'paradigm' that did not permit such an explanation.*[271]

Each person had started with a focus on a problem and the difficulties they were trying to overcome. When further conscious thought was deemed useless, that stopped, although each continued to carry the problem around with them. Then, suddenly, while doing something quite different—shaving, cooking, or engaged in simple conversation—an idea just popped into thought.

Thus, *the unconscious appears to play an essential role in creativity.* This is affirmed by the ICALS neuroscience finding in the area of the unconscious: **The unconscious plays a big role in creativity**. This idea has been emphasized by researchers in the area of creativity, in particular in the work of Andreasen[272] and Christos.[273] During the process of creating new ideas and approaches to solving problems or making decisions—as occurs in the abstract conceptualization mode—the unconscious can be taken into account and utilized to its fullest.

Henri Poincaré suggests that creativity *tugs* on the unconscious.[274] He builds on the four stages of creativity identified by Gram Wallas in his book, *The Art of Thought,* published in 1926. These are preparation, incubation, illumination, and verification. The initial phase, *preparation,* is the conscious probing of a problem or an idea. The second phase, *incubation,* occurs while the conscious mind is focused elsewhere, and may last for minutes, months, or years. During this time, the unconscious mind may well be contemplating the challenge and trying to make sense of the situation. Poincaré credited this phase with the novelties denied through waking, rational thought. The flash of insight, or tug, comes in the third phase, *illumination,* when creative thoughts burst through the unconscious stirring of incubation into the conscious, where it can be explored and tested in the fourth phase, *verification* or *validation.* Underlying all this, of course, is the openness to new ways of doing things (see Chapter 11 on humility).

Poincaré visualized the creative thought mechanism of the unconscious as similar to the workings of the atom.

> *Figure the future elements of our combinations as something like the hooked atoms of Epicurus. During the complete repose of the mind, these atoms are motionless, they are, so to speak, hooked to the wall ...; On the other hand, during a period of apparent rest and unconscious work, certain of them are detached from the wall and put in motion. They flash in every direction ... [like] a swarm of gnats, or, if you prefer a more learned comparison, like the molecules of gas in the kinematic theory of gases. Then their mutual impacts may produce new combinations.[275]*

The role of this preliminary conscious work is to mobilize certain of these atoms,

> *... to unhook them from the wall and put them in swing ... After this shaking up imposed upon them by our will, these atoms do not return*

to their primitive rest. They freely continue their dance. Now, our will did not choose them at random; it pursued a perfectly determined aim. The mobilized atoms are therefore not any atoms whatsoever; they are those from which we might reasonably expect the desired solution.[276]

Then, there is a bursting forth. As Tchaikovsky describes,

Generally speaking, the germ of a future composition comes suddenly and unexpectedly.... It takes root with extraordinary force and rapidity, shoots up through the Earth, puts forth branches and leaves, and finally blossoms. I cannot define the creative process in any way but this simile.[277]

There were three findings related to Creativity in the ICALS research which specifically relate to the unconscious. These are:

Conscious and unconscious patterns are involved with creativity. Creative insight is the result of searching for new relationships between concepts in one domain with those in another domain.[278] It creates a recognition and understanding of a problem within the situation, including the how and why of the past and current behavior of the situation. It is often the result of intuition, competence, and the identification of patterns, themes and cue sets.[279] Insight may also provide patterns and relationships that will anticipate the future behavior of the situation.

The unconscious produces flashes of insight. As was introduced previously, spurious memories can generate new ideas that combine in different ways to make new associations.[280] This relates to uncontrolled intuition.

Volleying between the conscious and the unconscious increases creativity. "New ideas are generated through the process of shifting from conscious to unconscious as the mind contemplates and searches for solutions."[281] The process of shifting between conscious and unconscious thinking makes use of the memories and knowledge in the unconscious and the goals and thinking of the conscious mind. This increases the chances of associating conscious ideas to create new ones, which supports innovation. While this occurs in the transition from the waking state to the sleeping state (and vice versa), to do this while awake requires some level of control and discipline in implementation. The body asleep, mind awake approach achieved through hemispheric synchronization is an example.

Ultimately, Sternberg, who takes a confluence approach, that is, integrating all the other approaches, says that *creativity is a choice*. In his words, it is a decision which has three parts: "The decision to be creative, the decision of how to be creative, and implementation of these decisions."[282] Thus, he agrees that *creativity can be developed*. Sternberg forwards an investment theory of creativity, requiring a confluence of six interrelated resources: intellectual abilities, knowledge, styles of thinking, personality, motivation and environment. His bottom line is that while *creative intelligence* is a part of human creativity, more is needed. As he describes that "more",

> *Creativity also involves aspects of knowledge, styles of thinking, personality, and motivation, as well as these psychological components in interaction and the environment. An individual with the intellectual skills for creativity but without the other personal attributes is unlikely to do creative work.*[283]

There is a state of *extraordinary creativity* which can be developed. Extraordinary creativity relates to certain individuals who have the capacity to repeatedly create novel ideas or processes that push the limits of understanding or application. These are the truly creative people who keep an open mind, maintain a high curiosity, and investigate many paths toward new possibilities. They have particular mental capability and capacity for challenging the status quo and seeing things from unique perspectives[284] while exhibiting humility. These are people who have flashes of insights and moments of inspiration. And no doubt those characteristics describe some of you who are reading this text.

The experience of extraordinary creativity does not appear to be designed or purposed; in fact, to labor or work toward it can move us in the opposite direction. However, English Professor Robert Grudin thinks that we can practice *deserving it* through embedding habits in our daily lives.[285] Describing these habits as a demanding and integral code, Gudin collectively calls these habits the *ethics of inspiration*. This utopian code includes: love of one's work, fidelity, concentration, love of problems, a sense of the openness of thought, boldness, innocence, an uncensored mind, civility, gentleness, and liberty. This is a good set. As Grudin describes, "It is a garden of mind, recalling Eden. Rousseau's vision of philosophy as a recapturing of nature, and Milton's idea that the purpose of education was to rebuild the ruins of the Fall."[286]

A most remarkable integration of creativity and education in the mainstream of large-scale school systems was the result of the work of Edward de Bono. He died in 2021 at the age of 86 having lived a life somewhat akin to the endeavors of a "renaissance man". He was born in Malta where he achieved a medical PhD, from which he branched out internationally, achieving status with faculty positions at leading universities and popularity with considerable following at an endless number of large corporations and organizations. One of his prolific endeavors was documenting and writing especially about his approach to creativity. In total his books numbered 86, and have been translated into 46 languages. Part of the vast effort with translation was due to his creative thinking techniques which are being used in school curriculums in more than 20 countries. The depth of his work is not only evidenced by the popularity of his programs and writings, it is also noted in his naming on the short list for the Nobel Prize for Economics in 2005.

It is very interesting that de Bono's early work in the late 1960's was characterized by describing how the brain's nerve networks functioned as an information system. From this pioneering work, drawing from neuroscience, he formulated his concepts with natural thinking, logical thinking, mathematical thinking, and lateral thinking. Lateral thinking is the most popular and may be generally described as looking at a problem, situation, question, or challenge from a different viewpoint, you might say *looking at it laterally or sideways*. As you refer to his books you will find that he developed rather detailed approaches to cause users to shift their thinking.

One of the findings in the area of Creativity from the ICALS research is **extraordinary creativity can be developed**. A basic operation of the brain is that of associating patterns within the mind to create new patterns (thoughts, ideas, and concepts), what is referred to as ordinary creativity. (Associative patterning was introduced in Chapter 3.) As forwarded above, everyone possesses ordinary creativity, the creation of new ways of doing things in their daily lives through discussing new ideas and developing insights and deeper understanding. In this sense *all learning is an act of creation*.

Not only can "ordinary" people become highly creative—noting that each person is unique and thus there is little agreement for the term "ordinary"—but their creativity can become an inspiration to others. As Green writes:

> *When we think about inspiration, what inspires us most are **ordinary people who have done extraordinary things** [emphasis added]. We appreciate when someone has the ability and willingness to be selfless, creative, innovative, or just dares to be different .... The beautiful thing about inspiration of this kind is that "ordinary" part .... [E]ach of them came from backgrounds of great poverty and difficulty. Each of them faced giant mountains to climb. They managed to reach the summit of those mountains not simply because they were great leaders, but because they were not afraid to be who they were. They were authentic.*[287]

There are many techniques, both individual and group processes, that serve to open the mind to new possibilities and stimulate creative thinking. Such processes use all forms of learning with some being from concrete experience, dialogue, or social interaction; others from internal reflection and comprehension. In either case, if learners understand the dangers of assuming they already "know" the answer, they may deliberately keep an open mind and improve both the efficiency and the effectiveness of their learning and creativity. (See Chapter 10 on humility.)

Becoming aware and understanding that every healthy mind is capable of creativity, and that the rules and practices of creative thinking are available to anyone, opens the door to enhanced individual and organizational creativity and learning capacity.

Innovative Thinking

Creativity comes exclusively from people, a capacity to see new ideas from associating internal and external information, a capacity. Some experts believe that creativity should also have utility and lead to a new product or process, that is, innovation. Innovation means the creation of new ideas *and* the transformation of those ideas into useful applications; thus, the combination of creativity and contribution as operational values promote innovation.

As an idea generator, knowledge is the currency of creativity and innovation, and knowledge cannot exist without information. As information flows freely and is generally available to all—and as people recognize the power of and creative potential in the mind/brain and learn to tap into that power and creativity to create change for the greater good—each individual has the ability to be an innovator, that is, providing a "new" product or process in their domain of focus.

The relationship between information, knowledge, creativity and innovation is entangled. Knowledge is effectively applying information (in terms of producing the expected result). Innovation is effectively applying creativity (in terms of a useful process or product). While knowledge comes from the past and creativity requires knowledge, both knowledge and creativity are capacities which can be applied in the present (actual) or engaged in the future (potential). Further, they both emerge from the associative patterning process of the brain, that is, the unique complexing of external and internal information (organized patterns). See Figure 8.

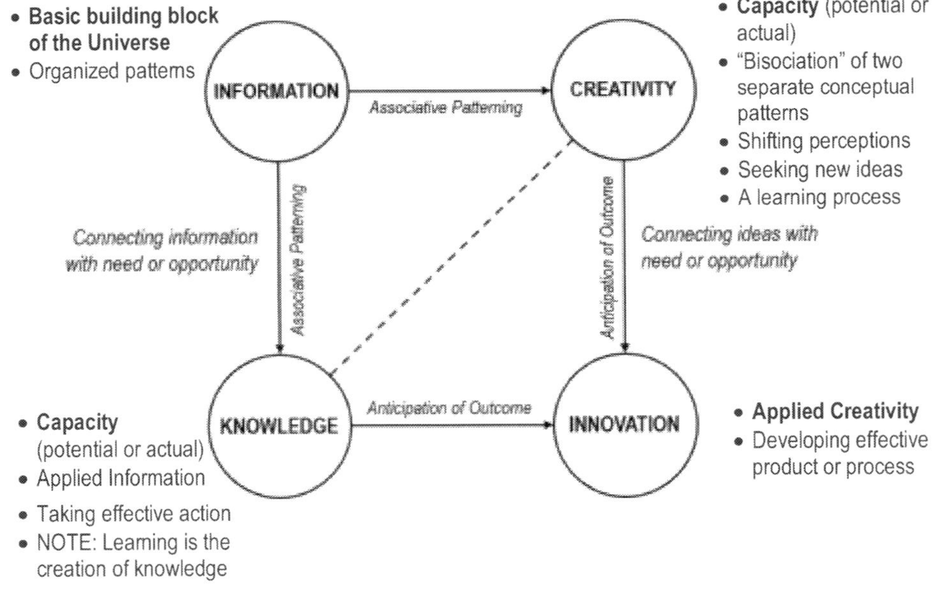

Figure 8. Exploring the relationships among information, knowledge, creativity and innovation.

In Figure 8 there is a dotted line between knowledge and creativity, which when combined lead to innovation. Note that innovation is not necessarily an immediate result. As Fritz Machlup said in the early 1960's in his seminal work on the knowledge economy, "We shall have to bring out clearly that this is not a simple unidirectional flow from one stage to the next, from inception to development, to eventual adoption, but there are usually cross-currents, eddies, and whirlpools."[288]

Past experiences, feelings, knowledge, goals and the situation at hand all influence how creative an individual will, or can, be. It is the context of the activity or situation at hand (need, challenge, etc.) that triggers the putting things together (bisociation) in an unusual way to create (and recognize) something that may be new and potentially useful (innovation). Thus, knowledge—context sensitive and situation dependent—serves as an action lever for innovation.

In Summary

Here's a few points we have learned from these findings. Since both *conscious and unconscious patterns are involved in creativity*, it just makes sense that *volleying between the conscious and unconscious increases creativity*. Further, it makes sense that the environment in which you are conscious is enriched, while, simultaneously, meditation is used to quiet the mind, that is, *moving from an outer enriched state to an inner enriched state*.

Recognizing that we are multidimensional and that *the unconscious brain is always processing*, randomly mixing patterns to create new patterns, it can be seen that the experiences to which we expose ourselves and the people with whom we interact directly impact our creativity. Even when we are unconsciously unaware of our surroundings and the activities underway, information is being received and processed. The flow of our everyday lives really counts!

Chapter 6
SELF as the Ground of Learning

The Growth of Self ... The Power of Belief ... The Subject-Object Relationship ... The Role of the Unconscious ... The Fickleness of Memory ... Forgetting ... In Summary

TOOLS: (3) Self-Belief Assessment; (4) Sleep on It; (5) Letting Go.

Regardless of our frame of reference—DNA, experience, culture, family, thought patterns, beliefs, values, or emotions and feelings—each and every human being is unique. As introduced in Chapter 1, and worthy of repeating here, at the core of all this difference is the *self*, with a subjective mind, exploring the world from the inside out, a protagonist ready for action. Yet in the magical environment of the womb, the conscious self does not exist. The thinker of the future that we become (the self) is a fresh slate, even as energies entangle to create the unique web of associations and responses that will help ensure survival of the budding human, what we describe as the "personality".[289]

Self is an emergent quality of the human which moves beyond biological drives and cultural habits. As a working definition, self is defined as the totality of the conscious and unconscious mind, the brain and the body. While there is a close relationship of self to our understanding of consciousness, self is inclusive of the personality, and—as our consciousness expands and becomes co-creator of our reality—inclusive of every aspect of what it is to be human. As Csikszentmihalyi says, "Inside each person there is a wonderful capacity to reflect on the information that the various sense organs register, and to direct and control these experiences."[290] This is rather like a figment of our imagination, "something we create to account for the multiplicity of impressions, emotions, thoughts, and feelings that the brain records in consciousness."[291] *Self is who you choose to be*, although the mind can be difficult to control. As introduced in Chapter 1, Csikszentmihalyi forwards, "Nowadays learning to control the mind may have become a greater priority for survival than seeking any further advantages the hard sciences could bring."[292] We agree.

Self has many facets. We often perceive ourselves as the roles we play, the identities we take on, but those are just the *way we choose* to manifest self. Self is so much more. As American psychologist William James argues, "Although the self might feel like a unitary thing, it has many facets—from awareness of one's own body to memories of one's self to the sense of where one fits into society."[293] Thus, as noted earlier, there is no single point within the mind/brain/body complex where we could situate self. It is the interactions among *all* of these neuronal patterns, firings and connections—with a focus on conscious thought and choice—that create the self.

Similarly, contemporary neuroscience does not identify a separate neurological function or structure where self—or consciousness, which enables the recognition of self—exists. Self is the quantum leap that occurred after the emergence of self-reflective consciousness, a distinct self that could take charge of the domain of consciousness, and determine which feelings or ideas take precedence. "Having had this experience of something inside us directing consciousness we gave it a name—the self—and took its reality for granted. And the 'self' became an increasingly important part of human beings."[294]

Self—including self-referential memory, self-description, self-awareness and the personality—coevolves with, and is the creator of, its immediate environment, with physical life a process that occurs between the individual and its environment. Our self-awareness—which represents the unique ability to reflect on the past and potential future of ourselves, our world and the Universe—is possible *because* of the relationship between ourselves and our environment. This is an important area of neuroscience findings—that of epigenetics—that emerged in the ICALS research focused on experiential learning.

The Growth of Self

Lipton forwards that science is in the process of shattering myths and rewriting the belief that we are controlled by our genes. Epigenetics is the study of mechanisms by which the environment influences gene activity. As Lipton posits:

> *Genes are simply molecular blueprints used in the construction of cells, tissues and organs. The environment serves as a 'contractor' who reads and engages those genetic blueprints and is ultimately responsible for the character of a cell's life. It is a single cell's*

'awareness' of the environment, not its genes, that sets into motion the mechanisms of life.[295]

Thus, the genome (an organism's genetic material) is more fluid and responsive to its environment than was previously thought.[296] In addition, the neural organization of our brain is not set in stone at birth.[297] This relates specifically to the following two neuroscience findings in the area of epigenetics emerging from the ICAS research.

Genes are not destiny. The environment can change the actions of genes via non-expression. This area on epigenetics does not necessarily directly impact many aspects of the adult experiential learning model. However, it plays a significant role in alerting learners to the influence they have on their own learning, which can significantly enhance learning efficiency. What has been discovered is that gene expression is significantly influenced by the local environment of the cells, which to some extent is under the control of individuals. As Lipton describes, "The belief that we are frail, chemical machines controlled by genes is giving way to an understanding that we are powerful creators of our lives and the world in which we live."[298] And, as C.A. Ross proposes, "Genes for brain growth and development are turned on and off by the environment in a complex, rich set of feedback loops."[299]

All brains are made of the same chemicals, the same types of neurons, the same myelin sheaths, and the same proteins. Thus, we might wonder why there is such variation in individual learning capacities. As neuroscience removes one myth after another it becomes clear that, aside from disease or physical damage, *the role of nurturing and the environment have significantly more impact on the learning capacity of the individual than previously thought*. This introduces a second neuroscience finding in the area of Genetics from the ICAS study.

Genes are operating options modulated by inputs from the environment, resulting in behavior. The environment referred to here is predominantly the local environment of the cells within the body. However, the environment of the cells is, in turn, determined by the environment of local subsystems of the body. These subsystems (such as the amygdala relating to the emotions) may play a significant role in conditioning learners to understand and take action on the influence and capability they have in determining their own success.

Environmental influences such as nutrition, stress, and emotions, released into the bloodstream, can modify genes through chemicals without changing their basic blueprint. This finding also emphasizes the importance of good nutrition, stress control, and emotional management in maximizing learning efficacy. And those modifications can be passed on to future generations![300]

Further, as introduced in Chapter 1 emerging from our understanding of plasticity, what we think and believe can determine what we do. It is our actions that determine our success, not our genes.[301] In other words, belief is a powerful force of self, as supported by the neuroscience finding in the area of epigenetics on the power of belief.

The Power of Belief

What we believe leads to what we think leads to our knowledge base, which leads to our choices and actions, which determines outcomes. While this has been previously forwarded based on observation, neuroscience findings now confirm that if we believe that we cannot do something, our thoughts, feelings, and actions will be such that, at best, it will be much more difficult to accomplish the objective. If we believe we can accomplish something, we are much more likely to be successful, and this results from choice, not genes. During the last century, Henry Ford is credited with saying "If you think you can or your think you can't, you're right!" He was right. As Lipton points out:

> *Henry Ford was right about the efficiency of assembly lines and he was right about the power of the mind ... Your beliefs act like filters on a camera, changing how you see the world. And your biology adapts to those beliefs. When we truly recognize that our beliefs are that powerful, we hold the key to freedom ... we can change our minds.* [302]

From the neuroscience perspective, this finding is derived from several areas of thinking among diverse researchers, presenting a chain of logic that ties our beliefs to our actions and our successes—or failures. Our beliefs heavily influence our mindset or frame of reference—the direction from which we perceive, reflect, and comprehend an external experience or situation. Thus, beliefs influence how we interpret and feel about the information that comes into our senses, what insights we develop, what ideas we create, and what parts of the incoming information of which we are most aware. From these reflective observation and abstract conceptualization processes, we create our understanding and meaning of the external world. How we *see* the external world and how we emotionally feel about external events drives our actions and reactions. How we act and react to our external environment influences whether we are successful or not, that is, whether we achieve our goals or not. All these latter concepts may affect many of the sub-elements of the adult experiential learning model.

When introducing the ICALS theory in Chapter 4, we forwarded that *the beliefs and attitudes of the learner will affect learning and can be under the control of the learner if he or she so chooses.* This idea of "choice" holds a close connection to the discussion of epigenetics, that is, *it is how genes are expressed that determine our future.* Similarly, it is our choice of beliefs—and attitudes and emotions—that determine our actions. And now we recognize that environmental influences (which include stress and emotions largely based on beliefs) can modify our genes without changing their basic blueprint with, in some cases, those modifications passed on to future generations.[303] Recognizing that environments can *and do* change the actions of genes,[304] we can consciously and actively manage our feelings, attitudes, and mind-set through meditation and other mental exercises.[305] A first step, which makes sense, is *understanding our beliefs.* See TOOL 3 below.

****

TOOL 3: Self-Belief Assessment

Beliefs can separate us from others. From the viewpoint of openness to change, take an inventory of your current beliefs. This is an awareness-raising exercise in which you can explore your beliefs. Remember, change begins with awareness, and only you can change yourself.

STEP (1) Find a place where you can be comfortable and uninterrupted. Have a pad of paper and pen in front of you. Draw three columns on the paper, with the one on the left the thinnest and the other two equal in size.

STEP (2): One-by-one, bring each of the following concepts into your mind, putting the concept in the left-hand column, and quickly jotting down your feelings about each of them in the middle column, catching the first thoughts and feelings that come to mind. Here are the concepts: Voice, Accent, Spirit, Egotism, Judgment, Gratitude, Skin Color, Education, Diversity, Life, Arrogance, Experience, and Connection.

STEP (3): Consider other concepts that will trigger beliefs that you have, and repeat Step 2, bringing each of those concepts to mind. Push your edges, your comfort zone, including concepts based on old beliefs.

STEP (4): Review your thoughts/feelings.

STEP (5): Think about a person that you respect and admire, someone who has high moral standards. Imagine you are that person. Now, go back and review your thoughts/feelings from *that individual's perspective*. This will require a longer period of time as you reflect on how (as that individual) you feel about these items. Write your responses in the third column.

STEP (6): Now compare the responses in column 2 and column 3. Where the two responses are different, *Ask*: Are there any beliefs that caused me to respond in this way? Is there a belief I have that is not helping me be the best person I can be? Can I improve my response? How would I go about doing that?

STEP (7): As appropriate, change your response, and act on your thoughts.

****

Lipton points out that positive and negative beliefs not only impact our health, but every aspect of our life.

> *Consider the people who walk across coals without getting burned. If they wobble in the steadfastness of their belief that they can do it, they wind up with burned feet. Your beliefs act like filters on a camera, changing how you see the world. And your biology adapts to those beliefs. When we truly recognize that our beliefs are that powerful, we hold the key to freedom ... we can change our minds.*[306]

This emerged as a finding from the ICALS research: **Positive and negative beliefs affect every aspect of life**. This finding has widespread application throughout experiential learning, that is, positive and negative beliefs not only impact our health, but also most aspects of our lives. Remember: *When we recognize how powerful our beliefs are, we hold the key to freedom and we can change our minds.*[307]

A related but interesting emergent finding links beliefs and stress, specifically, **Belief systems can reduce stress through reducing uncertainty**. Since an individual's belief systems create a perspective on the world that is held to be correct, this introduces a level of perceived certainty that reduces stress and increases repeatability in an individual's life. Conversely, belief systems may also narrow perspectives and thereby reduce learning. This finding impacts sensing, feeling, understanding, and meaning in the adult learning model.

Lipton forwards that, "your beliefs act like filters on a camera, changing how you see the world. And your biology adapts to those beliefs."[308] To understand the connection between beliefs (patterns in the mind) and the physiology of the brain, consider the following rather dramatic scenario: You have just received a phone call from the local police telling you that your daughter has been killed in an automobile accident. Now envision your feelings, emotions, and behavior, the changes in your body, your actions, and so on. These are all real and can be observed and measured. Ten minutes later you receive a second call from the same policeman who tells you that there was a mistake and it was not your daughter who was killed. Now imagine the change in your behavior and feelings. All of this was created by your *own* thoughts and feelings based on the *belief* that your daughter had died, yet your daughter was perfectly fine and healthy the entire time. All of your bodily changes were created by beliefs and thoughts within your own mind. Thus, beliefs and biology are not independent; they are intimately connected through the relationship of patterns of the mind (in the brain) and the physiology of the brain. Patterns of neuronal firings and changing synaptic strengths can create and release hormones that can change the body. Since creating and associating thoughts in the mind are the domain of learning and these thought patterns exist in the physiological structure of the brain, learning and neuroscience are irrevocably interrelated.

In summary, considering what we believe and how we think determine what we do, it is our actions that determine our success, not our genes,[309] and our beliefs and thoughts and feelings are what will ultimately determine our ability to successfully navigate change, uncertainty and complexity in changing times.

The Subject-Object Relationship

The distinction of self is an important aspect of the ICALS expanded experiential learning model. At the highest level, self represents the perception of separation, with boundaries encompassing a set of physical, mental, emotional, and spiritual characteristics and beliefs. Self is a learned pattern and, while the understanding of self in a "self-world" distinction has shifted, nonetheless there has been considerable development through the years of the "I-as-subject" concept. For example, the substantive self-consciousness thesis considers the self as a persisting object with self-consciousness, the recognition of self as an object. Gibson forwarded that the self and the environment are co-perceived,[310] that is, our perception of self co-evolves with the perception of the environment as we move from infancy into adulthood. This awareness of self as a persisting object—which is the substantive self-consciousness thesis— supports connecting the self to the physical body and the perceived boundaries of the body as separate from our environment and other "selfs" within that environment, although it does not explain whether the self as an object is a mental or physical self.

As one object in a world of perceived objects, we can now consider the relationship between subject and object, bringing in the "I-as-Subject" Thesis—with "I" as the self-conscious subject of thought (me)—which includes experiences producing knowledge that help build the idea of self. From the viewpoint of associative patterning, we would say that interactions in the environment have provided incoming information that, when complexed with internal information, produced new patterns of thoughts and feelings. These patterns—part of a continuous information stream—produce our ever-changing internal map of self and the world within which we live, all of which we now understand is occurring in the cortical columns of the neocortex. A network of neurons mimics the structure of the body parts to which they belong and literally maps the body, creating a virtual surrogate of it, what Damasio refers to as a neural double.[311]

As IImmordino-Yang and Singh describe, "Humans are born with the propensity to impose order, to classify and organize our environment in accordance with our individual ways of theorizing about and acting in the world."[312] Each individual considers the world and their place in it from a different perceptual framework, creating and recreating models as a continuous stream of input comes in through the senses. As introduced earlier, recent findings in neuroscience have identified reference frames within the cortical columns, which are similar in all regions of the neocortex, that enable the ability to *relate place and movement to self* and incoming sensory information.[313] In brief, neurons are about the body, and this "aboutness", this relentless pointing to the body, is the defining trait of neurons, neuronal circuits, and brains. When the body interacts with its environment, changes occur in the body's sensory organs, such as the eyes, ears, and skin; those changes are mapped, and thus the world outside the body indirectly acquires some form of representation within the brain.[314]

Let's further briefly explore this important learning about the creation of reference frames in the cortical columns throughout the neocortex. Not only do these reference frames enable tracking objective locations, but "by defining an object using a reference frame, the brain can manipulate the entire object at once."[315] These reference frames—both in terms of grid cells and place cells—are also needed to *plan and create* movement of the physical self.

There is considerable precedence for considering the information processing system of self as subject. The subject of experiences is linked to the point of view of the self, that is, a specific way of looking at a field of consciousness. Taking a reductionist approach, this would mean that "perceptual states are conscious just when they are representations from the subject's point of view."[316]

In literature, as in life, the awareness of events from a specific point of view is in their associations through comparison or contrast. This question of association is a basic and indispensable principle in the description of any system. The process of selection depends on comparison and contrast, similarities and differences. Thus, the importance of thinking tools such as Knowledge Capacities (see Chapter 5). When the "I" as subject perceives the "things" around the "I" as objects, it is through the unconscious lens of association with and comparison to other objects in the environment. The illusion of self as separate, with boundaries, is what enables this process, with the self being the observer and the objects outside the self being that which is observed.[317]

Damasio forwards that *conscious minds begin when self comes to mind*, that is, when there is awareness of who we are, the choices we have, and the ability to bring the three parts of time (past, present and future) together in support of our decisions and actions. Damasio describes three distinct steps to achieving self, starting with the *protoself*, that is, the generation of primordial feelings. The *core self*, all about action, is the next step, which "unfolds in a sequence of images that describe an object engaging the protoself and modifying that protoself, including its primordial feelings."[318] Along with its primordial feelings, the protoself and the core self are what constitute the material me, the physical me. The *autobiographical self* is the third and final step and includes biographical knowledge that pertains to the past, as well as anticipation of the future, bringing the three parts of time together. The higher reaches of the autobiographical self embrace "all aspects of one's social persona ... a 'social me' and a 'spiritual me'."[319] The combination of the core and autobiographical selves construct a *knower*, another variety of subjectivity.

Mulvihill notes that during the initial processing and linking of information from the different senses, "it becomes clear that there is no thought, memory, or knowledge which is 'objective', or 'detached' from the personal experience of knowing."[320] While we agree, conscious awareness of the autobiographical self moves the self into the position of both the observed *and* the observer, an expansion of self-conscious awareness, *enabling the living of life more fully, expanding learning while increasing our engagement with choice.*

In quantum physics there is an Observer Effect, recognition that the act of observing or measuring some parameter changes that parameter, that is, the observer affects the observed reality. This description emerged out of research by the Weizmann Institute of Science noting that, when observed, particles can also behave as waves.[321] The import of this phenomenon shifts the focal point of the subject/object relationship back to the observer (as subject). As can be seen, the overlay of self in the ICALS model provides the foundation for experiential learning and directly affects all five modes of the ICALS model.

The Role of the Unconscious

The unconscious part of the brain plays a significant role in our thinking and feeling, and while self is developed through conscious choices, much of the model of self comes from the unconscious. This is a finding in the ICALS study: **The model of the self comes mostly from the unconscious**. This model creates the idea of *who* we are and the image we build up of ourselves.[322] During active experimentation, the self interacts with objects and the external world, and hence is typically very aware of the perceived boundaries between the individual and that external world. However, much of the processing of what is learned is occurring at the unconscious level, with new learning based on conscious decisions subject to association with what has already been learned (the process of associative patterning). Thus, the model of self is continuously being updated at the unconscious level.

As introduced in Chapter 5, most of our thinking is unconscious, as is most of the processing of the emotional meaning of stimuli.[323] As discovered in the ICALS study in the area of the unconscious: **The unconscious brain is always processing**. Because of this, it could be said that the unconscious affects all five modes of the adult experiential learning model. However, we consider several sub-elements in the learning model to provide an indication of the unconscious brain's potential impact on learning. Note that the brain uses 80 percent as much energy when sleeping as it does when awake.[324] This is consistent with the finding that the high rate of energy usage by the human brain is not changed much by mental activity.[325] It is also known that when under deep anesthesia 50 percent of the brain's energy consumption is used for maintenance. Specifically,

> *Deep anesthesia abolishes electrical activity and reduces the metabolic rate of the whole brain by 50 percent, suggesting that, on average, energy consumption is equally divided between neural signaling and maintenance.*[326]

Building on these findings, it appears that in terms of the sub-elements of concrete experience (sensing, feeling, awareness, attention, and intuition) that our unconscious mind uses about 30 percent of the brain's energy for processing during sleep, that is,

If, 50% of the brain's energy consumption is used for maintenance;

And, 80% of the brain's energy is being used when we sleep;

Then, 30% of the brain's energy is used by the unconscious mind during sleep.

This would be much higher if the individual were awake. The unconscious would then be processing a great deal of information from the external environment. These incoming signals often show themselves as feelings or intuitions and guide our reactions to future experiences. As Pert has noted, the emotional system works through the generation and transmission of chemicals throughout the body, which in turn impact neuron activity.[327] These chemical changes represent the emotional tag of the amygdala. (The emotions are expounded in Chapter 8.)

Playing a key role in the survival of the individual, the unconscious part of the mind can detect patterns without the individual's conscious awareness, and individuals may act upon those patterns without being aware they are doing so. As Bargh explains, "An individual's behavior can be directly caused by the current environment, without the necessity of an act of conscious choice."[328] Such implicit behavior, and learning, is widespread. From a learning perspective the key is realizing that *a great deal of learning occurs in the unconscious part of the mind*. As a result, our unconscious can influence our thoughts and emotions without our knowing it.

This supports an additional finding in the ICALS study: **We may act for reasons we are not aware of**. This aspect of the unconscious clearly influences many sub-elements of active experimentation, and is an aspect of social engagement in terms of unconscious learning. Because of the dual roles of the conscious self with its stream of consciousness and the large and continuously operating unconscious, our conscious mind cannot and does not control all aspects of our actions. It is important to recognize that the unconscious significantly *affects our thinking and actions, and therefore our learning process*.

There is also a great deal of activity in the unconscious mind during sleep. Such activity may include dreaming, problem solving, or creating new ideas. Several researchers have concluded that during sleep the unconscious mind is evaluating and filtering the previous day's incoming information, keeping what is important and discarding what is not important.[329] One of the ways to direct this process from a point of consciousness is through inner tasking and lucid dreaming. One author touts this as his pathway to ideas. "It has been my experience from the beginning, as far back as I can remember, being able to state a problem in my head and then within a short period of time, definitely overnight, having a variety of solutions available." See TOOL 4 below.

****

TOOL 4: Sleep on It

Sleeping on a question or problem can yield an answer the next morning. This is a particularly powerful way to access tacit knowledge.

STEP (1) Prime your conscious mind. Early in the evening, prior to going to bed, take a focused period of time to "brainstorm" with yourself. *Ask* yourself a lot of questions *related* to the task at hand. Reflect carefully on the questions and be patient. This is the process of active reflection.

STEP (2) Before going to bed, ensure that you have a pen and paper available right beside the bed, accessible without you getting up. Write the specific problem or question you want to address on this pad of paper.

STEP (3) Tell yourself, as you fall asleep at night, to work on that specific problem or question. Then, release all thought from your head, allowing yourself to fall into sleep.

STEP (4) When you wake up the next morning, but before you get up, lie in bed and ask the same question, listening patiently to your own quiet, passive thoughts. Frequently, but not always, the answer will appear.

STEP (5) Write the answer down quickly before it is lost from the conscious mind (as with dreams).

Another aspect of this approach is useful when a group or team is tackling a difficult problem. It has been found that the answers from the team can be improved if, rather than acting on the quick responses, *let the team sleep on the problem* and review the answers they come up with in the morning. What happens is that while you sleep your unconscious mind is processing the information taken in that day, keeping the valuable information and discarding that which doesn't make sense or is not important to you. It is also working on solutions to issues that have come up that day. When the team gets back together the next day, there will be new ideas and thoughts, and a clearer vision of the best way ahead.

****

From a learning perspective, it is recognized that a great deal of past learning resides in long-term memory and the unconscious.[330] The challenge for learners is to build up this knowledge base over time, recognize its value, and *be able to retrieve what they know when it is needed.* This "retrieval" may take the form of intuition, heuristics, judgment, or inference.[331] One approach is to put yourself into situations which will trigger potential responses in the direction you desire. Another related useful technique for triggering the surfacing of that which is stored in the unconscious is to learn to ask yourself the right questions, and carefully listen to your answers, engaging in conversation with your self. Dialogue with a trusted colleague is an additional way to bring out information and knowledge that individuals may not know they have.[332]

You can also "open the brain" through intentional focus on an object (referential) or on a state (nonreferential) through "just thinking."

Andreasen says that an example is the state of "random episodic silent thought" (REST), which is a way of "achieving a more creative state of mind."[333]

Sound can also play a role in opening the brain, specifically, in opening the connection between the conscious and the unconscious mind. Pinker notes that neuroscience has slowly begun to recognize the capability of internal thoughts and external information such as sound to affect the physical structure of the brain—its synaptic connection strengths, its neuronal connections, and the growth of additional neurons.[334]

An example is hemispheric synchronization, which is the use of sound coupled with a binaural beat to bring both hemispheres of the brain into unison.[335] Binaural beats were identified in 1839 by H.W. Dove, a German experimenter. In the human mind, binaural beats are detected with carrier tones (audio tones of slightly different frequencies, one tone to each ear) below approximately 1500 Hz.[336] The mind perceives the frequency differences of the sound coming into each ear, integrating the two sounds as a fluctuating rhythm and thereby creating a beat or difference frequency. This perceived rhythm originates in the brainstem, is neurologically routed to the reticular activating formation, also in the brainstem,[337] and then to the cortex where it can be measured as a frequency-following response.[338] This inter-hemispheric communication is the setting for brain-wave coherence which facilitates whole-brain cognition, assuming an elevated status in subject experience.[339] What can occur during hemispheric synchronization is a physiologically reduced state of arousal while maintaining conscious awareness,[340] and the capacity to reach the unconscious creative state described above through the window of consciousness.

Kandel forwards that a fundamental principle of the unconscious is that it never lies, *but it could be wrong!* This basic principle emerged as a finding in the ICALS study. Specifically: **The unconscious never lies**. This may seem odd, but when it is realized that the unconscious is a part of the overall living individual, "lying" to self would not make sense.[341] This idea relates to the truth and how things work under reflective observation. Although the intuition coming from our unconscious may not be right, it is still what our unconscious *perceives as truth*. The same phenomenon would occur with problem solving in the abstract conceptualization mode. Being aware of, and listening to, our unconscious can improve our learning scope and rate.

As can be seen, the unconscious part of the mind/brain plays a significant role in experiential learning. Since the unconscious receives information from the environment of which the conscious mind may not be aware, an enhanced awareness of the unconscious may show itself through a sense that something is right or wrong (sensing), or a good or bad feeling about a situation (feeling), or perhaps through intuition. These represent efficient ways of learning and are in addition to, rather than in place of, conscious reflection and comprehension. Also, the unconscious may enhance learning effectiveness by developing understanding and insight over time.

For example, we often take action based on our intuition that comes from associating previous experiences in our unconscious. Recognizing the power of the unconscious, and understanding to some degree how it works, can help learners build and more fully engage their unconscious capability. As reviewed above, part of our memory is unconscious, that is, we do not have direct access to it upon conscious request. Nevertheless, a large amount of what we have learned in the past still resides in memory. Recognizing this enables a learner to build knowledge that may not be immediately apparent or consciously beneficial but may pay dividends later as individuals learn to more effectively access their unconscious. This is a contribution to long-term sustainable learning. We now specifically address memory.

The Fickleness of Memory

As Tennessee Williams wrote in *The Milk Train Doesn't Stop Here Anymore*, "Has it ever struck you … that life is all memory, except for the one present moment that goes by you so quickly you hardly catch it going?"[342] And even memory—a natural product of experience—can pass quickly! Humans have three distinct forms of memory storage capabilities: sensory memory, short-term memory (including working memory), and long-term memory. Sensory memory is quite short, generally a few seconds in endurance, and refers to information received through the senses.

Short-term memory takes over when information is transferred from sensory memory to our consciousness.[343] While short-term memory lasts longer than sensory memory, it is still initially very limited, engaging approximately 5 to 9 bits of information for approximately 30 seconds or so.[344] However, a second phase of short-term memory is working memory, which occurs when material is kept in conscious focus for a longer period of time,[345] and can happen when we are studying, repeating or rehearsing, focusing (for a period of time) on a core issue, etc. Because short-term memory can only handle 5 to 9 bits of information, displacement occurs when it is full and a new bit of information enters. An example is trying to remember a phone number and the last couple of digits dropping out of memory. However, as emerged in the ICALs research: **Working memory is** [also] **limited.** This item could impact learning in the aspects of concepts, ideas, and logic as well as problem solving and building models and theories. While working memory is essential for having a conversation and comprehending what we are reading, it can get overloaded, and when it does it may cause confusion and mistakes.[346] As well as impacting concepts, ideas, and creativity, working memory also affects conceptual thinking.

Because working memory is limited to roughly 7 plus or minus 2 bits, it is very helpful to chunk patterns to improve recall.[347] Chunking is bringing ideas and concepts together to create understanding through the development of significant patterns useful for solving problems and anticipating future behavior within a domain of focus. A study of chess players concluded that "effortful practice" was the difference between people who played chess for many years while maintaining an average skill and those who became master players in shorter periods of time. The master players, or experts, examined the chessboard patterns over and over again, studying them, looking at nuances, trying small changes to perturb the outcome (sense and response), generally "playing with" and studying these *patterns*.[348] In other words, they use working memory, pattern recognition and chunking rather than logic as a means of understanding, memory recall and decision-making.

Long-term memory is relatively permanent and can be thought of in terms of declarative memory and nondeclarative memory. Declarative memory—which can be tacit or explicit—includes the areas of semantic memory, factual knowledge such as meanings, concepts and math,[349] and episodic memory such as events and situations.[350] Nondeclarative memory includes those acts and habits which are done by rote due to extensive practice and conditioning. An example would be riding a bicycle, which involves embedded tacit knowledge that has become part of the structure of cellular memory, what we refer to as embodied tacit knowledge. Long-term memory has no fixed limits on how long the information can be stored and retrieved.[351] Factors that influence an individual's length of memory recall are repetition during learning, frequency of recall, and the emotional level when first experienced.

Memory as part of the learning cycle. Memory is involved in all phases of the learning cycle. For example, in grasping experience through apprehension an individual may use sensing, feeling, awareness, attention, and intuition. If the current experience is similar to a past experience (either good or bad) memory will influence the level and nature of the incoming information as it relates to the reflection and comprehension processes. This influence may significantly change the nature and interpretation of the experience.

Reflection creates understanding and meaning by associating and integrating incoming information with information stored within memory (associative patterning). Abstract conceptualization or comprehension works with concepts and solves problems by recalling ideas and past experiences from memory and creating new ideas and solutions to problems. Here both working memory and long-term memory will likely be actively involved. Working memory refers to representations of information currently in active use, small amounts of information for short periods of time. If not rehearsed, the information will be forgotten within a few seconds.[352]

Most action we take is an implementation of tacit knowledge residing in memory[353] with which both working memory and long-term memory become involved. Imbued with emotional relevance, as action is implemented, rapid feedback information is combined with past memories to offer a quick reaction guide to new actions.[354]

A good, accurate memory that can be recalled when needed is clearly an important aspect of experiential learning. We consider the following findings from neuroscience that relate to memory. Note that memory is represented in the brain as an extremely large number of patterns, each composed of a large number of neurons with many synaptic junction connections and with many different junction strengths. These patterns exist throughout the brain and a single pattern (concept, object, thought) may be spread out over many regions of the brain.[355] This is consistent with our new understanding of cortical columns throughout the neocortex and their entanglement. When memory is recalled, the temporal sequence of patterns representing the memory is brought into consciousness by assimilating the multiple parts of the memory that were stored in different areas of the brain.

A related finding from the ICALS study is **memory is scattered throughout the entire cortex; it is not stored locally**. This phenomenon primarily impacts unity and creativity. Because memory is scattered physically throughout the cortex there may be errors in its reconstruction. On the other hand, from a creativity viewpoint, such errors—which build on the ability to assimilate memory from many sources—may create new ideas and concepts in combination with continuous sensory input.

A related finding is the recognition that **memories are re-created each time they are recalled and therefore never the same**. The environment influences the re-creating process. This item is shown as impacting attention, understanding, truth/how things work, and integrate/look for unity. All of these factors in the adult experiential learning model would be very sensitive to the accuracy of memories. For example, if a specific memory was recalled as an inaccurate representation, it could significantly change the interpretation (and therefore the meaning) of the original event. Note that since we as humans are "verbs", not nouns, continuously changing and adapting to our environment, representations will never be exactly the same.

There are several other factors that influence the integrity of the memory as it is recalled. One of these is the overall health of the brain. Both nutrition and exercise can significantly influence the biological health of the neurons, their chemicals, and their patterns.[356] Exercise is discussed in Chapter 7 as part of managing self.

A second effect is whether the pattern has been recalled recently. Repetition of use strengthens neuronal memory patterns, thereby making them more easily recalled. An affirming finding in the ICALS finding is **repetition increases memory recall**. This item would impact control, rigor, and discipline in abstract conceptualization and focus attention in active experimentation. This is a well-known phenomenon and goes back to Hebb's rule that *neurons that fire together wire together*. Thus, the more a given pattern is repeated the stronger the neuronal connections are and the easier a specific memory can be recalled.

A third factor is the emotional tag of the memory. If there was a strong emotional context when the memory was first created its recall will be much easier.[357] There is an exception to this that occurs when the emotional impact is extremely negative and the memory is buried within the unconscious mind and is not available to consciousness. See Chapter 8 for a deeper discussion of emotions.

A fourth factor is the rate at which neurons are dying, or perhaps being born within the various regions that contain related memory patterns.[358] However, as discovered in the ICALS study, **memory patterns decay slowly with time**, depending upon the emotional and cognitive importance of the memory. When coupled with the potential errors during memory recall, it suggests that learning may be enhanced by evaluating specific material or experience for its relevance and importance to the learner. One example while reading is to underline important passages while reading and "judge" the value (to the reader) of these passages, marking them with one to three stars in the margin. This presses the reader to *consider how important* the specific information is to his or her own interests and goals. And, as we know, repeatedly focusing attention on critical knowledge also helps solidify past knowledge, and it provides a good mechanism for reviewing the marked passages in the future.

One of the somewhat surprising results from neuroscience is that typically an individual retains only a part of the incoming information, for example, from an image. To create the whole concept (in the image example, a whole picture), the rest of the incoming information is filled in by the individual when they describe or try to remember the image.[359] Edelman calls this reconstructing memory.[360] What this means is that the mind does not store memories in the same sense that a computer does. Rather, the mind stores the most important part of incoming information, that is, its meaning or *invariant* form, and fills in the missing parts during recollection. Understanding this helps the learner to be aware of the imprecision and possible irregularities in their memories, and to treat them carefully during the learning process.

This is an important finding emerging from the ICALS study. **The brain stores only a part of the meaningful incoming sensory information. The gaps are filled in (re-created) when the memory is recalled**. Two related findings are that **memory stores invariant forms (used to predict the future)** and **the mind uses past learning and memories to complete incoming information.** These items primarily relate to the sensing aspect of concrete experience in the adult experiential learning model. It makes the point that when we sense or feel our environment not all of the incoming information goes into memory. In fact, at best, only things that are meaningful to us are usually remembered. Memory storage, recall, and recognition all occur at the level of the invariant forms, which are distributed throughout the brain, what Hawkins describes as "a form that captures the essence of relationships, not the details of the moment."[361] *This means that individuals cannot fully trust the details of their memory.*

Further, an individual's reaction to an experience may be based on past, misinterpreted information that in turn may color the individual's immediate experience and therefore impact learning. This could also impact the quality or validity of an individual's intuition. What this says to experiential learners is to be cautious of their own and others' memories in terms of accuracy of recall. There is still much that is unknown, or perhaps misinterpreted, and one should be cautious in one's interpretations of specific research findings. This is not to impugn the research so much as to recognize the complexity and interrelationships among the subsystems of the brain. This suggests the value of keeping an open mind, both during conversations as well as during internal reflection and comprehension, as the conscious mind communicates with its memory.

Memories as stories. Since memories are stored as temporal sequences of associated patterns, information is much more easily remembered and recalled if put in a story-like form. We might go so far as to say that human memories themselves are story based. You need to finish one part of the story before moving on to the next. Whether written, oral or visual, the narrative is conveyed in a serial fashion. As Hawkins describes:

> *All memories are like this. You have to walk through the temporal sequence of how you do things. One pattern (approach the door) evokes the next pattern (go through the door), which evokes the next pattern (either go down the hall or ascend the stairs). ... Truly random thought doesn't exist. Memory recall almost always follows a pathway of association.*[362]

Consciousness itself is a single stream of thought. While consciousness has always been a difficult concept to pin down, it can be considered a state of awareness and a private, selective and continuously changing process; a process, a sequential set of ideas, thoughts, images, feelings and perceptions and an understanding of the connections and relationships among them and our self.[363] Similarly, with memory, one part of the story is associated with the next part of the story in a linear telling: "It's almost impossible to think of anything complex that isn't a series of events or thoughts."[364]

Stories come with many indices, multiple ways that the story is connected to memory. "The more indices, the greater the number of comparisons with prior experiences and hence the greater the learning."[365] In order for memory to be effective, it must have not only the memories themselves (events, feelings, etc.) but memory traces (or labels) that attach to previously stored memories. These indices can be decisions, conclusions, places, attitudes, feelings, questions, etc.

For example, a simple story can convey deep emotion and compassion. In 1997 when there was a mid-air explosion of Silk Air M1 185 in Palembang, a Singapore Armed Forces helicopter was second to arrive at the site. Reflecting Singapore's commitment to humanitarian efforts, one participant in the rescue effort shared the following story:

> *What we saw was unexpected and is difficult to describe. Pieces of twisted metal and mangled passenger seats half submerged in the murky waters of the Musi River. Luggage and body parts floating with the current. A lifejacket and teddy bear entangled in tall grasses at the river's edge. Our teams dove into the grisly work of search and salvage, hoping beyond hope that the river-soaked teddy bear had a live owner. It didn't.*[366]

A related finding from the ICALS research is: **Memory recall is improved through temporal sequences of associated patterns, that is, stories and songs**. This item would affect attention, understanding, and meaning in that if the information were difficult to recall it could distort any of these aspects of learning. On the other hand, if the memory had been stored in story form it would be easier to recall and have less chance of recall errors. For example, compare recalling a story to recalling a random sequence of numbers. Hawkins states that,

> *Written, oral, and visual storytelling all convey a narrative in a serial fashion and can only be recalled in the same sequence…You can't remember the entire story at once. Your memory for songs is a great example of temporal sequences in memory…You cannot imagine an entire song at once, only in sequence.*[367]

Bloom points out that progress in neuroscience involving aspects of the "movie-in-the-brain" is leading to increasing insights related to the mechanisms of learning and memory.

> *In rapid succession, research has revealed that the brain uses discrete systems for different types of learning. The basal ganglia and*

cerebellum are critical for the acquisition of skills—for example, learning to ride a bicycle or play a musical instrument. The hippocampus is integral to the learning of facts pertaining to such entities as people, places or events. And once facts are learned, the long-term memory of those facts relies on multicomponent brain systems, whose key parts are located in the vast brain expanses known as cerebral cortices.[368]

Memory patterns are also very resilient to neuron death. For example, as many as 40 percent of neurons composing a memory pattern may be lost and the pattern still be recalled.[369] From an evolutionary view, this takes into account the continual death of neurons in the brain.

Forgetting

An important related finding from the ICALS study is: **Memory patterns cannot be erased at will**. This item is significant because in problem solving, building models, and using theories we frequently rely on past knowledge and memory. However, as new knowledge is created, we do not forget what was learned before and therefore must be careful not to use outdated knowledge to solve new, different problems. In addition to solving problems, the brain also operates by storing solutions to problems, and while often very efficient it can fail when new challenges present themselves and the old solutions no longer apply.[370]

In both the conscious and unconscious mind, traumatic events—especially those charged with strong emotions—can cause blockages, impacting our thoughts, feelings and actions. Recall that the mind/brain is an associative patterner, continuously processing incoming information from the environment, specific to the situation and context at hand, and complexing it with internally-stored information. Of course, much of this is happening below our conscious awareness, that is, in our subconscious, orchestrated by and very much based upon the preferences and desires of our personality.

While the majority of our discussion on forgetting will be focused from the viewpoint of the mind/brain, we first address the element of consciously choosing to forget. In a study performed by The Scripps Research Institute, scientists identified a molecular biology process focused on active forgetting, which is most likely regulated. Through studying the memory function in fruit flies, found to be highly applicable to humans, it was discovered that "a small subset of dopamine neurons actively regulate the acquisition of memories and the forgetting of these memories after learning, using a pair of dopamine receptors in the brain."[371] Dopamine, a neurotransmitter, plays an important role in the brain's reward and pleasure centers as well as memory, learning and cognition. The study suggests that, prior to being consolidated, new memories include an active dopamine-based forgetting mechanism, and that those new memories begin to erase unless they have some importance attached to them. Since forgetting is the fastest shortly after learning, that is, 15-20 seconds after a thought has presented itself in short-term memory, it appears that this may be part of the initial selection process orchestrated by the subconscious.

What would serve as indicators of importance or significance as part of this selection process? As with people, memories are not created equal, nor would we want them to be. Indicators for strong memories would include our conscious focus (repetition and rehearsal), our mental models (including values and beliefs and the strength of those values and beliefs), and the emotional tag attached to incoming information, perhaps a connection to a significant event in our lives. Memories and the emotional tags that gauge the importance of those memories are activated automatically without any conscious effort and become part of an individual's everyday life.[372]

The question becomes, **can/does the human mind really forget?** To "forget" has to do with the inability to remember: to leave behind unintentionally, to fail to mention, to disregard or to cease remembering.[373] Forgetting can be either spontaneous or a gradual process, and can involve apparent partial or total loss of a memory or modification of that memory. In his memory studies, Connerton forwards there are seven types of forgetting: repressive erasure, prescriptive forgetting, formation of new identity, structural amnesia, annulment, planned obsolescence and humiliated silence.[374] Each of these descriptive terms well represent the state of each.

Perhaps the earliest work on the process of forgetting is the forgetting curve hypothesis forwarded by Hermann Ebbinghaus.[375] The forgetting curve shows that there is a decline in memory over time, purporting that half the memory of newly learned knowledge is forgotten in a matter of days or weeks unless it is consciously reviewed.[376] Ebbinghaus further hypothesized that the speed of forgetting was dependent on a number of factors, including the difficulty of materials, its meaningfulness to the individual, the way the materials were represented, and physiological factors such as sleep and stress. An affirming ICALS finding is that **less than seven hours of sleep may impair memory**. However, given enough sleep, Ebbinghaus concluded that these factors could be overcome through basic training in mnemonic techniques, which can produce over-learning and slow forgetting.

The trace decay theory of forgetting,[377] says that memories in short term memory leave a trace in the brain that, if not rehearsed, automatically fades away in 15 to 30 seconds. In other words, according to this theory it is the length of time between a memory and the need to recall that memory that determines whether the information is retained or forgotten. There are, however, other theories, and recent neuroscience findings, that may help our understanding of the concept of forgetting, which will be interjected throughout this discussion.

Perhaps one of the best-known theories related to forgetting comes out of the field of psychology. Freud argued that quite often threatening or anxiety-provoking material cannot gain access to conscious awareness; in other words, it is repressed and gets stuck in the subconscious. There is considerable controversy regarding the idea of repression. For example, in looking at reports of recovered memories, Andrews and his colleagues discovered that some of the recovered memories were false.[378] Further, Lief and Fetkewicz found that people can be misled into believing events that didn't happen,[379] and Ceci found that preschool children had difficulty distinguishing between real and fictitious events, which can continue into adulthood![380]

We now also understand that reliving events, whether in therapy or other conversation and especially when connected with deep emotion, embeds those memories deeper into long-term memory much like purposeful rehearsing, making them more impervious to any possible "forgetting" process. Ultimately, from the mind/brain perspective, *that which does not get attention eventually does go away* (use it or lose it). Thus, a good process for "forgetting" is *inattention*, and the very best way to avoid attending to some memory is to have a stronger, more significant memory replace it. This would infer stronger links than to similar stimuli in the older memory. For example, the best way to "forget" a traumatic relationship breakup is to have a new relationship. As a second example, if there was a negative memory of coal in a stocking for Christmas, the very best way to overcome that effect is to have a positive memory of a reward for a good deed performed in a following Christmas stocking, perhaps acknowledging a reversal of behavior. However, this would prove quite difficult for deeply embedded memories with traumatic emotional tags, highly dependent on an individual's ability to develop positive events of greater significance to replace these older connections.

Other theories of forgetting include Interference Theory (assuming that what is remembered can be disrupted by previous and future learning); Consolidation (focused on the biological processes of consolidation or lack of consolidation over time); Retrieval Failure Theory (the inability to access information stored in long-term memory); and State Dependent Cues (the ability to access information stored in long-term memory is dependent on the physical or psychological state).[381]

Both learning and "letting go"—whether in terms of *freeing* or *filing away* (putting away on the bookshelf)—are the primary processes through which we change and grow. Since humans have limited processing capability and the mind is easily overloaded and clings to its past experience and knowledge, letting go becomes as important as learning to facilitate the free flow of energy and prevent blockages. In an article titled "Forgetting is Key to a Healthy Mind", Wickelgren confirmed that letting go of memories supports a sound state of mind, a sharp intellect, and superior recall.[382] In this context, we are referring to memories that are in conscious awareness and subject to focus and choice.

Letting go or releasing is the art of being able to let go what was known and true in the past or to let go of concepts, beliefs and mental models that inhibit your growth and expansion. It is moving out of mind and thought into direct experience, *allowing yourself to be that which you already are without bringing along events and perceptions of the past*. As forwarded in the earlier discussion on mental models, being able to recognize the limitations and inappropriateness of past assumptions, beliefs, and knowledge is essential before creating new mental models and for understanding ourselves as we grow. It is one of the hardest acts of the human mind because it threatens our self-image and may shake even our core belief systems. See TOOL 5.

Note that "forgetting" and "letting go" are different concepts than that of "unlearning". As expressed above, new learning can, over time and especially when emotions are involved, replace old learning in terms of unused connections and neuronal firings ("use it or lose it"). However, it is difficult, if not impossible, to unlearn the truth of who you are. Even when this knowing is not at the conscious level, it lies beneath, ready to emerge in response to triggering and conscious choice.

****

TOOL 5: Letting Go

The Sedona Method, featured in the movie and book based on *The Secret*, is a technique developed for eliminating blocked energies that hold you back from being and doing what you choose. The method includes three ways to approach releasing: (1) by choosing to let go of the unwanted feelings; (2) by welcoming the feelings; and (3) by diving into the very core of the feeling. This unique and powerful technique consists of a series of simple questions that you ask yourself, which leads to expanding awareness of your feelings and enabling a choice to let them go.[383]

Let us follow another scenario where a letting go sequence does *not* occur. You are a senior executive in a large engineering firm, and have been in the job for five years. The first year, it was new and exciting and you knew you could make a difference. And you did, with revenues soaring. The second year you put what you had learned in place, developing and stabilizing long-term processes. The third year you did much activity by rote, and just focused on challenges and opportunities. The fourth year the company was bought up by a larger conglomerate and their processes and approaches became standards of behavior right there beside the rote stuff that kept your (now department) running like clockwork. By the fifth year the paperwork had become nearly unbearable, with additional levels of review and approval, the necessity to only use "approved" products that came from other parts of the organization or their partners, and increased reporting and audit procedures. Somehow, the fun has dissipated, the work has become repetitive and boring, and there is no time for—or interest from management in—creative solutions. While you are financially thriving, your energy is stuck and you are stuck. The letting go that needs to occur here is choosing to leave the job; yet you remain. This is a choice.

The biggest barrier to learning and letting go arises from our own individual ability to develop invisible defenses against changing our beliefs or behaviors. These self-imposed mental defenses have been eloquently described by Chris Argyris,[384] who concludes the mind creates defense mechanisms to support belief systems and experience. These mechanisms are invisible to the individual and may be quite difficult to expose in a real-world situation. They are a widespread example of not knowing what we know, representing invisible barriers to change.

****

In Summary

By now it may be quite clear that the self is an emergent, individualized part of each of us which reflects the agency of what it is to be human. Simultaneously, it is clear that humans are not a "single" thing—not just a body, not just a soul, not just a self—but all of these and more, an integrated complex adaptive learning system, interconnected yet perceived as separate, individualized yet each critical to the larger whole, a tiny spark which reflects the light of unconditional love, the animated spark of life, with the opportunity to co-create ourselves through learning.

UNLEASHING Litmus Test #2*

*After reading **Chapters 4-6**, and considering earlier chapters, reflect on each question for one minute (Reflective Observation) prior to answering silently, verbally or in writing.*

1. **AWARENESS** means something has come to your attention, it is perceived, it has been mentally engaged.
   Ask: As the ICALS theory reveals learning from the inside out, how does that shift my insight about how I should learn?

2. **UNDERSTANDING** includes your perception of the situation—the who, what, where, when, and why, and the anticipated results. As the situation becomes more complex, you need to re-create your understanding.
   Ask: What new single thinking process or technique can I use regularly to create a difference in my daily life?

3. **BELIEF** means you accept what you are aware of as true and understand it exists. Beliefs which dominate other patterns are prominent. Strong patterns are created by experiences and are closely related to emotions.
   Ask: How will increasing my thinking skills enable me to learn more efficiently and effectively?

4. **FEELINGS** are the foundation of learning—positive feelings make actions important to you and worthy of your efforts. Reason cannot operate without emotions.
   Ask: In what way can I increase my enjoyment of learning?

5. **OWNERSHIP** implies a personal commitment for you to take responsibility and act.
   Ask: How could my commitment to learning impact the lives of others?

6. **EMPOWERMENT** refers to self-empowerment, that is, having the *knowledge* to make the necessary change and the *courage to act* on what you have learned.
   Ask: In what ways can I continuously learn how to unleash my mind?

*These questions are based on the Individual Change Model.[385] CHANGE comes from within, that is, unleashing your mind is YOUR choice.

Chapter 7

Managing Self

Attention and Intention Create Our Reality ... Stress Focuses Attention ... Intention is the Source of Action ... The Capacity of Presencing Accesses the Field of the Future ... Physical and Mental Exercise Promote Learning ... The Aging Self Can Be Managed ... An Enriched Environment Expands the Mind ... In Summary

SUCCESS FACTOR: Attention/Intention

SKILL SETS: Managing Self, Presencing

FIGURES: (9) Attention and intention weave thoughts and feelings into reality; (10) An optimal level of stress facilitates learning; (11) The harnessing of anxiety as energy? (Cartoon by Jackie Urbanovic).

TOOL: (6) Holding Neurovascular Reflect Points.

First, let's explore the concept of "managing". The old organization model of management as planning, organizing, commanding, coordinating and controlling the work of others is now obsolete in organizations, giving way to leadership roles focused on the customer's needs and wants, thinking strategically, decisions made at the point of action, and learning continuously.[386] Further, Davenport proposes that managers in the future must adapt their activities to the new challenges they face.[387] Bennis, Parikh, and Lessem describe the *new managerial role* in terms of knowing enough about the organization, and the internal dynamics of the organization, to be able to create harmony between the two.

> *As such a new-paradigm manager, firstly you possess general insights into people, things, ideas, and events, as well as particular professional and managerial know-how. Secondly, you have greater insight into environmental forces and trends than is required for everyday business and management functions. Thirdly, you have in-depth insights into your own inner dynamics, covering the functioning of your body, mind, emotions, neurosensory system, and states of consciousness.*[388]

And that professional and managerial know-how includes things like overseeing work to doing it; organizing hierarchies to organizing communities; imposing work designs and methods to understanding them, hiring and firing workers to recruiting and retaining them, building manual skills to building knowledge skills, evaluating visible job performance to assessing invisible knowledge achievements, ignoring culture to building a knowledge-friendly culture, supporting the bureaucracy to fending it off.[389]

As we now understand, "self" is an emergent quality of living as a human, the part of us that makes choices and carries out those choices through actions (if we choose). And as part of a complex adaptive learning system, those choices affect every aspect of who we are as well as the relationships among those aspects, with "no choice" representing a choice! Thus, when referring to self, we consider managing in an active and emergent context, not so much as a sense of controlling as a sense of *choosing*, that is, asserting our individuated agency in choosing the thoughts we think, the feelings we have, the beliefs we hold, and the actions we take. This means moving beyond the concept of "total freedom" connected to and perceived by an ego-focused "I" to a deeper understanding that your self is part of a larger energy field to which you are always connected and with which you are continuously interacting (see Chapters 9 and 10). This is an underlying theme throughout the *Unleashing the Human Mind* text.

Recognizing that "self" is in the decision-making role, perhaps we can use what we've learned from managing our organizations to better understand how to manage ourselves. Clearly planning, organizing, commanding, coordinating and controlling make some sense, and, indeed, "planning" remains a critical skill for the human (see Chapter 10). Yet there are many autonomic systems within the body such as our heartbeat, food processing, and everyday breathing which we would hardly want to spend our time commanding and controlling. So, there may be something to this new model of managing.

Re the "firstly" idea in the Bennis, Parikh, and Lessem quote, at a conscious level—and as the primary contributor to "preference"—through experience the self has developed and therefore possesses general insights into people, things, ideas, and events of interest. And with a little imagination, we could apply most of those "professional and managerial know-how" items to the various aspects of self, which could prove a fun thinking and learning exercise for those of you who enjoy challenges. Re the "secondly" idea, looking at "life" and the environment from the highest vibrational level of the human (rather than the level of a system or cell, etc.), certainly the self can think strategically both in terms of running the overall system (in terms of preferences such as eating, exercising, sleeping, attitudes, learning, etc.) as well as applying and responding to environmental forces (choice).

Re the "thirdly" idea, each of us has in-depth insights into our own inner dynamics, covering the functioning of our body, mind, emotions, neurosensory system, and states of consciousness. It appears that what we are being asked in this new role of managing self is to engage our preferences and insights—as well as our professional and managerial know-how—as we move through the various roles of planning, organizing, commanding, coordinating and controlling. This means thinking strategically while simultaneously listening to the *feel* of our inner dynamics—our body, mind, emotions and soul (which is defined as the animating principle of human life in terms of thought and action, specifically focused on its moral aspects, the emotional part of human nature, and higher development of the mental faculties) (see Chapter 10).

A number of years ago (and this shows our age), there was a management video of someone trying to herd a bunch of cats. It was quite a spectacle, as well can be imagined! However, it's not far off from managing our thoughts and feelings. For example, tell yourself NOT to think about something and sure enough that's exactly what you WILL think about. Similarly, somewhere along the way, we began to recognize that the way we tell children what to do really counts. For example, "Stay on the sidewalk" is much more effective than "Don't go in the street." In the first instance, the visual pulled up is of the child on the sidewalk; in the second instance, the visual pulled up is of the child in the street.

Visualization is a powerful tool for managing self. For the majority of people, vision is the main way they process the world around them. And, because of ever-increasing social media, radio stations, web pages, cable television networks, video gaming, and on and on, the brains of the younger generations are literally wired for faster programming and more visual programming. Approximately 70 percent of the neurons in the cortex are devoted to vision, with, interestingly, two out of three of these neurons focused on *inner vision*, that is, imagination, dreaming, etc., with "the ability to visualize our past, present, and future to explore and imagine new ideas and creations … one of our greatest human assets."[390] Indeed.

There is a vast amount of literature available on various visualization approaches, imagination, dreaming and visioning. It is not the intent to cover the sphere (which would be impossible) or to present an approach that is purported to be "the" approach to managing self. There is no such thing. Because each person is unique in terms of past (heritage, personality, experiences), present (self, needs, passions) and future (desires, dreams), the only person capable of understanding and managing your "self" is you. However, a greater understanding of self in terms of learning—which, whether at the level of cells or self, or anywhere in between, is what enables us to adapt to the environment—can facilitate choices that head us in the direction we desire.

In Chapter 3 we introduced a graphical representation of the future consciousness expansion developed by the Universal Knowledge Guild, which was tied to the human abilities needed during each phase of expansion. An important concept introduced was that of *thought attendance*, which was coupled with an openness to new ideas and learning with a rising awareness that form follows thought—a consciousness of our own consciousness and the thoughts and feelings emerging from self. It is with this in mind that we address several areas related to managing self which are supported by neuroscience findings from the ICALS research.

Attention and Intention Create Our Reality

Attention and intention are tools of the self that are directly related to consciousness and lay the web for interaction with the world in which we act. There are many ways that information (any organized or non-random pattern) comes to the attention of self, ranging from existence in the external environment to a direct interaction with others, to a frightening event, to the internal experience of dreaming, meditation, mindfulness, reading, or reflecting. All of these are ways that information can come to the attention of the mind, and—with a strong emotional tag attached, perhaps caring or passion—thereby interact and influence the self's thoughts and perceptions of the world.

Attention can only occur in the NOW, and sustained attention is a series of NOWs. For example, the eyes work much like an old-time movie reel, with a fixation occurring approximately three times a second as the eyes make a small, quick movement (a saccade) and then stop. However, we perceive it as a continuous flow of movement. Just as the eyes shift their focus from object to object, attention shifts driven by our thoughts and feelings. When these thoughts and feelings are focused and consistent accompanied by intelligent actions, we become the effective co-creators that we are.

Similar to how the body is modeled in somatic regions of the neocortex, Graziano proposes that attention is modeled by a particular region.[391] This modeling is what leads us to actually believe we are conscious, much like we believe we have an arm or a leg. In other words, we create a model of attention, and as we now know, that is being created in multiple cortical columns across the neocortex as different sensory streams come into mind.

Planning—the mental activity of forethought focused on achieving a specific goal—can help clarify and solidify intention. See Figure 9. Planning uses patterns from the past to predict the future, and intention is all about the future. Planning, an existential skill which is an important part of the human past, present and future skill set, is addressed in more depth in Chapter 10.

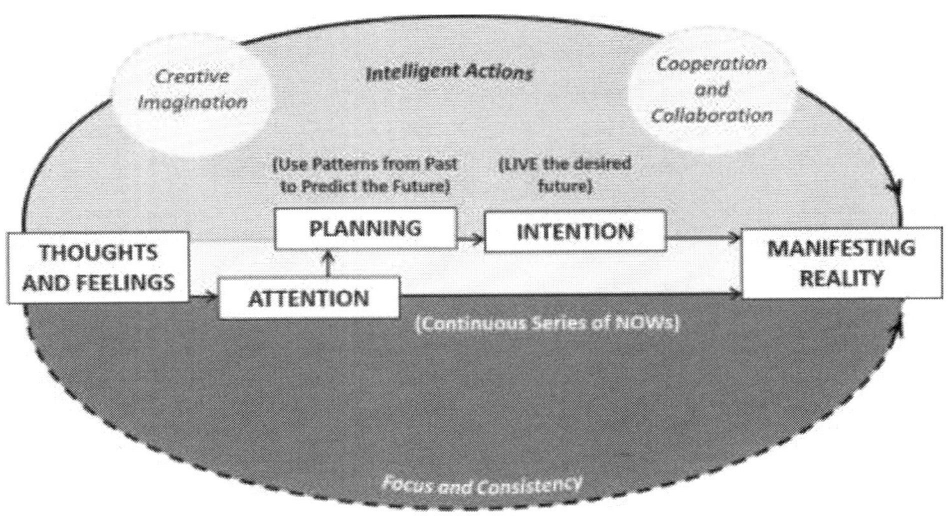

Figure 9. Attention and intention weave thoughts and feelings into reality.

There are many ways to purposefully capture attention and set intention. For example, some 20 years ago in the U.S. Department of the Navy Acquisition System, change was afoot, with dozens of initiatives simultaneously pushed down from the Department of Defense that impacted every aspect of the acquisition process for multi-year, million-and-billion-dollar programs. While a myriad of resources from both the public and private sectors were available in support of this change, we were faced with how to quickly move the system from within. The Inspector General teams served as a pivot point. To facilitate and accelerate implementation, the *inspection teams* were trained as knowledge brokers, with access in terms of availability and *understanding of* the change needed and support materials available, coupled with an understanding of future program office needs. These teams immediately scheduled "pre-audits" of every program office, with the checkpoints representing 100% implementation of each of the change initiatives. When each pre-audit was completed, the team sat down with program office leaders, provided resources for every gap that was noted, and scheduled a full audit of completed implementation six months later. This process gained the immediate *attention* of leadership and set the *intention* in terms of expected quality and timeliness for full implementation.

Let's take a few minutes to focus on attention and intention separately. Before the turn of the century, Portante and Tarro argued that attention had become the scarce resource of the information economy.[392] They referenced Lanham, a professor of rhetoric at UCLA, who was focused on the implications for information technology of human attention management.[393] A primary point was that black-and-white text could never hold the attention of a generation stimulated by color and movement, brought up with the singing and dancing of numbers and letters on *Sesame Street.*

As the information age has exploded and the availability and accessibility of that information increased, the gap between the attention of individuals and organizations and the information that needs to be attended to has widened. Davenport and Beck describe attention as a slippery intangible asset, and begin their book *The Attention Economy* with a focus on the current attention deficit.[394] For example, they describe an organization's attention deficit in terms of organizational ADD. The symptoms are the increased likelihood of missing key information needed for decisions; diminished time for reflection on anything but simple information transactions such as e-mail; difficulty holding others' attention; and a decreased ability to focus when necessary.[395] This work includes an in-depth treatment of attention and approaches for improvement. A fascinating finding is the recognition that time management and attention management are *not* linked.

At the unconscious level, the human senses know how to process vast amounts of information effectively. As introduced in Chapters 5 and 6 and affirmed by the finding that **the unconscious brain is always processing**, it is our senses that are the information processors providing the information for our decisions and actions. And since consciousness is a part of self, then "it follows that what we pay attention to over time will shape that self."[396] Even the *objects and activities* that we pay attention to become part of our self. For example, when we are asked who we are, the response may be "a housewife" or "a dancer" or "a mechanical engineer", each a label that we have attached as a part of our self. In one context, the self could be thought of as a hierarchy of goals, because *it is the purpose and goals of the self that focus the largest amount of attention*. Not only *what* we pay attention to, but *how* we pay attention is important. For example, we might consider a person who continuously worries about getting hurt as neurotic, or a person who avoids eye contact and stays relatively quiet as shy.

In terms of mental development, it is the frontal lobes that help an individual pay attention and ask good questions. As Amen[397] describes, a more developed frontal lobe allows you to take better advantage of new knowledge, to know what to focus on, and to relate it to life experiences so that it has more useful value to you. The eighteen-year-old may be able to memorize facts more easily, but his frontal lobe isn't as good at selecting which facts to memorize. A "more developed frontal lobe" occurs through mental exercise. Two neuroscience findings emerging in the ICALS research are pertinent to this discussion. The first is in the area of mental exercise, and the second relates to aging.

Exercise increases brainpower. There is a direct connection between exercise and blood flow and oxygen to the brain. As Medina explains, consistent exercise reduces the lifetime risk for a number of elderly mental problems such as dementia and Alzheimer's.[398]

At any age, mental exercise has a global positive effect on the brain. It has been shown that as long as individuals use their brain and continue to actively think, barring age-related diseases, their brain capacity can maintain a good capability. While neurons continue to die as one ages, in several areas of the brain they can also continue to be created so long as the brain is active. For example, the hippocampus produces new neurons after birth and continues to do so well into old age.[399]

Spatial attention allows humans to selectively process visual information through prioritization of an area within the visual field. A region of space within the visual field is selected for attention and the information within this region then receives further processing. Spatial attention to a specific thing, person or event increases the intensity of the related neuronal firings, which in turn affects the conscious experience of focus, amplifying the contrast in the experience and making it less faint and more salient.[400] This translates into increased memory and recall. A related emergent finding is that **repetition increases memory recall**. This item impacts control and discipline in abstract conceptualization and focused attention in active experimentation. This well-known phenomenon goes back to Hebb's rule that neurons that fire together wire together. The more a given pattern is repeated, the stronger the neuronal connections are and the easier a specific memory can be recalled. This affirms what most good students learn along the way, that repetitive focused attention is a conscious tool for managing memory and recall.

Further, there are potential long-term impacts. As Jensen says, "It is now established that contrasting, persistent, or traumatic environments can and do change the actions of genes."[401] Staying in an environment—whether negative or positive—is going to affect not only the present but the future.

Similar to spatial attention, focusing attention on a particular non-spatial stimulus feature such as color increases its representational precision resulting in more concrete conscious experiences.[402] Thus, attention directly impacts the breadth and depth of neuronal connections in the short and long-term. Attention is also addressed in contrast to the mind "at rest" in Chapter 10 under the heading "The Mind Requires Time for Inner Reflection".

Stress Focuses Attention

We have discovered through neuroscience that while one situation may engage our full attention, another similar situation may be filled with monkey chatter or emotional distractions. Stress plays a large role in arousal and attention. Surprisingly, this can be negative or positive. One emergent finding is that **stress focuses attention**. Stress takes precedence. This item would clearly impact the magnitude and area of one's attention relative to adult learning. For example, if stress is too high, it may result in intense fear and the inability to act, which reduces learning.

Consider the amygdala, the part of the brain where incoming sensory input is continuously screened for potentially dangerous situations. If a threat is sensed, the amygdala immediately sends a signal that sets in motion a quick action (such as the fight or flight response) before the cortex is even aware of what has happened. As Zull details, "Our actions will not be controlled by our sensory cortex that breaks things down into details, but by our survival shortcut through the amygdala, which is fast but misses details."[403] The negative impact of this is that excessive levels of cortisol (the substance the adrenal glands secrets during stress) can cause permanent damage to the locus ceruleus, which is important for selective attention.[404] Begley also notes that attention, one of the parameters of successful learning, also *pumps up* neuronal activity. She says that, "Attention is real, in the sense that it takes a physical form capable of affecting the physical activity [and therefore the structure] of the brain."[405] Let's explore that more fully.

As described earlier, all signals of information coming into experience are tagged with an emotional factor (see Chapter 8). As Akil and his colleagues state,

The stress system is an active monitoring system that constantly compares current events to past experience, interprets the relevance (salience) of the events to the survival of the organism's ability to cope.[406]

Similarly, a finding from the ICALS study is that **stress is an active monitoring system that constantly compares current events to past experiences**. Three sub-elements of the adult learning model that could be affected by this item would be anticipation, act on environment, and object-based logic. For example, in anticipating the outcomes of some planned action developed from abstract conceptualization, an individual would think about past examples of similar situations and estimate the problems and potential issues, that is, what stress may occur as a result of the action. Or, if a trusted other was involved, there would be a reduction of stress and fear.

If the emotional content of incoming information is one of strong fear or uncertainty to the individual, stress is created and may significantly limit any learning involved. Conversely, if there is too little arousal/stress involved then there may be no desire for learning. Thus, for each individual there exists some *optimal level of arousal/stress*.[407] Note that low levels of stress are often referred to as arousal. Thus, learning is highly dependent on the level of arousal of the learner. Again, too little arousal and there is no motivation, too much and stress takes over and reduces learning. Maximum learning occurs when there is a moderate level of arousal, and there is an optimum level of stress for each individual that facilitates learning. There are two related findings from the ICALS study.

First, **there is an optimum level of stress for learning (the inverted "U")**. This level is somewhere between a positive attitude and a strong motivation to learn (arousal), and some level of fear of learning or the learning situation. See Figure 10. Among sub-elements, this level can impact problem solving, creativity, and anticipation in the abstract conceptualization learning mode. And as noted above, the sub-element of reducing stress and fear under social engagement in terms of social support comes into play with this finding.

Second, **maximum learning occurs when there are moderate levels of arousal–thereby initiating neural plasticity**. Moderate levels of arousal (a state of high attention without debilitating anxiety) initiate neural plasticity by producing neurotransmitters and growth hormones, which literally build connections. As Cozolino and Sprokay confirm, "learning is enhanced through dopamine, serotonin, norepinephrine and endogenous endorphin production."[408] While having an effect in all five modes of experiential learning, maximizing learning in the NOW directly affects the sub-elements of problem solving, creativity, and anticipating the future as well as enhancing understanding, meaning, truth and how things work. Further, Cozolino and Sprokay offer that, "We appear to experience optimal development and integration in a context of nurturance and optimal stress.[409] Nurturance is explicated in Chapter 9.

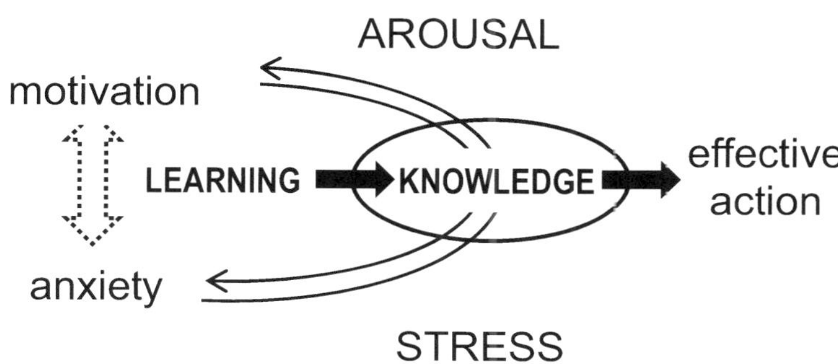

Figure 10. An optimal level of stress facilitates learning.

Excitement can serve as a strong motivation to drive people to learn, but cannot be so strong that it becomes high stress moving to anxiety. For example, Merry sees adaptation not as a basic transformative change, but as having a *new range of possibilities*. When people face growing uncertainty and stress, their resilience allows them to find novel forms of adaptation to the changing conditions.[410] In other words, with a stressful external environment, people will naturally tend to find ways of reacting and adapting to that environment.

Controlling Stress. It is possible for individuals to control their perception of stress by recognizing its existence and understanding that stress is created inside the body and can therefore be understood and managed.[411] Note that when we talk about controlling or managing stress, we refer to an *excess* of stress, which has a negative impact on aspects of self. As stated earlier, patterns in the mind *can and do* influence neurons in the brain, so that mind over matter is possible, and therefore an individual can consciously change his or her *level of response to stress*, just as levels of emotions can also be shifted (see Chapter 8). Stress is a function of individual perspective and interpretation, how we perceive a situation, or the uncertainty versus control that we see in a situation.[412]

Let's take a step backwards and actually clarify our understanding of exactly WHAT stress is. By understanding the origin of stress and learning to manage stress we can minimize its deleterious effects. So, according to Hans Selye, a recognized guru in harnessing stress, stress is defined as "the state manifested by a specific syndrome which consists of all the nonspecifically-induced changes within a biologic system" or, from a simpler point of view, "Stress is the nonspecific response of the body to any demand."[413] While stress may have its own characteristic form and composition, this definition does not attribute a particular cause, which is fully dependent on the *perception of the individual self*.

A confirming neuroscience finding from the ICALS study states that **stress depends on how we perceive a situation**. As Thompson explains, "The extent to which situations are stressful is determined by how the individual understands, interprets, sees, and feels about the situation."[414] From a learning perspective, this item would impact awareness, attention, understanding, and meaning as well as, perhaps to a lesser degree, other sub-elements of the experiential learning model.

Figure 11. The harnessing of anxiety as energy?
Cartoon by Jackie Urbanovic.

Since stress is a result of the perception of an individual in a given situation, an individual can learn to control or minimize it. The cartoon drawn by artist Jackie Urbanovic, conveys this dilemma! One approach is to perceive things from a different frame of reference, which is what is suggested in Figure 11. Is it indeed possible to harness the energy of anxiety? As shown in the discussion above, the evidence would suggest yes, to a certain degree.

Another way to reduce stress is to not judge a situation by the possibility of its outcomes. In other words, *most stress is caused by anticipation of future events* that offer a perceived threat of some nature. A tool for interrupting stress[415] is Holding Neurovascular Reflect Points.

****

TOOL 6: Holding Neurovascular Reflect Points

Ideally, we would benefit from optimal functioning of the forebrain, the thinking part of our brain, to make prudent/wise choices under stressful circumstances. This simple yet invaluable energy technique can interrupt the stress response, as well as reprogram the way your body responds to stress.

STEP (1): Find a quiet place where you will not be caught up in the everyday sounds and movement of life for a few minutes. Close your eyes and take a few deep breaths.

STEP (2): Using your palm or fingertips, hold or touch the main neurovascular reflect points above your eyes, with your thumbs on your temples. The neurovascular points or frontal eminences are the raised areas on the forehead directly above the eyes. The main neurovascular reflex points allow us to shift our physical body's autonomic response and assist us to meet stress with a high functioning thinking brain. They are often referred to as the "Oh my God" points which you often intuitively hold when shocked by an alarming event. By simply holding or touching the main neurovascular reflex points with your palm or fingertips, you boost blood and oxygen flow back up into the forebrain, allowing for clear thinking while shifting energetic patterns to calm and re-center emotionally.

STEP (3): Remaining in this position, breathe deeply for 1-5 minutes. This simple and gentle pressure instructs the primitive brain that the crisis is not a real physical threat that needs to be met with a fight-or-flight response flooding toxic stress hormones into the bloodstream. Considering that stress reactions are physical, mental and emotional responses, this process is the foundation for reprogramming what becomes an emergency response loop.

By placing fingertips over the "Oh my God" points, thumbs on your temples and breathing deeply for 1-5 minutes while feeling stressed or focusing on a stressful memory, your mind will clear and your emotions will calm as you free yourself from your memory's emotional grip.

STEP (4): When you reach completion, thank your body for its response, take several cleansing breaths, on the out-breath releasing any tension still remaining, and open your eyes.

****

Situations occur throughout life that help us understand who we are and how well we handle stress. When these situations become closer and closer, it is paramount that we shift our frames of reference to consider them from an external viewpoint. Emotions are wonderful indicators of our preferences, and without them we would have no judgment and would not be able to make decisions. That is exactly the role they need to play in a CUCA environment, when things are not in our direct control and the future is unknown.

This was driven home to two of the authors several years ago when taking a prop jet from Washington, D.C. to New York. Here's what they write about the experience:

As what was to be a short hop dragged beyond an hour and a half, we began to wonder what was happening! We were no longer circling LaGuardia Airport (where we had been scheduled to land) but were over JFK International Airport, making a low pass with a dozen emergency vehicles (lights flashing) following along under us. The pilots announced that while our landing gear was down and they felt everything was fine, the light didn't show that it was locked. We continued circling after that announcement, clearly getting rid of any excess fuel. As panic took hold in the cabin, some people were struck silent, others couldn't stop talking, and many were praying while tightly gripping the arms of their seats. But, surprisingly, two people on board were smiling as they exchanged looks. There was nothing we could do to change whatever was going to happen.

We were both in a state of wonderment, thinking excitedly—and simultaneously saying out loud to each other—"What's next?"

Intention is the Source of Action

As early as the hunter-gatherer, we see the beginnings of structure and dedicated efforts to meet objectives through intention, planned action, and individual roles. *A choice of the self, intention relates to the world.* Intention is the source with which we are doing something, the act or instance of mentally and emotionally setting a specific course of action or result, a determination to act in some specific way. The spiritualist Ramon relates intention directly to the consciousness of an individual, describing consciousness as the "energized pool of intent from which all human experience springs."[416]

Searle believes that people have *mental states*, some conscious and some unconscious, which are intrinsically intentional. From his viewpoint, these are *subjective* states that are biologically based, that is, both caused by the operation of the brain and realized in the structure of the brain, with consciousness and intentionality "as much a part of human biology as digestion or the circulation of the blood."[417] Intentionality is not a description of action, rather it is in the *structure of action*. We look to Searle's theory of intentionality for a baseline definition: "Intentionality is the property of many mental states and events by which they are directed at or about or of objects and states of affairs in the world."[418]

Thus, if you set an intention, it is an intention to *do something*. However, states such as those represented by beliefs, fears, hopes and desires insinuate intention, and they are *about something*. The relationship of intent and action has two schools of thought. The ideomotor model of human actions contends that human intentions are the starting point of the actions associated with those intentions.[419] Conversely, the sensory-motor model of human actions identifies sensory stimulation as the origination of actions.

Recall that the mind is an associative patterner that can be triggered by external events. While the direction of a belief would be from the mind to the world, the direction of a desire would be from the world to the mind, triggered by an external event. Does this insinuate a causal relationship between the external event and desire? Searle is careful with his response. He contends that causality is generally considered a natural relationship between world events, while *intentionality is not a natural phenomenon*, not part of the natural world, rather something transcendental, something that stands over or beyond.[420] For purposes of this conversation, we defer to the concept of *deliberate intent*, that is, *choice at the conscious level* which, once direction is set, is supported at the unconscious level as we live out our day-to-day lives.

As part of the Princeton Engineering Anomalies Research (PEAR) program active over a 25-plus year span, Jahn and Donne studied anomalous human/machine interactions, which focused on the effects of consciousness on physical systems and processes, and remote perception, the sending and receiving of information over distances.[421] Consistent with quantum biology proposed by Pribam[422] and Popp,[423] they discovered that "the unconscious mind somehow had the capability of communicating with the subtangible physical world—the quantum world of all possibility. This marriage of unformed mind and matter would assemble itself into something tangible in the manifest world."[424] Thus, through the power of intent, of human wishing and will, we can create order. *This touches the very essence of human creativity.*

In our normal everyday life, as we move in and out of familiar environments and situations, we are in a relatively low state of attention, and intention takes the form of anticipated outcome in response to our thoughts, words and actions. We are certainly taking in the sights, sounds and smells around us, at least unconsciously. Then, something occurs. It might be an unexpected incident, or maybe hearing a beautiful passage of music, or an unexpected election result, or watching a dramatic moment in the theatre or at a movie, or some thought triggering the excitement of possibilities. *You are at a point of peak intensity, fully attending to the instant* (thought attendance). And in that instant, a desire may emerge to do something with this; perhaps share that music with a loved one, or perhaps take that creative idea and act on it, or become an advocate to right some wrong. This is a birthing of intent.

As can be seen, attention and intention are interrelated, and both are necessary to balance current priorities with future opportunities and guide you in your personal Intelligent Social Change Journey toward intelligent activity. Your thoughts and actions gravitate toward what you pay attention to; and what you intend requires your attention. In the words of a popular Broadway song from some year ago, *you can't have one without the other*. And both can be the result of conscious choice.

The Capacity of Presencing Accesses the Field of the Future

To a large extent, presencing has very much to do with mindfulness. Perhaps the most straightforward definitions of "presencing" are offered by Brown, Ryan and Creswell when they simply describe mindfulness as "being attentive to and aware of what is taking place in the present"[425] and as a "process of openly attending, with awareness, to one's present moment experience."[426] Since "attention" and "awareness" are the two key components of mindfulness, as a high-level description this certainly makes sense, and they loosely include the concept of "being focused upon". Digging a little deeper, Kabat-Zinn adds the elements of "on purpose" and "non-judgmentally" and relinquishing personal attachment to the experience that is unfolding moment-by-moment.[427] Other researchers see this as a transformational concept,[428] a "becoming aware with respect to one's habitual and automatic patterns nested within one's larger sense of being in the world."[429] There is much literature on the process of mindfulness which would support the concept of presencing. This is not to be confused with the extended cognition of routines and habits, with these, embedded in the unconscious, reflecting more a nature of mindlessness rather than mindfulness.

Intent is also necessary for presencing, a core capacity needed to access the field of the future. Presencing, or "letting come" is consciously participating in a larger field for change. As Senge and his colleagues describe:

> *Genuine visions arise from crystallizing a larger intent, focusing the energy and sense of purposefulness that come from presencing ... Crystallizing intent requires being open to the larger intention and imaginatively translating the intuitions that arise into concrete images and visions that guide action.*[430]

Similar to Kabat-Zinn's description of mindfulness, for Senge intent is based on purpose. Purpose is the reason for an individual or an organization to exist, that is, asking: *Why are we here?*[431] Thus, change is purposeful, and at the very core of change is learning, learning to do new things and learning to do things for a different reason.[432] When this deep source of intention is tapped into, people experience synchronistic events. Synchronicity is a coincidence of events that seem related, yet with no obvious connection one to the other.

Looking from the larger perspective, without presence we would become our environment. We are chameleons, and brilliant at imprinting what we see and bringing it into our own energy. When we are conscious, we are present. But while presence and consciousness are interrelated, and presencing is core to consciousness, the idea of presence goes beyond an awareness of the NOW to include deep listening and the ability through awareness to move beyond the way we've done things in the past, that is, it brings with it the freedom of choice. As Hawkins explains,

It is this accessibility of the past—the ability to jump back in time and slide forward again to the present—that gives us our sense of presence and awareness. If we couldn't replay our recent thoughts and experiences, then we would be unaware that we are alive.[433]

Presencing is a core capability of the future, a way to access the living fields that connect us and *that which is seeking to emerge*.[434] These living fields are, in the larger sense, what could be referred to as the consciousness field, the quantum field or the God field. The concept of connecting to "that which is seeking to emerge" is certainly consistent with the quantum field as a probability field containing all possibilities.

Senge and his colleagues introduce seven capacities that are foundational to see, sense and realize new possibilities: suspending, redirecting, letting go, letting come, crystalizing, prototyping and institutionalizing.[435] Each of these capacities enables various activities which serve as a gateway to the next capacity. This theory has become more popularly known as the *Theory of U*. This theory was introduced by Scharmer—what he calls the "social technology of presencing"—which forwards that the awareness, attention and consciousness of people in a social system determine the quality of the system.[436] He says that we each have a blind spot that can keep us from being *present*, with presence necessary for profound systemic changes.

The idea of "letting come" is the process of *allowing*, whose importance cannot be overemphasized. As the adage goes, we are often our own worst enemies. As we realize the power of the mind/brain in terms of thought and feelings in the process of co-creating, we do not want to interject barriers to our desired progress forward. Similar to humility, presencing opens us to receiving and learning. "Over time, your appreciation for the question will become equivalent to your appreciation for the answer, and your appreciation for the problem will become equivalent to your appreciation for the solution. And in your newfound ease with what-is, you will find yourself in the state of allowing what you truly desire."[437] Despite this focus on self, presencing is not an individual concept, but rather considers the individual as a living system that is part of the whole. In other words, there is a sense of social identity present. In the context of the mind, this would mean *consciously* showing up, being aware of our inner connections to the larger Field.

Physical and Mental Exercise Promote Learning

Physical exercise plays a significant role in the mental and physical operation of the mind/brain. Exercise increases blood flow, bringing glucose as an energy source for neuron operation, and also provides oxygen to take up the toxic electrons.[438] Exercise stimulates neurogenesis, the creation of new neurons in certain locations in the brain, and exerts a protective effect on hippocampal neurons, thus heightening brain activity. The hippocampus is part of the limbic system and plays a strong role in consolidating learning and moving information from working memory to long-term memory.[439] During this process the hippocampus, "constantly checks information relayed to working memory and compares it to stored experiences."[440] We've previously introduced the process of associative patterning. This process integrates the various incoming information and creates higher-level images, concepts, understanding, and meaning. Stonier addresses this process of creating meaning as follows: An incoming piece of information (pattern) is combined (associated) with information (patterns) from memory. When these patterns are linked successfully, the incoming information becomes meaningful, that is, the incoming information is put into a context that represents broader understanding and meaningful information.[441]

Exercise boosts brainpower and stimulates the proteins that keep neurons connecting with each other. From the "use it or lose it" concept previously introduced, we can now see that to build learning capacity, the learner needs to exercise both body and mind on a regular basis. According to Medina, our brains were built for walking about 12 miles a day or the equivalent exercise. Exercise keeps neurons connecting and brings glucose for energy to the brain. As Medina states, "A lifetime of exercise can sometimes result in an astonishing elevation of cognitive performance, compared to those more sedentary."[442]

Another finding from neuroscience is that **volition is necessary for benefit. Forced exercise does not promote neurogenesis.** Volition is taken to mean, "the act of making a conscious choice or decision."[443] This item addresses the importance of physical exercise on the health of the brain since volition is very important for neuron activity to occur, and therefore directly affects learning effectiveness. Similarly, Begley offers that forced exercise does not promote neurogenesis, while voluntary exercise is marked by the absence of stress and does promote neurogenesis.[444] *Exercising the mind by mental activity and the physiology of the brain by physical activity provides optimum growth and health in both areas, leading to improved learning efficacy.*

Physical exercise plays a significant role in maintaining a healthy brain, which in turn can support a healthy mind. We can thus understand the duality of the challenge of maximizing learning. On the one hand, the material part of the brain, the neurons and chemistry, must be kept healthy and capable of doing their part. On the other hand, how the learner interfaces with the external world and the way he or she goes about learning, creating, and mixing patterns of his or her brain neurons will influence the state of the mind/brain during the life of the learner. A related neuroscience finding is **physical activity increases the number (and health) of neurons**. Physical activity has a generic effect throughout the brain and hence the improved health of the brain would have different effects on different sub-elements of the adult experiential learning model. Begley has offered that, "It has become a truism that the better connected a brain is, the better it is, period, enabling the mind it runs to connect new facts with old, to retrieve memories, and even to see links among seemingly disparate facts, the foundation for creativity."[445] There is a caveat, and that has to do with sleep.

Thus, the material brain can influence the creation, association, and exercise of the brain patterns while at the same time these patterns can influence the architecture of the brain. What patterns are created, how many, and how often they are utilized is influenced by the physical and mental environment within which the learner lives, and the learner's decisions and actions. We can thus perceive the self-organization of the environment-learner feedback loop as being perturbed by both environmental changes and the learner's actions and reactions. The bottom line of all of these findings is that both physical and mental exercise are essential to the health of the brain.

However, there is an important caveat. This cycle must begin somewhere, which means it follows the choices of self, who sits in the driver's seat in the journey of life.

The Aging Self Can Be Managed

During the past few decades, research has been conducted to understand the changes in the mind/brain that occur during aging. Today there is a relatively comprehensive picture of the changes occurring in the aging brain. Historically, there has been a widespread belief that as individuals grow older, their mental powers decrease due to the continuous loss of neurons. Recent research contradicts this myth. It is now clear that—barring diseases, accidents, or genetic issues—so long as individuals use their mind/brain and continue to actively exercise and think, their brain capacity will maintain a good capability in healthy bodies. Although neurons do continue to die as one ages, they also can be created in some parts of the brain as long as it stays active. It is reasonable to assume elderly people should be able to use their brains and minds as long as they keep them active throughout their lives.[446]

A related finding from the ICALS study is **physical exercise reduces cognitive decline and dementia in older people**. As stated earlier, exercise promotes blood flow, providing energy to the brain cells. In some respects, 50-year-olds will do better than 18-year-olds in academic studies because their frontal lobes, which include the prefrontal cortex, are better developed. It is the frontal lobes that help individuals pay attention and ask better questions.[447]

Marion Diamond has spent 30 years researching brain health using rats, which have the same basic brain structure pattern as humans. She repeatedly compared rats in an enriched environment with those in an impoverished environment and found that the rats living in the enriched environment showed an increase in cortical thickness in the frontal area compared to the impoverished rats. In the visual cortex, there was a seven percent difference in thickness between those living in enriched and those living in impoverished environments.[448] Other tests showed that if the rats lived together cortical thickness also increased compared to rats living alone, but not as much as rats housed together in an enriched environment, demonstrating the importance of the sociability factor. When the researchers held and stroked the rats every day their average lifespan increased from 600 days to 900 days, demonstrating the importance of care and nurturing.[449] An enriched environment is addressed further below.

Recall the use it or lose it idea previously introduced. No matter what the age, any set of neuron circuits that do not get used will tend to grow weaker. While older people may not be able to memorize new facts as easily as younger individuals, their frontal lobes are considerably more developed and as a result they know what to focus on and how it may be useful to them. Amen also explains that the best mental exercise is new learning, acquiring new understanding and knowledge, and doing things that they have never done before.[450] In one research program known as the Nuns study, researchers from the Rush University Medical Center in Chicago studied 801 nuns, priests, and other clergy engaged in mentally stimulating endeavors over a period of five years. The result was that those subjects who increased their mental activity during the five years "reduced their chances of developing Alzheimer's disease by one-third."[451] These subjects also reduced their age-related decline in mental abilities by 50 percent and concentration and attention span by 60 percent.

A supporting finding in the ICALS study is **the best mental exercise to slow aging is new learning and doing things you've never done before**. Exercising the mind in multiple areas of thinking strengthens the overall brain. New learning creates new connections, new patterns, and strengthens existing connections. No learning allows these connections to weaken.

As introduced in the discussion of attention as a finding, **mental exercise at every age has a global, positive effect on the brain**. Combine this with the finding introduced in Chapter 5 that **despite certain cognitive losses, the engaged, mature brain can make effective decisions at more intuitive levels**. The implications of these findings are significant because they open the door for older people to take advantage of their true mental capacity throughout life. An additional value of these findings is that anyone at any age, if they so choose, can begin actively using their minds and improving their capacity to learn, no matter what their past learning history has been. Now that we understand and realize that aging need not degrade mental capacities, *anyone who desires to continue learning throughout life can do so*. When this choice happens, the payoffs would seem to be potentially very large.

An Enriched Environment Expands the Mind

A neuroscience finding in the ICALS study was that **an enriched environment increases the formation and survival of new neurons**. An enriched environment can influence both the nature of the experience of the learner and his or her learning efficacy. Begley says that "exposure to an enriched environment leads to a striking increase in new neurons, along with a substantial improvement in behavioral performance."[452] A related finding introduced in Chapter 5 in the conversation on creative thought—and appropriate to repeat here—is that **an enriched environment can produce a personal internal reflective world of imagination and creativity**. This means an enriched environment entering the mind/brain/body through sensory experiences stimulates the mind to associate patterns and create new possibilities. Let's dig down a bit to explore just what is meant by an "enriched" environment?

Kempermann and his colleagues discovered that there was a significant increase in neurons in young adult mice which spent 45 days in an enriched environment.[453] For these mice, that environment was created to resemble the complex surrounding of the wild, including such things as wheels, toys, and tunnels. After the 45 days, the animals had undergone a dramatic spurt of neurogenesis. As Begley describes,

> *The formation and survival of new neurons increased 15 percent in a part of the hippocampus called the dentate gyrus, which is involved in learning and memory. The standard 270,000 neurons in the hippocampus had increased to some 317,000.*[454]

As Fred Gage (a member of the team) described this finding to the Dalai Lama, "It's not a small number: 15 percent of the total volume can be changed just by switching experience."[455] Indeed, as introduced above, Begley describes this as one of the most "striking" findings in neuroplasticity.[456] There are other changes that occur as well, specifically, thicker cortices are created, there are larger cell bodies, and dendritic branching in the brain is more extensive. These changes have been directly connected to higher levels of intelligence and performance.[457] This research on the effects of enriched environments on brain structure is both credible and well established.

As a second example, Skoyles and Sagan presented the results of research on adolescent monkeys that suggested prefrontal cortices respond better than other parts of the brain to an enriched learning environment. After a month of exposure to enriched environments, the monkey's "prefrontal cortices had increased their activity by some 35 percent, while those of animals not exposed to an enriched environment had slightly decreased their activity."[458] These authors go on to say that, "As the most neurally plastic species, we can choose to put ourselves in stimulus-rich environments that will increase our intelligence."[459] Again, *this is a choice of self.*

From these examples we're starting to build an understanding of this concept of "enriched environment". Only, what does it specifically mean in terms of human learning? In general, we could say this is when the surrounding environment contains many interesting and thought-provoking ideas, pictures, books, statues, etc. We might reflect on the use of space design and plants, art and music, furnishings and light to help create that environment. And much of that "enrichment" might be focused on social aspects, that is, engagement with other people. This certainly makes sense. People do not often learn in isolation, but are very much engaged in a continuous process of what is often referred to as social learning. This will be further explicated in Chapter 9 which focuses on social learning.

Now, if we had asked the question—"What does it mean in terms of human learning?"—20 years ago, we might have said that effective learning requires concentration; no physical, mental or emotional distractions. This is true; our minds cannot function if they are bombarded with disturbances. From the viewpoint of the formal classroom setting, this means surroundings that are quietly attractive, passive yet comfortable. If the surroundings provide safety, positive feelings and an aura of warmth and confidence, the desired learning will likely be easier and quicker, and the results remembered longer. We would also agree that the idea of an enriched environment might be highly sensitive to the specific learner. For example, some types of music have long served as learning aids. But in today's world, an enriched environment would also very much include technology augmentation, with computer technology, web services and social media all offering platforms that can stimulate and support learning. Appendix I provides a gentle short story which explores the potential joys of learning in the future.

In Summary

This chapter began the exploration of managing self—which itself represents the powerful force needed to address the uncertain future—with a focus on the success factor of attention/intention, the personal tool that creates our reality, and the skill set of presencing, which enables us to access the field of the future. We also touched on the power of stress, the role of physical and mental exercise in learning, managing the aging self and the mind expanding offered by an enriched environment, all supported by neuroscience findings in the ICALS study.

However, the complex adaptive learning system that is the human is just that, a complex adaptive system with a large number of interrelated parts that include nonlinear relationships, feedback loops, dynamic uncertainties, self-organization, emergent properties (which make the whole of the system very different than the sum of the parts), all operating at some level of disequilibrium in a continuous interaction with—and adaption to—the environment. As powerful as the conscious mind can be, the conscious high-level viewpoint from self which is at the helm of this system must make decisions based on the best understanding of the system—coupled with acknowledgement and trust in the unconscious workings of the system—which are very much a result of past experiences and decisions of the self. In like manner, the decisions made today will affect the workings of the system in the future.

With this in mind, the next four chapters continue with a focus on managing self, or self-management, each with a focus on a specific aspect of self. Chapter 8 focuses on emotions, with an emphasis on heart-mind entrainment, seeking coherence through the balance of the emotional and mental fields. Chapter 9 expands the perceived field of self, the social aspect of self, bringing in the understanding that we are in continuous two-way communication with those around us, affected by—and having the ability to affect—others. Chapter 10 further expands the boundaries of self, recognizing we are part of an even larger field—as noted, whether that is considered as a consciousness field, a quantum field or a God field—which begs us to consider the animating energy which enlivens us, exploring this energy in terms of spiritual learning, the process of elevating the mind as related to intellect and matters of the soul to increase the capacity for effective thought and action. And Chapter 11 reflects on the humanness of humility, and the powerful role it plays in learning.

Chapter 8

Heart-Mind Entrainment
(Feelings and Emotions)

The Heart and Mind Have a Strong Connection ... Emotions Offer Learning Possibilities ... Emotions are Sensitive to Meaning ... Truth is a Shifting Target Relative to Self ... Inducing Resonance ... Emotions Can Be Managed ... In Summary

SKILL SET: Managing Self (Emotions)

TOOLS: (7) Connecting through the Heart; (8) Releasing Emotions Technique.

Emotions—considered both a "mental state that arises spontaneously rather than through conscious effort and is often accompanied by physiological changes"[460] *and* "organized patterns of thoughts and behaviors that we actively construct in the moment and across our life spans"[461]—play a large role in what it is to be human. They involve mental and physical reactions to specific situations (and people) as well as *long-term drivers that consciously or unconsciously guide our lives.*

Emotions play a key role in experiential learning. As mentioned earlier, all incoming signals and information are immediately passed through the amygdala (the oldest part of the "old" brain), where they are assessed for potential harm to the individual. The amygdala places a tag on the signal that assigns it a level of emotional importance.[462] If the incoming information is considered dangerous to the individual, the amygdala immediately starts the body's response, such as pulling a hand away from a hot stove. As emerged in the ICALS study, **the entire body is involved in emotions and the body drives the emotions**. As LeDoux offers, *emotions are a slave to physiology*, not vice versa.[463] This aspect of our emotions may singularly impact how we act on the environment and how much control, rigor, and/or discipline we can exert. It also impacts how well we can focus attention and maintain awareness.

In parallel—but slower than the amygdala's quick response—the incoming information is processed and cognitively interpreted, although by that time anything of significance to the individual has an emotion or feeling attached to the thought. For example, when an individual thinks about recent occurrences like an argument or a favorite sports team losing in the Rose Bowl, feelings are aroused. Or, recall the internal responses to holding the hard copy of your first book, or your new born child. Thus, as introduced in Chapter 1, it can be understood that **emotions influence all incoming information**. This finding can potentially influence all aspects of adult experiential learning.

By now, we understand that the neocortex (the "new" brain) is the organ of intelligence and that as such it is objective. However, it is largely guided by the "old" brain through emotions, which relate to passions, values, and virtues. In other words, the neocortex is objectively processing incoming information that already has an emotional tag attached, that is, the organ of intelligence is objectively processing subjective information. We'll talk more about that below.

There are three specific aspects that are most likely to be subjected to strong emotional influence: understanding, meaning and truth/how things work. However, Haberlandt goes so far as to say that there is no such thing as a behavior or thought not impacted by emotions in some way.[464] *Even simple responses to information signals can be linked to multiple emotional neurotransmitters.* Similarly, reiterating this point, Mulvihill explains,

> *Because the neurotransmitters which carry messages of emotion are integrally linked with the information during both the initial processing and the linking with information from the different senses, it becomes clear that there is no thought, memory, or knowledge which is 'objective,' or 'detached' from the personal experience of knowing.*[465]

In summary, what our new understanding of the neocortex implies is that as the cortical columns continuously process incoming signals using various reference frames—ever adjusting their model of reality, striving for a higher truth—the incoming subjective emotions and feelings forwarded from the "old" brain become part of the objective model, at least until trumped by other stronger emotions and feelings or conscious intellectual choice. Thus, the emotions we feel—or choose to feel—affect every part of our lives, personal and professional, whether alone or in a group.

Ralston identified eight principles of emotions that provide unique insights into the *way* people behave and *why* they behave that way.[466] This is a good set that warrants repetition here. These principles are:

1. Emotional needs express themselves one way or another.
2. Anger is an expression of need.
3. Our feelings and needs are not wrong or bad.
4. Emotions are the gateway to vitality and feeling alive.
5. We can address emotional issues and still save face.
6. Immediate reactions to problems often disguise deeper feelings.
7. We must clarify individual needs before problem solving with others.
8. We need to express positive feelings and communicate negative ones.

If you've been around in life for a while, you no doubt have heard the expression, and probably felt, the *emotional rollercoaster of life*. This refers to the up and down ride between excitement and disappointment, emotional highs and lows. Emotions—context sensitive and situation dependent—form a dynamic flow, and humans seem to have a penchant for living from one extreme to another. However, as will be seen, as we ride this rollercoaster there are choices that can be made, with emotions serving as a guidance system for self, alerting and protecting individuals from harm, and energizing them to action when they have strong feelings or passions,[467] hopefully with those strong feelings and passions consistent with the individual's direction of choice.

The Heart and Mind Have a Strong Connection

A large body of research is emerging out of the HeartMath Institute, which has through science answered the question of how, and why, as an information-processing center the heart affects mental clarity, emotional balance, creativity, intuition and personal effectiveness. The heart communicates with both the brain and the body in four different ways: neurological communication through the nervous system, biochemical communication through hormones, biophysical communication through pulse waves, and energetic communication through electromagnetic fields,[468] with an electromagnetic field generated by the heart radiating outside the body, carrying information that affects the environment and others in the environment.

Coherent entrainment occurs when biological oscillators within the body are pulled into synchronization by the rhythms of the heart. The goal is to follow the process of entrainment—where a system or subsystem aligns its functional behavior with that of another system or subsystem—to a state of coherence, where these systems/subsystems become phase or frequency locked.[469] This cross-coherence, which is similar to what can occur among laser photons, is what scientists simply refer to as "coherence". From the viewpoint of physiology, McCraty and Childre say that cross-coherence is "when two or more of the body's oscillatory systems, such as respiration and heart rhythms, become entrained and operate at the same frequency."[470] Increased heart coherence, which can be measured as heart-rate variability or HRV, increases systemic harmony and "entrainment" of human biological-psychological systems.

Heart-mind coherence does *not* mean that the heart and mind are doing the same thing, acting the same way at the same time. What it *does* mean is that there is a "cohesion", a working together, a consistency. From an organizational viewpoint—recognizing that organizations are also complex adaptive systems—for one of the authors this was expressed as a "connectedness of choices", where different decisions and different activities from different parts of the organization—even when they appeared to be diametrically opposed—worked together to head the organization in the same direction toward a collective vision. As McCraty and Childre describe from the system level,

> *It can appear at one level of scale that a given system is operating autonomously, yet it is perfectly coordinated within the whole. In living systems, there are micro-level systems, molecular machines, protons and electrons, organs and glands each functioning autonomously, doing very different things at different rates, yet all working together in a complex harmoniously coordinated and synchronized manner.*[471]

This coherence is the optimum state for the working together of emotions and thought, achieving mental clarity and emotional balance.

The heart energy center, long associated with love and the recognition of Oneness and connections—and indeed home to our internal sense of connection—allows us to love life itself, developing an understanding of life and leading to expansion of our consciousness. There is a resilience here, supporting the continuous change in which our body thrives. When we are young, our Heart Rate Variability, that is, the distance from the peak of one heartbeat to the next, varies. As we get older and become more rigid in our behaviors and thinking, the distance from the peak of one heartbeat to the next becomes more even. This regularity limits our life span; for when the heart beats in an even fashion it wears out much faster. Resilience can be regained by establishing coherence between the heart and brain.[472] So, how is that achieved?

A quick methodology that supports this coherence starts with breathing a bit slower and deeper than usual while focusing on your heart. You can tap your fingers gently in the heart area to help that focus occur. Then, think and feel positive, perhaps focusing on someone special in your life, or on something for which you are extremely appreciative. It is that simple—and yet can be that difficult, for life is filled with past memories and current perturbations, all of which have a tendency to creep in and out of our brains as "monkey chatter". An exercise to connect to the larger energy field through your heart is provided in TOOL 7 below.

When fully functioning, the human has the power to comprehend and empathize with others (with the potential of that empathy moving into compassion), and to sense and understand the connection to a larger force, giving us an awareness of the larger eco-field beyond the perceived boundaries of our own life. The energy field of the human heart and the Earth's energy field are coherent, that is, *they are part of a unified field*. Unfortunately, in today's environment we are recognizing the impact of technologies on the Earth's energy field, and it begs the question: How are these technologies impacting the human heart?

****

TOOL 7: Connecting through the Heart

Neurons are not only located in the human brain, but in the heart, with the capability to think, feel and remember. Because these neurons are quite often not as actively engaged in everyday life as those firing within our heads, there are fewer mental models to limit the movement of our thoughts and feelings. This exercise is focused on connecting to the larger energy field through your heart.

STEP (1): Find a quiet place where you will not be bothered for a half hour. Make your body comfortable (sitting or lying down) and close your eyes.

STEP (2): Think about where your thoughts originate, and place your hand where you think that occurs. This will generally be on one side of the upper head, or on the forehead between the eyes. This "thinking spot" will serve as the starting point of your journey. Put your arm down and let your body relax. NOTE: This "thinking spot" will be perceived differently by different individuals.

STEP (3): Take several slow, deep breaths, breathing in through your nose and out through your mouth, consciously releasing any tensions or random thoughts with the out-breath.

STEP (4): Imagine an elevator inside the middle of your head next to the thinking spot where your thoughts originate. The door opens, and you enter the elevator.

STEP (5): The doors close, and you begin moving downwards. It is a beautiful elevator, with glass panels such that you can observe without from within. Slowly, imagine the downward journey of the elevator, moving through the throat, down the neck, into the upper chest, and ever so slowly down right behind your heart space. Take your time; there is no hurry. Keep all of your thoughts focused from the location of your elevator. If you have difficulty imagining this slow journey, take your hand and gently trace the journey down from your thinking spot. Down, down, down, until you reach your heart space.

STEP (6): The elevator doors open and you enter your heart. It is warm, softly beating, and quite welcoming. Keeping your focus in this place, imagine yourself growing larger and larger. You are energy, with no barriers. From your heart space, you expand outward, encompassing your entire body, then move beyond the body, expanding wider and wider beyond the confines of your body. You feel free, and continue expanding, wider and wider, until you fill the room or area immediately surrounding your body. Keep your focus outward from the inward core of your heart space, and continue expanding, expanding, wider and wider. You, as your heart, now fill the house or field within which your body resides.

Do not stop, continue expanding, periodically stopping to enjoy your growth, your freedom. Move beyond the local area to the larger geographical area of which you are a part. Keep expanding, more and more, wider and wider. You pass, and expand beyond, trees and houses, roads and the cars moving along them, up through the clouds, down into the Earth, ever-growing, expanding, expanding. The slow rhythm of your heart gently pulses through the field. You have no limits. You become as large as the Earth, and expand beyond, reaching the limits of the solar system, and, if you wish, expanding beyond. Continue expanding as far as you are comfortable. Then, pause and enjoy the feeling of this expansion. The lightness, yet the power of the energy of which you are a part. Observe what is beyond and, if you dare, continue expanding.

STEP (7): When you are ready to return, think about your body and where it resides, and slowly bring your awareness back to your body. You may repeat this process any time you choose.

NOTE: If you reflect on any issues or problems going on in your life after experiencing this journey, they may have diminished their level of importance. Experiencing this journey, understanding that we are so much more, provides us with a larger systems perspective from which to view local situations.

****

Emotions Offer Learning Possibilities

Understanding the role of emotions in experiential learning allows the learner to take control and turn initially negative emotions into learning possibilities. Both the excitement of learning and the fear of failure or embarrassment impact the efficiency and effectiveness of the learning process. A passion for learning opens the mind, creates possibilities, and expands the scope of interest and creativity. A fear of learning closes the mind to options and causes it to focus on safety and protection, which minimizes learning. Further, we have recognized that there is a strong link between an individual's emotions and his or her cognitive appraisals.[473] Remember, this is because the input of subjective emotions and feelings becomes input as part of the objective models of reality being continuously created and re-created in the cortical columns of the neocortex.

To develop understanding, see the meaning or significance of something, or recognize relationships, patterns, and probabilities requires an open mind that can explore different avenues of thought. This is difficult if fear due to excessive stress is in the background of learning.[474] This leads to the core finding that **emotions can increase or decrease neuronal activity**. The level of emotion that an individual experiences can influence not only *awareness* but also how well he or she can *focus attention.* Because of the presence and significance of emotions throughout the mind and the body, this item could well impact any of the sub-elements that are part of the adult experiential learning model.

Our *rate of learning* depends upon what we already know.[475] A generic fear of learning may create a potential long-term degenerative spiral, resulting in continuous non-learning, the exact opposite of what would be desired. A related finding, which we have previously touched on, is that **emotional fear inhibits learning**. Because emotional fear *spreads throughout the body*, it impacts many aspects of the adult learning model. For example, sensing, feeling, awareness, attention, understanding, and meaning would all be changed in some way by high levels of emotional fear. On the converse side, a positive attitude filled with excitement and energy would enhance all of these same characteristics. *The recognition and belief that the mind can influence the brain/body (the physical material world) is a key concept for enhancing experiential learning.*[476] We now recognize that the world as we each perceive it is a simulation, with the firing neuronal activity—our current thinking and perceiving—relative to the models we have created of the world rather than the physical world we perceive as reality. While input (electrical spikes or action potentials sent through sensory nerve fibers) comes into the brain, everything that we are perceiving is created by/occurring in the brain. Thus, "the brain only knows about a subset of the real world, and … what we perceive is our model of the world, not the world itself."[477]

Where learners can control their emotions by taking a positive attitude, they can also *enhance their learning*. As noted above, all of this is driven internally, within the learner, by chemicals generated by the perceptions and feelings of the situation and the material to be learned.[478] While the external environment and the material to be learned may have a significant role in the learning outcome, *the perception and interpretation by the learner significantly impacts the result*.

In the NOW, our unconscious mind is continually interpreting situations and these interpretations influence our emotional response. Since our emotional state affects learning and feelings about the information learned, recognition and control of feelings related to some learning situation can significantly enhance our learning capacity.[479] For example, if we voluntarily desire to learn, our emotions create less stress and better support, and that helps focus the learning process. On the other hand, if learning is involuntary, the emotional system may create more stress and thereby decrease learning. This is consistent with the finding introduced in Chapter 4 that *voluntary learning, as distinct from forced learning, has little or no stress attached to it and therefore fosters a much higher level of learning.*

A related finding is that **unconscious interpretation of a situation can influence the emotional experience**. As previously forwarded, the unconscious mind is always working and, through the release of neurotransmitters, it can create bodily emotional experiences. This process can impact the understanding and meaning aspects of an external experience during a learner's reflective observation.

Another aspect of the emotional system is the role it plays in individual memory. While this was previously addressed in our discussion of memory, in this chapter we are looking from the viewpoint of emotion. As you will recall, an overarching finding is that **emotional tags impact memory**. Since the amygdala checks all incoming information for danger and places a tag on it relative to the emotional impact, some level of emotion from zero to high is attached to our thoughts and memories. In her work that revolutionizes educational theory and practice, Immordino-Yang points out that interdependent neural processes *support both emotion and cognition.* Therefore, "It is literally neurobiologically impossible to build memories, engage complex thoughts, or make meaningful decisions without emotion."[480]

Similarly, situations that have a high emotional impact are much easier to recall, sometimes remembered throughout life. From a learning perspective, this means that (consciously or unconsciously) the learner is always evaluating the importance of incoming information and this process helps the individual to remember the information.[481] We remember things that are emotionally significant. This helps our understanding and the creation of meaning, as well as our ability to anticipate the outcomes of actions. Thus, this item supports both reflective observation and abstract conceptualization. The mind solves many complex problems by recalling the core (or meaning) of past solutions. The most significant past problems have emotional tags that aid in recall.

Emotions are Sensitive to Meaning

A fascinating aspect of the emotional system is that it is not concerned with the *details of the information.* Its primary concern is with the *meaning and impact* of the information.[482] As knowledge, the construction of meaning is relative, highly individualized and subjective, using our creative juices to move us toward the reality we desire, which can also be an intimidating responsibility![483] With this new frame of reference comes the understanding that events themselves, that is, things happening in the environment in which we live and interact, *do not themselves have meaning*. It is our self, emerging from all the experiences of life, that *assigns meaning to events*. In other words, we are the determinants of meaning. As Walsch describes,

> *Events are events, and meanings are thoughts.* **Nothing** *has any meaning* **save the meaning you give it.** *And the meaning you give to things does not derive from any event, circumstance, conditions, or situations exterior to yourself. The Giving of Meaning is entirely an internal process.* **Entirely.**[484]

A related finding in the ICALS study is that **emotions miss details but are sensitive to meaning**. Emotions exist to alert and protect individuals from harm and to energize them to action when they have *strong feelings or passions*. As introduced in Chapter 1, emotions represent public expressions of response and feelings represent the private mental experience of emotion. Passion indicates those desires, behaviors, and thoughts that suggest urges with considerable force.[485] However, emotions are concerned with the *meaning* of the information and not the details. Next time you feel an emotional response to someone or a situation—which will most likely occur several times throughout your day—ask yourself: *What is the meaning behind this response?* This focus on meaning may influence a number of aspects of the adult experiential learning model either positively or negatively, depending on circumstances. Specifically, awareness, open-mindedness, intuition, creativity, understanding, and meaning, as well as risk taking, resonance with people and ideas, and the evolution and sculpting of the brain, are all influenced by the emotional state of the individual.

Remember, it is the unconscious that is attaching the emotional tag to incoming information, which leads us to a related finding from the ICALS study. **The unconscious processes emotional meanings of stimuli**. As LeDoux notes, the unconscious is where much of the emotional action is in the brain.[486] Since thinking has emotional tags attached to it, this item would apply to most sub-elements of the adult experiential learning model. This also emphasizes the role of "meaning" relative to unconscious processes vice memorizing or simple understanding. *The mind/brain is a meaning-making system.*

The *amount* of meaning in life is directly related to an individual's level of consciousness. Primary consciousness is the ability to construct a mental scene, but with limited semantic or symbolic capability and no true language. An incoming scene, say an image, is immediately (within fractions of a second) evaluated by the brain's value systems and, through the interaction of the memory system with its previous experiences and incoming signal, *a meaning is associated with the perception*. This new perception may be put into memory, depending on its importance to the individual, which is expressed through the emotional tag connected with that incoming information. Remember, "perception" is a continuously shifting emergent property that is based on the "consensus" of the models of reality formed by the cortical columns of the neocortex. Note that while the quality of emergence denotes some level of stability, perceptions are highly relative, that is, context sensitive and situation dependent.

Levels of consciousness are very much related to emotions and feelings. Recall that in Hawkins' model (Chapter 2), levels of consciousness falling below 200 are weak attractors and represent negative energy. They are focused inward, centered around self. For example, consider level 150 (anger). We always say "I'm angry at …" or "I'm angry because of …" and give credit to external sources. However, the anger is within you; our emotions are *our* emotions.

As you move up the levels of consciousness scale, the increased awareness that comes with expanded consciousness offers greater choice as life unfolds. We are no longer victims to our environment in terms of our emotional responses, but can use those very human emotions as the action guidance system they were intended to be, thanking them, then engaging our mental faculties to consider our next actions and interactions. Feelings can also serve as indicators of your consciousness level. Feelings of boredom and loneliness indicate a lack of consciousness—occurring in the lower levels of the scale. As people move from the lower negative emotions into courage (200), increasingly the well-being of others becomes more important. By the 500 level—that of love—"the happiness of others emerges as the essential motivating force."[487]

A powerful aspect of Hawkins' scale, based on 20 plus years of research with thousands of participants, is the recognition that vibrational power advances logarithmically. Thus, a single individual at the highest consciousness level (1,000, an avatar) can counterbalance the negativity of all mankind. As Hawkins advocates,

Were it not for these counterbalances, mankind would self-destruct out of the sheer mass of its unopposed negativity. The difference in power between a loving thought ($10^{-35\,\text{million}}$ microwatts) and a fearful thought ($10^{-750\,\text{million}}$ microwatts) is so enormous as to be beyond the capacity of the human imagination to easily comprehend ... even a few loving thoughts during the course of the day more than counterbalance all of our negative thoughts.[488]

As can be seen, positive emotions and feelings such as love and joy offer us an incredible opportunity to change lives—our own and others. The further we move beyond the 200 level of courage—moving toward willingness, acceptance, reason, love, joy and peace—the greater our positive impact on the world. Our emotions and feelings are that powerful!

Truth is a Shifting Target Relative to Self

As introduced in Chapter 5 on thought and thinking—and just as critical to our understanding of emotions—**the unconscious can influence our thoughts and emotions without our awareness**. This affects our understanding of the making of meaning and a desire to find the truth and how things work as described under the reflective observation mode and enhanced in the social engagement mode, with the term "reflective" implying that it may take considerable time to integrate and understand the meaning of an experience. Meanwhile, the unconscious—processing somewhere around eleven million pieces of information per second—has been fast at work relating incoming information to the internal map (the combination of models throughout the neocortex) that has emerged from the experiences, feelings, beliefs, values, etc. culminating in the individual self, and assigning emotional tags. As a comparative point, the conscious mind processes around 40 pieces of information in that same timeframe.

Much like knowledge, *truth is a shifting target relative to our self*. As Cooper describes, "Truth is a living, dynamic awareness that grows in its meaning and value as our consciousness expands."[489] And because we are complex adaptive systems living in a continuously changing reality, an apparent truth and imagined truth may well accompany a perceived actual truth. Walsch likens these to three levels of reality,[490] that is, we can experience things at the level of distortion (imagined truth), the level of observation (apparent truth) or the level of ultimate truth (actual truth) which, of course, is only actual for the moment at hand, and is "actual" in terms of our perception!

Humans are truth seekers, with our individual truth locally structured based on the lower mental thought of logic, which can be connected to examples in the world. Conceptual in nature and mentally constructed—with "feelings" playing a large role in our perception—*truth is both internally discovered (learned) and created.* When we recognize truth as a subjective, relative value, it is also clear that it is highly contingent on an individual's level of consciousness and level of perception. Entirely subjective, perception is considered the result of using the senses to acquire information about a situation or the surrounding environment; the impressions, attitudes and understanding about what is observed. Since each individual is unique, both in terms of physical, mental, emotional and spiritual makeup, and in terms of experiences, individual perceptions also vary. Thus, the value of truth is *dependent on its meaning as translated by individual perception.* As Hawkins explains,

> *Truth isn't functional unless it's meaningful and meaning, like value, relies on a unique perceptual field. Facts and data may be convincing at one level and irrelevant at another.*[491]

Feelings change. Because truth is structured, it can be discerned in comparison to another truth and can be given a relative value regarding the *level of truth*, that is, a mathematical representation of how much a conceptual truth is true in a specific example. Further, because truth is relative, over time we are able to recognize situational value and determine contradictions, and under all of this discerning our "judging" is driven by our feelings. In other words, truth is found in our inner experiences, and we relate that to a clear factual knowledge which is applicable to our current situation.

The relationship of emotions and truth can be shown mathematically. Affirming *the transfer of emotions and intentions along with thought*—for example, through the concept of mirror neurons—mathematical research engaging transreal numbers shows that *knowledge by sentience gives us the possibility of knowing everything that is not propositional.*[492] Transreal numbers are an extension of real numbers developed by an English computer scientist, James A. D. W. Anderson, in 1997.[493] This "extension" allows division over zero, which is forbidden in the realm of real numbers.

Let's explore that a bit further. This "knowledge"—known through "sentient justification"—includes everything that is difficult or impossible to reduce into logical discourse. Examples include knowledge of material bodies, direct intuition, states of consciousness and perceptions regarding emotions. Worth detailing here, this expands our understanding of learning to include forms of apprehension of reality *not occurring through analytical means*. While "logic" says—as part of the extended argumentative structure—that only propositions that can be evaluated as true or false are factors suitable for deductive reasoning, "in a logic created from the transreals, it is possible to evaluate non-propositional objects such as sensations or feelings as **having truth value.**"[494] This mathematical discovery defines a new type of epistemic operator—a sentient one—working with an epistemic logic which includes "poetical discourse" based on images and an "immediate intuition of reality", that is, one that is both **sensory and mental**. This new understanding has ramifications for learning through technologies such as PEGs (Psychecology Games)[495] that introduce narratives involving sensations or feelings, which now can be recognized as *powerful contributors to argumentative structures that change our belief states.*[496] PEGs are further discussed in Chapter 12.

Inducing Resonance

As referenced earlier, a 2008 MQI research study explored ways to more fully—and by conscious choice—engage our tacit knowledge, that is, knowledge residing within that is difficult to consciously express. This understanding was described as moving from ordinary consciousness to extraordinary consciousness, beyond the usual or regular state of consciousness, which means acquiring a greater sensitivity to information stored in the unconscious. The challenge is to make better use of our tacit knowledge through creating stronger connections with the unconscious, building and expanding the resources stored in the unconscious, deepening areas of resonance, and sharing tacit resources among individuals. The resulting four-prong model includes surfacing tacit knowledge, embedding tacit knowledge, sharing tacit knowledge and inducing resonance.[497]

Resonance is a quality—a richness or significance—which evokes an association and/or strong emotion. All objects have simple wave functions that have unique wave signatures with a natural vibrational frequency. A resonance of thought occurs when written or verbally expressed ideas are consistent with the rhythm of your natural frequency, that is, vibrating at the same frequency.[498] Entrainment can occur when a strong frequency pulls a less-powerful frequency into its rhythm. Both entrainment and resonance can be used to alter your consciousness.[499]

The 2008 MQI research study found that through exposure to diverse, and specifically opposing, concepts that are well-grounded, it is possible to create a resonance within the listener's mind that amplifies the meaning of the incoming information, increasing its emotional content and receptivity. While it is words that trigger this resonance, it is the current of the search for truth flowing under that linguistically centered thought that amplifies feelings, bringing about the emergence of deeper perceptions and validating the recreation of externally-triggered knowledge in the listener.

The concept of truth has emerged throughout this book. Truth seeking is part of the Intelligent Social Change Journey. As a reminder, when we move from lower to higher mental thinking, from logic to concepts, we are looking at the relationships among things and how they work together, learning and ever creating larger and more complete concepts that expand our understanding of the larger world, i.e., higher truths. This is a living, dynamic awareness that expands our consciousness. Recall also that intelligence is defined as *an aptitude for grasping truths*. It is the neocortex, as the organ of intelligence, that is very much concerned with updating its models of the world based on a continuous stream of incoming sensory input. Connections are not fixed, and when we learn something, connections are strengthened. Conversely, when we forget, they are weakened. As previously noted, this is what we call Hebbian learning.

Inducing resonance is a result of external stimuli resonating with internal information, *bringing it into conscious awareness*. When resonance occurs, the incoming information is consistent with the frame of reference and belief systems of the receiving individual. This resonance amplifies feelings connected to the incoming information while also validating the re-creation of this external knowledge in the receiver. Further, this process results in the amplification and transformation of internal affective, embodied, intuitive or spiritual knowledge from tacit to implicit (or explicit). Since deep knowledge is now accessible at the conscious level, this process also creates a sense of ownership within the listener. For example, in a debate, the speakers are not telling the listener what to believe; rather, when the tacit knowledge of the receiver resonates with what the speaker is saying (and how it is said), a natural reinforcement and expansion of understanding occurs within the listener. This accelerates the creation of deeper tacit knowledge and a stronger affection associated with this area of focus.

An example of inducing resonance is a debate. Well-researched and well-grounded external information is communicated (made explicit) and tied to an emotional tag (explicitly expressed). The beauty of this process is that this occurs on both sides of a question such that the active listener who has an interest in the area of the debate is pulled into one side or another. An eloquent speaker will try to speak from the audience's frame of reference to tap into their intuition, their tacit knowledge. She will come across as confident, likeable and positive to transfer affective tacit knowledge, and may well refer to higher order purpose, etc. to connect with the listener's spiritual tacit knowledge. A strong example of this occurs in the U.S. Presidential debates. This also occurs in litigation, particularly in the closing arguments, where for opposing sides of an issue emotional tags are used to connect to the jurors and affect their judgment. To support unleashing the human mind, inducing resonance through paradoxical thinking is suggested in Chapter 12.

Emotions Can Be Managed

Since the emotional system is always operating, it continuously impacts all modes of the learning process. For example, research has shown that if the amygdala is damaged the individual is unable to make decisions. There is no "feeling" of which decision is best, no basis to consider one choice "better or worse" than another.[500] Enter the well-known widely-exampled case of Phineas Gage. An incident occurring on September 13, 1848, to the construction foreman of a railroad company laid the groundwork for this understanding. On that day, Gage placed an explosive charge into a rock with a tamping rod, which accidentally detonated, sending the three-foot, seven-inch rod through his skull. Incredibly, he survived, with physical recovery within ten weeks, except there was something that was fundamentally wrong. While Gage suffered no memory loss or apparent motor skill deficits—except the loss of his left eye and depth perception—his behavior had changed. He could no longer make decisions. The accident had destroyed the part of his frontal lobe that relates emotions such that he had a lack of emotions.

Since this incident, there have been many research studies that have involved traumatic brain injuries which have validated the significance of the frontal cortex emotional control center, which issues somatic (emotional) signals—highly dependent on neural substrates that regulate homeostasis, emotion and feeling[501]—that affect the decision-making process. There is no doubt today that emotions are critical to decision-making. Considering the role emotions play in human decision-making in everyday life, a high emotional event or situation can easily distract an individual's attention or focus it.[502] Thus, it is important that learners recognize and understand the role and significance of their emotions.

As can be discerned, emotional responses to life shift and change through experience and maturity. From a long-term perspective, learning can be enhanced by asking questions and deliberately assigning importance levels to what is learned. Further, by developing a positive attitude toward learning, individuals can improve their learning rate and their recall. Because emotions operate through chemical generation and flows, and consistent with the ICALS related finding, their impact may include the entire body.[503] How we feel about a situation or a goal is an overall state of the body rather than a specific set of patterns in the mind. With practice, an individual can learn to sense, and to some degree manage, internal feelings.[504] In that context, we address five "knowings".

A FIRST KNOWING in managing our emotions is the recognition that they ARE ours. Although we often perceive emotions as caused by external events or people, they emerge within ourselves. As a role play, imagine yourself as a young woman who has the opportunity of a lifetime to interview with a large public relations firm in New York City. You meet with several executives, you provide writing samples, they look at some of your layouts, and somewhere in between they feed you. You get the job at twice the salary you ever expected! Floating with happiness, you exit the office building and head toward your car, parked seven blocks away (assuming you were able to find a parking spot at all!) It is dark, and you notice two large men walking behind you. You speed up; they speed up. You turn the corner; they turn the corner. You feel a level of fear. You start running; they start running. Catching up with you, one of them reaches out and grabs your elbow, saying, "Excuse me, miss. You dropped your wallet."

Clearly, the feelings of fear were based on expectations driven by internal beliefs and perceptions, not by reality. The emotions building within you were YOUR emotions, your imaginings of possibilities. Nothing bad or negative happened other than in your perception of the situation, which is driven by the "old" brain whose job for millions of years has focused on survival. Quite to the contrary, something good was happening. And while you cannot run away from your fears, acknowledging them in context as a signal that you are out of alignment puts them in another perspective. It is through feelings (which are inwardly directed and private) that emotions (which are outwardly directed and public) begin their impact on the mind.[505] Thus, the feeling we get when we sense an external experience is an important part of our evaluation of that experience, and thereby affects our learning.

Further, values and beliefs—which drive strong emotions—trump reason. Values—as knowledge—are an individual choice and as such are situation-dependent and context-sensitive. For example, in our historic fear-motivated, belief-driven, scarcity-based economy it only made sense (in a general usage of the word sense) that societal values were highly materialistic. This frame of reference drives decisions and actions that prove highly destructive to both individuals and development of a global economy, with the individual's perspective, knowledge and knowing, hijacked by the allure of material goods and perceived societal norms. This misappropriated influence of materialism and misalignment with self leads to loss of confidence, confusion, and the inability to make intelligent decisions for the greater individual good or the greater good of humanity.

Self-balancing concerns important choices. For example, understanding that thoughts induce emotions and emotions affect thought, in this continuum there is a propensity to get caught up in a cycle of repetitive thinking. This is where beliefs come into play. A belief is a thought that you continue to think for an extended period of time. The longer you think the thought, the stronger the belief, with the belief potentially influencing you to behave in a delimited manner. This paradigm explains the prejudices and opinions that limit our creativity and the slow expansion of our consciousness.

A SECOND KNOWING in managing our emotions is an awareness of those emotions and what they are communicating. All emotions in the human emotion set are meant to serve as a guidance system, to *ensure that we are in alignment with ourselves.* Thus, when we are feeling higher order emotions such as joy and love, we are in alignment with ourselves. Conversely, when we are feeling lower order emotions such as anger and fear, we are NOT in alignment with ourselves.

Feelings come from the limbic part of the brain and often come forth *before* the related experiences occur. When that happens, *they represent a signal* that a given potential action may be wrong, or right, or that an external event may be dangerous, or joyful. Emotions assign values to options or alternatives, sometimes without our knowing it. There is growing evidence that *fundamental ethical stances in life stem from underlying emotional capacities*. These stances create the basic belief system, the values, and often the underlying assumptions that are used to see the world—our mental model.

While emotional processing can—and regularly does—take place outside of conscious awareness, once aware of these emotional responses, humans have the potential to consciously observe, influence and shift their emotions. Awareness often comes in the form of pain and, while this pain can be felt throughout the body, it is often felt as originating in the heart, the perceived center of emotions. Fortunately, thought has the potential to mitigate this pain. As Cooper describes:

> *Many of life's events can damage a person emotionally. The mind can help the heart overcome its hurt and pain by creating new thought forms that allow a different heart reaction to otherwise traumatic events. In this way, the mind and the heart work together to enact a different perception of traumatic events, allowing love and compassion to remain intact even when one is under attack.*[506]

As a rule of thumb, it takes only 17 seconds of focused feeling to shift from one emotional state to another. How might you do this? You do this through your thoughts and actions. (See TOOL 8 below.) In the body's complex responses to emotional arousal, there is an electrochemical pathway that moves from the brain through the limbic system and then throughout the body utilizing adrenal glands and the autonomic nervous system. The firing of specific nerve cells forms and releases amino acid neuropeptide chains which can activate or deactivate the biological process involved in both emotion and behavior. These chains are what Pert calls molecules of emotion.[507]

Fascinatingly, as emerged in an ICALS finding, **the brain can generate molecules of emotion to reinforce what is learned**. If positive, this phenomenon could provide control over attention during the concrete experience. It can also improve understanding and meaning, and seek truth/how things work, influencing activities in both the concrete experience and reflective observation modes of experiential learning.

What is missing from Pert's description of the chemicals and molecules that activate the body's responses to emotional arousal is an explanation of *how* thought activates these chemical messages in the first place. However, Pert suggests an interesting phenomenon that offers a clue. Building on Pert's work, Lambrou and Pratt contend.

> *The receptor sites on a nerve cell vibrate at a certain frequency. However, when the neurotransmitter locks onto a receptor site, the frequency changes. Something is going on at the energetic, or vibratory, level. Our thesis is that the energy flowing in the meridians activates certain cells to trigger the manufacture of the neuropeptides. This means that the energy of thought interacts with the meridian system, activating the electrochemical process and sending signals throughout the body.*[508]

This means that the energy of thought interacts with the meridian system, which is what activates the electrochemical process, sending signals throughout the body.

A THIRD KNOWING concerns the recognition that internal emotions are triggered by external situations and events—which can consciously be avoided or, conversely, set into motion. Physical bodily changes follow our perceptions, and the feelings that emerge ARE based on emotions.[509] Emotions can be expressed as explicit knowledge in terms of changes in body state. As Damasio notes, "Many of the changes in body state—those in skin color, body posture, and facial expression, for instance—are actually perceptible to an external observer."[510] Often, these changes to the body state represent part of an explicit knowledge exchange. Examples would be turning red with embarrassment or blushing in response to an insensitive remark

Further, "…once emotions occur, they become powerful indicators of future behavior."[511] While we cannot "will" our emotions to occur— understanding that emotions are things that emerge rather than things we consciously order to occur—we can set up and participate in situations where external events or internal thoughts or memories provide stimuli to trigger desired emotions.[512] We do this regularly when we go to the movies or visit an amusement park, or even when we consume alcohol or stimulate our palate with a gourmet meal.

Feelings as a form of knowledge have different characteristics than language or ideas, but they can lead to effective action because they can influence actions by their existence and connections with consciousness. When feelings come into conscious awareness, they can play an informing role in decision-making, providing insights in a non-linguistic manner and thereby influencing decisions and actions. For example, a feeling (such as fear or an upset stomach) may occur every time a particular action is started which could prevent the decision-maker from taking that action.

A FOURTH KNOWING urges us to release emotions no longer in service. Within the scientifically measurable energetic fields of which we are a part, or the energetic information fields with which we interact every instant of our lives—including both thoughts and emotions—energy can be captured, slowed or stopped, both intentionally and unintentionally, consciously and unconsciously. When that energy is retained in our system—whether physical, mental or emotional—it becomes stagnant, what we can describe as stuck, clogged, bounded, or gapped. When this occurs, it affects the energy flow within ourselves, as well as our interaction with the larger ecosystem of which we are a part, and it can have severe consequences.

Blocked or stuck energy is in stasis, that is, a state of no change or a motionless state. Remember, a complex adaptive system cannot exist long in stasis. A CAS is continuously either growing or declining, expanding or dying. In the body, stuck energy might mean preventing fluids from flowing normally through their regular channels. For example, a blockage to the circulation of air throughout our system might be caused by asthma, bronchitis or emphysema, or food lodged in our airway, and lead to death. A blockage to food and water intake or output can cause system collapse. And we all know the potential consequences of a blood clot. Similarly, while we don't generally think of our thoughts and emotions in terms of energy, they are part of a powerful energy force that has very much to do with both living and dying, and the quality of those experiences. From a knowledge perspective, stuck energy means we cease to learn, and when we cease to learn, over time, we lose the capability to interact and deal with changes in our environment. A deeper discussion of these concepts is included in Chapter 20, Part III, of *The Profundity and Bifurcation of Change: Learning in the Present*.

As the guidance system which punctuates the positive and negative aspects of our lives, our feelings and emotions need to be honored, and, as appropriate, considered in the making of our day-to-day decision. As positive emotions such as love and joy flow through our lives, they become more meaningful. Conversely, while we may choose to hold on to strong negative emotions to purposefully create a force to propel us on a course of action, once they have been honored in terms of recognition and understanding, it is indeed a good idea to release emotions when they are no longer needed. TOOL 8 is a technique for releasing emotions.

****

TOOL 8: Releasing Emotions Technique

STEP (1): Recognize and name the emotions you are feeling, fully acknowledging their presence.

STEP (2): Put your arms around yourself and, rocking in a motion from left to right and back in a self-embrace, and with gratitude for these emotions, *say out loud* "I am having a human moment."

STEP (3): Ensuring that you have learned all you need to learn from their presence, thank your emotions for their presence and for this learning.

STEP (4): Using your creative imagination, choose to release these emotions, visualizing them floating away in a balloon, or imploding into the air, or sending them to a junk yard for potential reuse. Have fun with this. The only limit is your imagination.

STEP (5): When a negative emotion departs it leaves a clear space that needs filling. To fill this space, spend a few minutes thinking about some happy memories, or engage in an activity during which you feel joy.

NOTE: It's useful to create a small card (wallet-size) with happy thoughts. This card can be used to spur happy thoughts in the technique above, or can be used to raise your vibration anytime you feel an emotional low.

****

The Dalai Lama uses the term *emotional hygiene* to describe the necessity for each of us to bring destructive emotions under control before going out and acting in the world. As any experiencer of life knows, these emotions can, and most likely will, cause harm to ourselves and others. Emotional hygiene is a good practice to remember, and where perhaps the tool above can be of service. Goleman suggests that having calm, clarity and compassion as we act on the world will result in the greatest good.[513] With reflection, no doubt we all agree.

A FIFTH KNOWING is the understanding that the brain can generate emotions through self-consciousness. Pert offers that through self-consciousness the mind can use the brain to *generate molecules of emotion and override the cognitive system.* As can be seen, this is directly related to the ICALS finding that **the brain can generate molecules of emotion to reinforce what is learned**. Through positive thought and direction, this phenomenon can provide control over attention during the concrete experience. It can also *improve understanding and meaning, and* *seek truth/how things work*, and influence activities in concrete experience and reflective observation of the adult experiential learning model.

A subtle but powerful factor underlying mental models is the role of emotions in influencing our perception of reality (recall the earlier discussion of perception). This has been extensively explored by Daniel Goleman in his seminal book *Emotional Intelligence*,[514] which is the ability to sense, understand, and effectively apply the power and acumen of emotions as a source of human energy, information, connection, and influence. It includes self-control, zeal and persistence, and the ability to motivate oneself. To understand emotional intelligence, we study how emotions affect behavior, influence decisions, motivate people to action, and impact their ability to interrelate. The large body of work on EQ provides a myriad of approaches for managing emotions, all of which contribute to managing self.

In Summary

Emotions are always part of learning processes and significantly modulate the learning process and its results. For example, in the concrete experience mode of the adult experiential learning model high positive emotions increase an individual's sensing, feeling, and awareness as well as support focused attention. Positive emotions not only excite an individual to learn, but also encourage exploration and flexibility in learning. When we are excited about something, our minds are more open and we look to building ideas and improving our understanding. On the other hand, negative emotions tend toward the opposite result. With strong negative emotions, the learner may start to close in and become more cautious and fearful, with the result of paying less attention to learning. As Pert details, "Emotional states or moods … [generate] a behavior involving the whole creature, with all the necessary physiological changes that behavior would require."[515] Although the amygdala may be the source of our emotions, these emotions quickly spread throughout the body.

As a reminder, *remember* that humans have the ability to balance their emotions. Our emotions are just that, *ours*. Thus, once emotions are acknowledged, embraced, and managed, meeting their purpose and responsibility as a guidance system, it is our choice how and to what extent we continue to "feel" and be governed by them.

Chapter 9
The Extended Reach of Self

Humans are Social Creatures ... We Seek an Attuned Other ... There are Embedded Mechanisms for Learning from Others ...Empathy is a Heart-Mind Tool of Connection ... We are Co-Creators ... Conversations Really Matter ... The Relationship Network Can Be Managed ... Social Networking Enabled a Shift to Idea Resonance ... In Summary

SUCCESS FACTOR: Social Engagement

SKILL SETS: Managing Relationships, Otherness

FIGURES: (12) There is a multiplier effect of ideas as they are shared, with all benefiting; (13) The movement from relationship-focused value built on trust and respect of people with whom you have personal relations TO relationship and idea-focused value built on trust of structure and people in that structure TO idea-focused value built on respect for, trust of, and resonance with, ideas.

TOOL: (9) Relationship Network Management.

Educators. Business Leaders. Politicians. Scientists. We've known about social learning, and actively engaged. We all learn from others. Isn't that what a classroom is all about? And sometimes we described it as learning by example, imitating, or copycatting. But it wasn't until the turn of the century that the language—a way to talk about it—and a fuller understanding emerged. Neuroscience and cognitive psychology helped with that—or, perhaps more accurately, the technological advancements enabling the creation and sophistication of brain measurement instrumentation helped. As previously introduced, these advancements included functional magnetic resonance imaging (fMRI), which is used to produce precise measurements of brain structures,[516] the electroencephalograph (EEG), which is another noninvasive technique that measures the average electrical activity of large populations of neurons,[517] and transcranial magnetic stimulation (TMS), which uses head-mounted wire coils that send very short but strong magnetic pulses directly into specific brain regions that induce low-level electric currents into the brain's neural circuits, turning on and off specific parts of the human brain.[518] This visibility into the workings of the mind/brain shows what actually happens *from the inside out*, providing the opportunity to seriously explore the unconscious mental life, expanding our knowledge and understanding about adult learning.

Humans are Social Creatures

While Kolb recognized the influence of the environment on internal learning, today we know that the environment is a critical part of and actively engaged in the learning process, with that understanding supported by neuroscience findings. We are social creatures. And while we as humans have no doubt intuited this for centuries. We now complete the quote that was introduced in Chapter 4. As Cozolino purports,

> *As a species, we are just waking up to the complexity of our own brains, to say nothing of how brains are linked together. We are just beginning to understand that we have evolved as social creatures and that all of our biologies are interwoven.*[519]

Now we know that there are physical mechanisms which have evolved in the brain to enable us to get the knowledge needed to keep *physically and emotionally* safe.[520] The related ICALS finding is **physical mechanisms have developed in our brain to enable us to learn through social interactions.** These mechanisms enable us to, "(1) engage in affective attunement or empathic interaction and language, (2) consider the intentions of the other, (3) try to understand what another mind is thinking, and (4) think about how we want to interact."[521] The mechanisms for this capability come from adaptive oscillators, which are part of our physiology, and also through mirror neurons, which are detailed later in this chapter. These oscillators are created by stable feedback loops of neurons, and can bring an individual's rate of neural firing into sync with another individual. This is when two people relate well to each other and learn to anticipate each other's actions.[522] Buzsaki calls this phenomenon mutual entrainment, meaning a measure of stability that oscillators have when they lock in with each other.[523] Such situations, when they occur, while supporting all the characteristics attributed to the social engagement mode, also enhance the sensing, feeling, and awareness in the concrete experience phase of the adult experiential learning model.

What we can learn from others should not be underestimated. A trusting relationship and a good holding environment "stimulate the brain to grow, organize and integrate."[524] During this process the learner's neurotransmitters in the frontal cortex are stimulated, leading to greater brain plasticity and increased neuronal networking, which in turn leads to meaningful learning.

Social interaction also directly influences the expression of genes. According to Bownds, social interaction influences this expression through the reaction and behavior of the learner. For example, gene expression in infants is influenced by sensing and behavior appropriate for its environment. Clusters of nerve cells in the brainstem distribute neurotransmitters (serotonin, acetylcholine) to areas of the cortex and influence both the growth of the brain during development and its behavior and plasticity in adults.[525] Which supporting neuronal pathways become permanent depends on the usefulness of the behavior in enhancing survival and reproduction.

The effects of social forces are often not in conscious awareness. LeDoux says that the present social situation and physical environment are part of what is connected. He finds that,

> *People normally do all sorts of things for reasons that they are not consciously aware of (because the behavior is produced by brain systems that operate unconsciously). And that one of the main jobs of consciousness is to keep our life tied together into a coherent story, a self-concept. It does this by generating explanations of behavior on the basis of our self-image, memories of the past, expectations of the future, the present social situation and the physical environment in which the behavior is produced.[526]*

Tallis suggests that unconscious learning most likely influences people's social preferences. In his words,

> *Human beings are constantly forming positive or negative opinions of others, and often have minimal social contact. If challenged, opinions can be justified, but such justifications frequently take the form of post hoc rationalization. Some of course are laughably transparent.[527]*

We Seek Attuned Others

Experiential learning is not just a function of the incoming information. The overall environment, a trusted other, and the conscious and unconscious state of the learner all have a role in the final efficiency and effectiveness of learning that occurs. A positive social relationship, and a corresponding environment, support individual learning through the creation of internal perspectives and feelings that facilitate pattern generation and the development of meaning by the learner. For example, the biological impact resulting from interactions with an *attuned other* can significantly enhance learning. During this process the neurotransmitters in the frontal cortex of the learner are stimulated, which leads to increased brain plasticity and neuronal networking, in turn enabling meaningful learning.

What is an attuned other? Attunement infers being in harmony with or resonating with another. It is a shared state, a vibrational entrainment—inclusive of understanding another's needs and feelings—that opens the door to learning. The notion of affective attunement is connected to Dewey's observations that an educator needs to "have that sympathetic understanding of individuals as individuals, which gives him an idea of what is actually going on in the minds of those who are learning."[528] A specific related finding in the ICALS study is that **social interaction mechanisms foster the engagement in affective attunement, consider the intentions of others, understand what another person is thinking and think about how we want to interact**.

As Johnson forwards, "According to social cognitive neuroscience, the brain actually needs to seek out an affectively attuned other if it is to learn."[529] Reflect back on your life and identify individuals with whom you have been affectively attuned. The limbic system, the primitive part of the human brain, and in particular its amygdala, is the origin of survival and fear responses. Affective attunement alleviates fear, which has been identified as an impediment to learning throughout the field of adult learning.[530] This concept—specifically, that **the brain actually needs to seek out affectively attuned others for learning**—became an ICALS finding. Being open to learning helps one integrate and look for unity as one reflects and opens up to sensory feedback during active experimentation.

The literature is extensive on the need for a safe and empathic relationship to facilitate learning. Cozolino says that for complex levels of self-awareness, that is, those that involve higher brain functions and potential changes in neural networks, learning cannot be accomplished when an individual feels anxious and defensive. Specifically, he says that a safe and empathic relationship can establish an emotional and neurobiological context that is conducive to neural reorganization. "It serves as a buffer and scaffolding within which [an adult] can better tolerate the stress required for neural reorganization."[531] Taylor explains that,

Adults who would create (or recreate) neural networks associated with development of a more complex epistemology need emotional

support for the discomfort that will also certainly be part of that process.[532]

A related ICALS finding is that **adults developing complex neural patterns need emotional support to offset discomfort of this process**. Drawing further on Taylor's research, Taylor suggests that this support is needed by individuals *developing complex knowledge*. Such emotional support enhances the feelings of an individual during concrete experience, and also aids in the creation and understanding of concepts and ideas during abstract conceptualization. This finding is an important understanding for those of us who return to school at an older age! Note that the term "affective" is attached to "attunement", which occurs in much of the related research. This emphasizes the emotional elements necessary to the process of attunement.

Another ICALS finding is that **affective attunement contributes to the evolution and sculpting of the brain**. Cozolino posits that the two powerful processes of social interaction and affective attunement contribute to "stimulate the brain to grow, organize and integrate."[533] As these new patterns are created in the mind, they in turn impact and change the structure of the brain. All of these influence sub-elements of the five modes of the adult experiential learning model.

Affective attunement involves a mentor, coach, or another significant individual who is trusted and capable of resonance with the learner. When attunement happens, a dialogue with such an individual can greatly help the learner in understanding, developing meaning, anticipating the future with respect to actions, and receiving sensory feedback. Cozolino and Sprokay believe that, "the attention of a caring, aware mentor may support plasticity that leads to better, more meaningful learning."[534]

While the role of a mentor or trusted other is significant, as is the environment, Caine and Caine emphasize that the role of the individual in this relationship is also critical:

In order to adequately understand any concept, or acquiring any mastery of a skill or domain, a person has to make sense of things for himself or herself, irrespective of how much others know and how much a coach, mentor or teacher tries to help. We have also argued that although there is an indispensable social aspect to the construction of meaning, there's also an irreducible individual element.[535]

This is consistent with the earlier discussion in Chapter 8 forwarding that it is the self, emerging from all the experiences of life, that *assigns meaning to events*.

Physical and mental exercise emerged as important factors, along with social bonding, for stimulating learning. The specific related ICALS finding is that **physical and mental exercise and social bonding are significant sources of stimulation of the brain**. Amen, whose research supports this statement, highlights the importance of physical exercise for the health of the brain, mental exercise for the health of the mind, and social bonding.[536] This was introduced in Chapter 7. Physical exercise provides oxygen and increased blood to the brain. Studies have shown that a higher concentration of oxygen in the blood significantly improves cognitive performance in learners. For example, tests demonstrated that individuals were able to recall more words from a list and perform visual and spatial tasks faster, with their cognitive abilities changing directly with the amount of oxygen in their brains.[537]

For Sousa, social bonding carries with it a positive, *trusting* relationship that allows the learner to take risks and not be concerned with mistakes made during learning.[538] It also encourages an open mind and willingness to listen and learn from a trusted other. (Note that the concept of trust emerges again and again throughout this research.) During this process, the learner's neurotransmitters in the prefrontal cortex (dopamine, serotonin, and norepinephrine) are stimulated and lead to increased neuronal networking and meaningful learning.[539] While these would apply to many sub-elements of the adult experiential learning model, we highlight the following: awareness, attention and concepts, ideas, and logic. Enriched environment, social bonding, and affective attunement are all connected to social support, which is one part of the social engagement mode.

It becomes clear that the nature of social interaction plays an important role in determining learning. The specific social interaction that influences the neural structure, the environment in which that occurs, the significance of the event to the individual in terms of meaning, and the stress level of the individual—as well as the level of physical and mental exercise engaged by the individual—all affect the nature and amount of learning that results.

From the learner's perspective this phenomenon suggests that the learner create and manage positive and beneficial social interactions wherever possible. *Being aware of this phenomenon is a first step toward effective management of the social learning process.* By being aware of these factors, learners may be able to change the local physical environment, improve communications with others, or perhaps position and adjust their own internal feelings and perspectives to maximize their learning.

There are Embedded Mechanisms for Learning from Others

Social learning includes *learning through and beyond our experience*, that is, not only engaging our own experiences but replicating the experience of others as our own, embedding it in our mind and body, and acting accordingly. Educators might refer to this as apprehension and discuss the power of intention and attention; cognitive psychologists might talk about how the neocortex cortical columns store patterns in invariant form, internal representations of the external world using reference frames to map and make sense of our experiences; and neuroscientists might bring up mirror neurons and, possibly, Heisenberg's recognition that energy and matter are indefinite, and that thought affects energy. Yes, all of these insights help more fully address this phenomenon.

It is not a surprise that humans are experiential learners. Nor is it a surprise that an important facet of learning can be watching someone else doing what you want to learn. This is validated by the ICALS finding that **what we see we become ready to do**. Because we may have instant recognition and understanding of an external phenomenon via mirror neurons, reaction time can be significantly enhanced as well as our learning. This relates to heightened boundary conditions in the sense of providing instant awareness and understanding of objects and boundaries, permitting faster responses when needed.

Mirror neurons? Let's explore that phenomenon. Our "self" is in continuous interaction with the environment, largely in the unconscious, and that interaction works both ways. Not only does the observer affect that which is observed, but that which is observed affects the observer. One element of that can be described in terms of mirror neurons, a recently discovered phenomenon in the brain. Research began with experiments with Macaque monkeys in the early 1990s, noting that activation of subsets of neurons in the brain-motor areas appeared to represent action. The initial experiments had one monkey grasp an object (an orange) while the experimenters monitored what went on inside an observer monkey's brain. Many variations of this were used to verify that an observer's neurons fire (or mirror) the actor's neurons, with testing moving from monkeys to great apes to humans.

Non-invasive measurement techniques such as fMRI enabled the experiments on humans to be greatly expanded. By 2006, research in humans included the discovery of mirror neurons located in the frontal lobe and in the parietal lobe, which includes the Broca's area, a key area for human language.[540] As Rizzolatti describes, "subsets of neurons in human and monkey brains respond when an individual performs certain actions and also when the subject observes others performing the same movements."[541] In other words, *the same neurons fire in the brain of an observer as fire in the individual performing an action.* These neurons provide an internal experience that **replicates that of another's experience**, thereby experiencing another individual's act, intentions, and/or emotions.[542]These researchers also found that the mirror neuron system responded to the intentional component of an action as well as the action itself.[543]

There is a caveat. Immordino-Yang's extensive research indicates that the network of neurons we call mirror neurons only fire when the goals of the others carrying out actions are understood, creating a resonance between the sender and the receiver. As she contends,

> *While this internalization of another's situation can be automatic, the representation of another's situation is constructed and experienced on one's own self in accordance with cognitive and emotional preferences, memory, cultural knowledge, and neuropsychological predispositions—the "smoke" around the mirrors.*[544]

In other words, at the unconscious level the self is very much involved in the process of mirror neurons.

Given this caveat, there are several related ICALS findings. First, that **neurons create the same pattern when we see some action being taken as when we do it**. This affects sensing, focusing, attention, and sensory feedback to the brain. Understanding this phenomenon allows the learner to make better use of learning opportunities. It also explains why our intuition can be right.

Carrying this further, Iacoboni proposes that mirror neurons facilitate the direct and immediate comprehension of another's behavior without going through complex cognitive processes, which makes the learning process more efficient because it can *instantly transfer not only visuals but emotions and intentions* as well.[545] This leads to a second ICALS finding that **mirror neurons facilitate the rapid transfer of tacit knowledge (bypassing cognition)**. By creating the same neuronal pattern in your mind that is in the mind of another person, the need for cognitive thinking is bypassed and tacit knowledge may be immediately transferred. This would affect understanding and meaning in reflective observation, and speed up the learning process.

While this phenomenon serves to explain how *actionable tacit knowledge can be transferred between individuals*—as well as the potential of mimicry to facilitate learning—it also serves as a warning to be aware that when we are mentally simulating another's behavior, we are not modulating that simulation through our own internal evaluation and judgment. The capacity to re-create feelings, perspectives, and empathy with people by reliving their experiences can greatly aid learning, providing we understand what is happening and its potential for misinterpretation.

A third related finding concerns creation of a neural resonance, or stated more clearly, **mirror neurons facilitate neural resonance between observed actions and executing actions**. By neural resonance we mean a positive, mutually reinforcing relationship between two people interacting with each other. An individual may interact with another individual more efficiently and effectively because of the understanding and affection developed from the mutually reinforcing relationship between the two. For example, we may be able to understand another's feelings and/or intended actions because (through mirror neurons) we generate similar feelings and/or intended actions (or reactions). Mirror neurons can facilitate a quick, positive relationship between the observer and the observed.

Mirror neurons also serve as a means of learning through imitation, which is "a very important means by which we learn and transmit skills, language and culture."[546] Notice that when it does occur, the exchange occurring in this process—consistent with the transfer of sensations and feelings proven mathematically through transreal numbers—includes intention, emotions and cultural norms. As Zull notes, mirror neurons are a form of cognitive mimicry that transfers active behavior and *other cultural norms*.[547] This is one of the findings in the ICALS study. When we see something happening our mind creates the same patterns of neurons that we would use to enact the same thing ourselves. While part of social engagement, grasping through direct comprehension, and therefore accelerating learning, this would also be related to the sensing, feeling, and attention in the concrete experience sub-elements of the adult experiential learning model.

As Immordino-Yang says,

Notably, a person need not directly experience in the environment every action or perception; instead, he can mentally conjure these experiences based on memory or imagination. Of necessity, these internally derived activations in the form of memories or imaginings will reflect the biological predispositions and previous subjective experiences of the learner.[548]

This iterative reconstruction of both perceptual and motoric experiences—whether imaginative or experiential—creates dynamic feedback loops between individual perception and acting, thinking and feeling.[549] This concept of mental conjuring based on memory and/or imagination provides a foundation for understanding the "existential" experience explored in Chapter 10.

Empathy is a Heart-Mind Tool of Connection

Related to the mirror neuron phenomenon, an ICALS finding was that **through mental reliving we recreate the feelings, perspectives and other phenomena that we observe**. Nelson and colleagues offer that we understand other people's behavior by mentally simulating it.[550] That is very close to a closely-related finding in the ICALS study, specifically, that **we may understand other people's behavior by mentally simulating it**. In addition to relating to all of the social interaction characteristics which are a subset of the social engagement mode, this relates directly to understanding, meaning, and truth/how things work in reflective observation, and to heightened boundary conditions in active experimentation. As Stern proposes, "This 'participation' in another's mental life creates a sense of feeling/sharing with/understanding the person's intentions and feelings",[551] which is very supportive of learning.

Empathy is the concept of experiencing the inner life of another, which Cozolino credits to a combination of visceral, emotional and cognitive information, a "muddle of resonance, attunement, and sympathy."[552] Riggio describes three different types of empathy: (1) *perspective-taking*, a cognitive-based form for seeing the work from someone else's frame of reference; (2) *personal distress*, literally "feeling" another's emotions, caused by what he terms as "emotional contagion"; and (3) *empathic concern*, the recognition of another's emotional state and feeling a resonance with it. Note that "feeling" is a key proponent of empathy.[553]

As an important element of individual development, in the Intelligent Social Change Journey (ISCJ) introduced in Chapter 1, empathy, which is required to co-evolve in a dynamic environment (phase 2 of the ISCJ), is part of a continuum based on an increasing depth of connection. The movement through this journey is from sympathy to empathy to compassion, and, for most humans sometime in the future, to unconditional love. This continuum represents the connection growth path in the Wisdom Model (see Appendix B).

Interestingly enough, from a neuroscience viewpoint, the physiological basis for empathy is so inherent in brain function that it has been extensively documented in scientific experiments with other tested primates. For example, Masserman reported that in a study of rhesus monkeys when one monkey pulled a chain for food, a shock was given to that monkey's companion.[554] The monkey who pulled the chain refused to pull it again for 12 days, that is, the primates would literally choose to starve themselves rather than inflict pain on their companions. Cozolino says that the insula cortex and anterior cingulate link hearts and minds and that this is best demonstrated when watching others experience pain.[555] These two regions become activated either when we experience pain, or when a loved one experiences pain. The degree of activation of these two structures has been shown to correlate with measures of empathy.[556] As deWaal sees it, *empathy is nature's lesson for a kinder society*.[557]

Although the neurobiology of empathy is still in its early development, the insula—described as the limbic integration cortex lying beneath the temporal and frontal lobes—appears to "play an important role in both the experience of self and our ability to distinguish between ourselves and others."[558] Beyond basic sensations, the left insula is involved in the evaluation of eye gaze direction, the response to fearful faces, and the observation of facial expression of the other.[559] Further research has found that the insula mediates the extreme limits of emotions, ranging from severe pain to passionate love.[560] More recently, the dorsomedial prefrontal cortex—which produces a representation of self in relation to others—is thought to play an important role underlying the function of empathy in feeling emotions about others' mental situations.[561]

These findings suggest that through feelings there is an active link between our own bodies and the minds and the bodies and minds of those around us. Thus, the feelings that we each perceive in the course of our daily living may be affected by, *or even belong to*, those around us.

We are Co-Creators

We as a humanity are in a continuous cycle of co-creation such that every moment offers the opportunity for the emergence of new and exciting ideas. Note that in every aspect of creating and learning, there is a "co" element. This is because ideas do not happen in isolation.

As introduced in Chapter 5, at some level everyone who is living is creating, and two related ICALS findings were introduced. The first, that **free-flow and randomly mixing patterns create new patterns**, is a way to describe the interaction among neuronal networks and patterns to create new ideas, and the second, that **accidental associations can create new patterns**, makes the point that creativity can happen by accident within an active mind that plays with ideas, connections, and their relationships. Much of the co-creating process is happening in the unconscious as humans interact with their environment and others in their environment.

Recall the definition of knowledge as the capacity to take effective action (justified true belief). As inferred by the law of relativity (knowledge is context-sensitive and situation dependent), all knowledge is incomplete. When knowledge is focused inward, that is, bounded and *not* shared, it has diminishing value as others continue to connect with the ever-changing and expanding reservoir of knowledge. An individual with bounded knowledge has ceased learning, with that knowledge over time losing any value it may have had in terms of taking effective action, and thus becoming a knowledge artefact as information. Further, there is a diminishing of consciousness and meaning that accompanies the cessation of learning (see Chapter 8). We can't stand on the sidelines. As complex adaptive systems, when we cease creating, when we cease learning, we enter a downward spiral that is characterized by the loss of consciousness, the loss of meaning, and, eventually, the loss of life.

The greatest meaning of life comes with the expansion of co-creating, sharing knowledge that facilitates the creation of new ideas and potential innovation. Both consciously and unconsciously—as demonstrated through our earlier discussion of mirror neurons—as we interact with others, we develop a deeper understanding of others and ourselves and an appreciation for diversity, creating a collaborative advantage. We are able to gain the advantage of other's thinking at or above our personal level of thinking, while simultaneously creating in a way that is uniquely ours, concurrently individuated and one.

We now understand from neuroscience findings that the mind is an associative patterner, ever creating and recreating knowledge for the moment at hand. Simultaneously, we understand that creativity is the bisociation of two or more ideas to create a new idea or apply an idea in a new context. Thus, there is a multiplier effect of ideas as they are shared. The more we participate in cooperative and collaborative experiences, the more opportunity for the bisociation of ideas. See Figure 12.

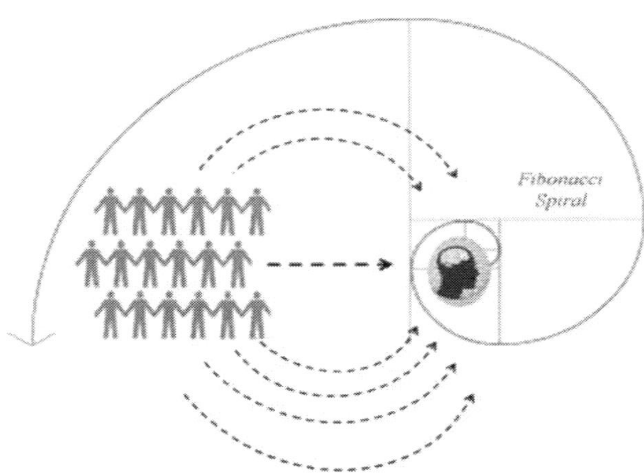

Figure 12. There is a multiplier effect of ideas as they are shared with all benefiting.

As a humanity, we are action-oriented and knowledge-driven. Waxing eloquent, *just as a winding stream in the bowels of the mountains curve and dip through ravines and high valleys, so, too, with the knowledge of our minds.* In a continuous journey towards intelligent activity, context-sensitive and situation-dependent knowledge, imperfect and incomplete—just as all humans are—experientially engages a changing landscape in a continuous cycle of learning and expanding. It is the context of the activity or situation at hand (need, challenge, etc.) that triggers the putting things together in an unusual way to create (and recognize) something that may be new and potentially useful (innovation). *We as a humanity are in a continuous cycle of knowledge creation such that every moment offers the opportunity for the emergence of new and exciting ideas, all waiting to be put in service to an interconnected world.*[562] And conversations—with ourselves and with others—is where that learning occurs.

Conversations Really Matter

At every point in the journey of self we are in the process of becoming something more or something else, with vastly different perspectives emerging from one activity to the next as to where we are at any point in time! And what is known to one is unknown to others, or perceived very differently. What seems real to one person can be totally unreal to another. Your truth is considered just a perception by others, and vice versa. For example, one person's terrorist is another's war hero, an unfortunate reality that we collectively face in challenging times.

Recall the description of intelligent activity as a state of interaction where intent, purpose, direction, values and expected outcomes are clearly understood and communicated among all parties, reflecting wisdom and achieving a higher truth. Okay, that sounds well and good, only how do you get a bunch of people to understand—much less be able to communicate—intent, purpose, direction, values and expected outcomes?

When engaged in dialogue—equally sharing and listening—conversations can surface our beliefs, values and mental models as we verbalize our thoughts on a specific issue at hand. Sharing our experiences and stories and listening to the experiences and stories of others not only clarifies self-understanding and increases empathic appreciation of others but, because our mind is an associative patterner, triggers new thoughts and connections. Arthur Shelley, a capability development guru and originator of *The Organizational Zoo*, believes that a good way to start is using images designed to stimulate Conversations That Matter.[563] We've seen the effectiveness of his approach, which is featured in Chapter 12 and detailed on the organizationalzoo.com website, where you find the metaphors, figures, and even the entire website for international collaboration in over a hundred languages.

Whether in dialogue, discussion, debate, or casual conversations, face-to-face or virtual, we can learn a great deal by talking and listening to ourselves and others. This helps keep learners mentally and physically safe in the sense that interaction with others may provide an atmosphere for relaxed yet challenging conversations, thereby supporting learning.

A related finding in the ICALS study is that **language and social relationships build and shape the brain**. This item could significantly impact the sensing aspect of concrete experience and the concepts, ideas, and logic of abstract conceptualization in the adult learning model. Good social relationships—including the language that is used—enhance learning through a reduction of stress, a shared language, and the use and understanding of concepts, metaphors, anecdotes, and stories.

The Relationship Network Can be Managed

Considering the continuous social interactions of the human, both on the conscious and unconscious levels, it makes sense that our everyday conversations *lay the groundwork for the decisions we will make in the future*. Since time is a scarce resource, it is critical to choose our relationships and interactions wisely. The relationship network is a matrix of people that consists of the sum of an individual's relationships, those individuals with whom the individual interacts with—or has interacted with in the past—and with whom there is a connection or significant association.[564] In short, all those with whom you have had repeated and comfortable conversations.

Whether virtual or face-to-face, relationships are ultimately about people and the way they interact with each other over periods of time. The fundamental principles of success in relationships parallels Sun Tzu's fundamental principles of success in warfare, i.e., know thyself, know the other, and know the situation. Principles of Relationship Network Management (RNM) start with the individual (what the individual brings to a relationship in terms of values, ability to communicate, expertise and experience, ideas, and willingness to share and learn). Then we move into understanding the situation (virtual or face-to-face, open or guarded communication, content of exchange, purpose of interactions, etc.), and the other (trusted or unknown, values, communication skills, frame of reference, expertise and experience, and willingness to share and learn).

There are five basic concepts that successful Relationship Network Management is built upon. These include interdependency, trust, openness, flow and equitability, all of which overlap. For example, interdependency includes a state of mutual reliance, confidence, and trust. Note that interdependence does not translate into freedom for individuals to do as they choose. Interdependence means a world of greater constraint or greater conflict, or both, "greater constraint in so far as individuals accept the demands of interdependence; greater conflict in so far as they do not." [565]

Trust is based on integrity and consistency over time, saying what you mean, and following through on what you say, and openness is directly related to trust and a willingness to share. From a neuroscience perspective, trust in a relationship is very important in enhancing learning. As discussed earlier in this chapter, when a secure, bonding relationship in which trust has been established occurs, there is "a cascade of biochemical processes, stimulating and enhancing the growth and connectivity of neural networks throughout the brain."[566] This process promotes neural growth and learning.

As can be seen, these qualities facilitate the free flow of data, information and knowledge among individuals, across teams and organizations, and around the world. Equitability in terms of fairness and reasonableness means that all those involved in the sharing gain something of value out of the relationship. These qualities are consistent with the definition of intelligent behavior.

****

TOOL 9: Relationship Network Management

A self-empowering tool in social networking is Relationship Network Management. (See the RNM Assessment Chart in Appendix F.) There are five steps to managing your relationship network.

STEP (1): Recognize the value of your network. When we recognize the value of our relationships, we can learn to consciously manage it, and provide the level of grounding needed to operate in the world of ideas.

STEP (2): Identify the domains of knowledge (areas of passion) that are important to you and what you want to achieve in life.

STEP (3): Identify the people with whom you regularly interact, both in your personal and professional life. Note how often you interact with them, the quality of the interaction, and whether they can depend on you and you on them to respond to questions with honest (and valued) opinions. *Ask*: What is at the root of this relationship? How do we complement each other? What do I learn from them? What do they learn from me? Is this relationship knowledge expanding? Consider the principals of Relationship Network Management and assure that each relationship exists within the bounds of the RNM principals.

STEP (4): Carefully compare the list developed in Step (2) with the Network and understanding developed in Step (3). Then, consciously choose to develop, expand, and actively sustain those positive relationships in terms of thought, feelings and actions. Where gaps are identified, that is, where you have no exposure to the domains of knowledge (passion) which are important to you, prepare a plan that will bring that knowledge into your awareness and experience. For example, taking a college class related to that knowledge area will open the door to networking with people with similar interests.

It is critical to choose your network wisely. At some point in the future, you will make a decision based on a conversation you had today or last week. Although you may or may not remember the conversation, the resonant content of that conversation is linked into your unconscious to associate with future thought. Thus, *your everyday conversations and reflections on those conversations serve as grounding functions for future decisions and actions*.

STEP (5) By choice, stay open to sharing and learning through your relationship network. See the tool of Humility provided in Chapter 11.

****

Social Networking Enabled a Shift to Idea Resonance

Global connectivity and the internet have brought—and continue to bring about—new modes of social networking, demanding a shift in our perceptions, and facilitating a shift from relationship-based interactions to idea-based interactions, *with affective attunement developing through virtual relationships* based on the resonance of ideas. Virtual networking primarily relies on this resonance of ideas to develop a level of trust. While this is quite different than the personal relationships or connections built up over time among personal and work interactions, those that connect continuously *do* build up a level of trust based on the responses of those with whom they interact, howbeit personal trust and the trust of ideas can be very different perceptions. See Figure 13.

As has been established throughout this chapter, from a neuroscience perspective, *trust in a relationship* is very important in enhancing learning. Recall that when a secure, bonding relationship in which trust has been established occurs, and (as noted earlier) there is "a cascade of biochemical processes, stimulating and enhancing the growth and connectivity of neural networks throughout the brain" which promotes neural growth and learning. Interestingly—and important enough to remember—the perception of trust achieves this same process.

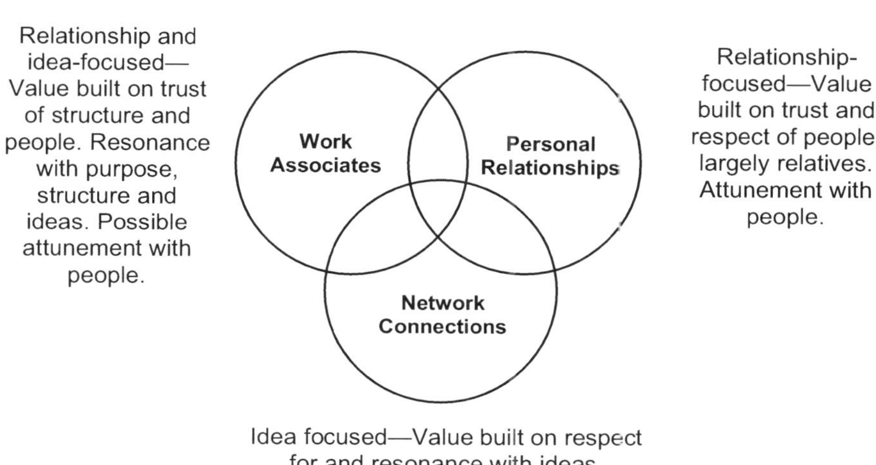

*Figure 13. The movement from relationship-focused value built on trust and respect of people with whom you have personal relations TO relationship and idea-focused value built on trust of structure (the workplace, partners) and people in that structure TO idea-focused value built on respect for, trust of, and **resonance with ideas.***

In the past, trust meant that you could rely on the integrity, ability or character of a person, thing, or process. Trust was built up over time. However, in today's world, where everything happens quickly and often over the internet, virtual trust is not only trust among individuals and groups communicating virtually—and trust of the ideas being communicated, that is, the idea resonance introduced above—but also a trust in the technology, the hardware and software used to communicate (security, reliability, accurate transmission, etc.) and the information being exchanged.

The trustworthiness of information can be considered from a number of viewpoints: (1) its relevance to specific objectives; (2) its quality (accuracy); (3) its timeliness; and (4) its completeness. Of course, trusting information also means trusting the *source* of that information, whether it comes from a person, organization, database, webpage, book or social network. Since *trust is a feeling*, trust of self (as noted) is also paramount. Feelings and emotions are different. Remember, we consider feelings as private, inwardly directed; while emotions are public, outwardly directed. While the *feeling* of trust may or may not be connected to emotions, it will undoubtedly *affect* emotions, especially when trust is broken.

While the system of self is robust and has a high level of trustworthiness in relation to our personal desires and beliefs, it's not perfect. As we have discovered, much of what affects our individual feelings and decisions is unconscious, where biases and beliefs may reside of which we are unaware. Further, the uniqueness of context and content of a situation—coupled with the complexity of a situation and uncertainty of the environment—pose the danger of us oversimplifying and relying on those largely unconscious beliefs from past experiences which are no longer applicable. As time speeds up and the world continues to change more rapidly, with unexpected twists and turns, old beliefs and theories become inappropriate and outdated and so, in other words, untrustworthy.

A potential pitfall of trusting relationships built on the resonance of ideas is the perception that *we are our ideas*. This is not the case. Ideas, regardless of the subject matter, can only represent a part of who we are. As complex adaptive systems, there are so many aspects of self that would not be reflected in the exchanges occurring through virtual social networking. Surprises can occur even in long-time face-to-face relationships.

In Summary

In Chapter 8 we introduced the strong connection between the heart and the mind, and said that the heart communicates with both the brain and the body in four different ways. Further, that there is an electromagnetic field generated by the heart radiating outside the body, carrying information that affects the environment and others in the environment. Then, in this chapter, we have touched on mechanisms created in the mind/brain to enable learning from others as well as conscious social practices that facilitate learning through the flow of knowledge.

As can be seen, learning is part of the social experience of life, with each minute of every day affecting in some way—shifting, changing, expanding, reducing—the complex adaptive learning system that is the human. The overall environment, a trusted other, and the conscious and unconscious state of the learner all have a role in the final efficiency and effectiveness of learning that occurs. By being aware of these factors, learners may be able to change the local physical environment, improve communications with others, or perhaps position and adjust their own internal feelings and perspectives to maximize their learning.

What is quite clear from this neuroscience-supported understanding—looking from the inside-out—is that we are all in this together. Although each self is individuated, no self is separate from its environment. We are each embedded in a social network of learning, co-creating and adapting to the emergent challenges and opportunities in the flow of life.

UNLEASHING Litmus Test #3*

*After reading **Chapters 7-9**, and considering earlier chapters, reflect on each question for one minute (Reflective Observation) prior to answering silently, verbally or in writing.*

1. **AWARENESS** means something has come to your attention, it is perceived, it has been mentally engaged.
   Ask: What comes to mind when I think of managing my SELF?

2. **UNDERSTANDING** includes your perception of the situation—the who, what, where, when, and why, and the anticipated results. As the situation becomes more complex, you need to re-create your understanding.
   Ask: When have negative emotions impacted one of my learning experiences?

3. **BELIEF** means you accept what you are aware of as true and understand it exists. Beliefs which dominate other patterns are prominent. Strong patterns are created by experiences and are closely related to emotions.
   Ask: What is my most significant belief about the experience of learning?

4. **FEELINGS** are the foundation of learning—positive feelings make actions important to you and worthy of your efforts. Reason cannot operate without emotions.
   Ask: What is something I really enjoy doing that I learn from?

5. **OWNERSHIP** implies a personal commitment for you to take responsibility and act.
   Ask: Who would I like to help enjoy their learning experiences more?

6. **EMPOWERMENT** refers to self-empowerment, that is, having the *knowledge* to make the necessary change and the *courage to act* on what you have learned.
   Ask: What is one thing that I would like to accomplish in the next 10 to 15 years and what learning opportunity will enable me to achieve it?

*These questions are based on the Individual Change Model.[567] CHANGE comes from within, that is, unleashing your mind is YOUR choice.

Chapter 10
The Expanded Human

Learning Itself is Spiritual in Nature … Virtues are Spiritual Characteristics in Form … The Mind Requires Time for Inner Reflection … There is a Brain-Heart/Mind-Soul Learning Continuum … Synthesis Allows Us to Map the Past and the Future … Planning is Core to Existential Learning … The Concept of Existentialism Takes on New Meaning … Through Music We Connect the Self and the Soul

SUCCESS FACTORS: Attention, Morality

SKILLSETS: Synthesizing, Planning

FIGURES: (14) The Brain-Heart/Mind-Soul Learning Continuum; (15) The human experiential state of acting-reacting-synthesizing; (16) The human soul existential state of being-reflecting-planning.

TOOLS: (10) Quieting the Mind; (11) Thinking Patterns; (12) Scenario Building.

As introduced in Chapter 1, "soul" represents *the animating principle of human life in terms of thought and action*, specifically focused on its moral aspects, the emotional part of human nature, and higher development of the mental faculties. From the philosophical aspect, it is the *vital, sensitive or rational principle in human beings*.[568] Our own conviction that this is in service to "others" was confirmed by one of our favorite researchers, Csikszentmihalyi, who says that "an enduring vision in both work and life derives its power from soul—the energy a person or organization devotes to purposes beyond itself."[569] Thus, our focus on the expanded human has a great deal to do with inner reflection—the inner self—and the development of morality and virtue, as well as connection to the larger field of which we are a part, again, whether you call that a consciousness field, a quantum field, or a God field. This serves as the context for this chapter.

As a point of reference, in a 1990-1993 World Values Survey, 93 percent of U.S. responders, 85 percent of Canadian responders, 81 percent of Swiss responders and 75 percent of West German responders expressed belief in a soul. Only 6 nations of the 38 nations surveyed had less than 50 percent of responders express that belief.[570] More recently, Haider found that more than 60 percent of *all* people believe in the soul.[571]

But what is the relationship of the soul to the mind/brain? Remember that the "brain" consists of an atomic and molecular structure and the fluids that flow through this structure, while the "mind" represents the patterns created in the brain's 86 billion neurons and their firings across neurotransmitters and trillions of synaptic interconnections. It is these patterns that encompass all our thoughts (see Chapter 1). However, as forwarded by Beauregard and O'Leary, "Your thoughts and feelings cannot be dismissed or explained away by firing synapses and physical phenomena alone."[572]

It is of importance to this line of thought to understand that the spaces between neurons (the synapses) direct signals using ions (parts of atoms), which function according to the rules of quantum physics, not classical physics. This means that we are connected to a Universe of probabilities, not laws, a field that forms the ultimate foundation of self. In this field there is what is called a quantum Zeno effect. It was discovered that when an unstable elementary particle is continuously observed it does not decay, while when unobserved decay occurs. Since the mind/brain is a quantum system, when you focus on a specific idea, you are holding the pattern of connecting neurons in place, that is, *the idea does not decay* which would occur if it were ignored.[573] This is driven by YOUR thought, YOUR free will. A physical connection to the brain is not present, and the neuronal firings that result from your thoughts and feelings certainly form the pattern, but do not sustain it. YOUR intent, YOUR attention, YOUR attending sustain the pattern. From this point of view, we move closer to understanding "soul".

As the animating principle of human life in terms of thought and action, the soul is considered as giving life. Animate comes from *animatio/animare*, which is the state of being lively, full of spirit and vigor. In our definition this reference to animating principles of human life is specifically focused on its *moral aspects*, the emotional part of human nature, and higher development of the mental faculties.

Morals are the result of learning experiences. They are concerned with what is perceived as right—beliefs or standards of behavior about how to act, largely held in the unconscious as part of the model of self, which create the idea of *who* we are and the image we build up of ourselves. This is consistent with the ICALS finding introduced in the Chapter 6 discussion of the role of the unconscious, specifically, that **a model of the self comes mostly from the unconscious**. We recognize now that this model is continuously being updated through the cortical columns of the neocortex as sensory input occurs during the course of living.

While not directly addressed, the importance of "moral aspects" emerges in the discussions of the Intelligent Social Change Journey and the wisdom model (Chapter 1), are alluded to in the exploration of the shift underway (Chapter 3), inferred in the brief mention of civilization (Chapter 5), assumed in the discussion of the effect of beliefs and attitudes on learning (Chapter 6), valued in relation to the higher-order emotions such as love and joy while exploring the levels of consciousness (Chapter 8), and touched in the discussions of a safe and trusted environment, social bonding, affectively attuned others, and the need for empathy in learning (Chapter 9). Also, throughout this book, the issue of trust emerges. Baier sees trust as "the very basis of morality".[574] Thus, the soul—as thought and action focused on moral aspects, emotions and higher development of the mind—weaves itself throughout the discussion of self.

Moral aspects can also be part of cultural norms, or part of an individual's intentions and emotions, and as such can be actively transferred through the cognitive mimicry process of mirror neurons when—as introduced in the previous chapter—the goals of the others carrying out actions are understood, creating a resonance between the sender and the receiver. This is consistent with the ICALS finding previously introduced, that **cognitive mimicry can transfer active behavior and other cultural norms.**

Learning Itself is Spiritual in Nature

Spiritual means *pertaining to the soul*, or "standing in relationship to another based on matters of the soul".[575] From this viewpoint, per the definition of "soul", spiritual pertains to human life in terms of thought and action, and as introduced above, specifically focused on its moral aspects, the emotional part of human nature, and higher development of the mental faculties. Not surprisingly, an alternative definition of spiritual is *of or pertaining to the intellect* [576] (intellectual, the capacity for knowledge and understanding, the ability to think abstractly or profoundly) and *of the mind* [577] (in terms of highly refined, sensitive, and not concerned with material things). Bringing these concepts together, our working definition of spirituality is *the elevation of the mind as related to intellect and maters of the soul reflected in thought and action.*

Since the definition of knowledge is *the capacity (potential or actual) to take effective action*, learning is considered *an increase in the capacity for effective action*. Thus, **spiritual learning is defined as the process of elevating the mind as related to intellect and maters of the soul** (that is, specifically focused, again, on its moral aspects, emotional human nature and higher development of the mental faculties) **to increase the capacity for effective thought and action.** And, as defined, elevating the mind in relationship to intellect and soul is the lofty goal of all learning.

In a formal 2008 study based on the definitions above, researchers at the Mountain Quest Institute explored the contribution of characteristics spiritual in nature to the learning process. This cross-discipline study engaged literature in learning, education, spirituality, psychology and knowledge.

Various learning models were applied during the research process, including exploring the three types of learning, which are consistent with the Levels of Learning I, II and II introduced in Chapter 1. In this treatment, the types were described in terms of:

Type 1, developing skills which require learning and practicing new ways of doing something;

Type 2, developing knowledge in a field (single loop learning), which requires studying and practicing better ways of taking actions, developing new processes, tools and methods, and applying new management ideas; and

Type 3, changing the basic theory and belief about how a system works (double loop learning).

The domain of spiritual learning resides largely in Type 3, and also moves beyond double-loop learning to what might be described as Type 4 learning, that which has been called intuition, or the "ah ha!" experience (consistent with Level IV learning), or what might be attributed in spiritual literature to unconscious streaming or channeling. Considering the workings of the mind in terms of associative patterning and associative attracting, this could be described as the ability to connect with (and tap into) the larger consciousness field. Type 4 learning emerges unconsciously as a form of knowing, with insights often taking the form of transformative knowledge. This is consistent with the ICALS finding that **the unconscious produces flashes of insight** which was introduced in our earlier discussion of creativity.

Based on hundreds of texts in each field which were part of the study, characteristics that were developed both in the learning literature and the spiritual literature, could be grouped into five general themes:

SHIFTING FRAMES OF REFERENCE: Abundance, awareness, caring, compassion, connectedness, empathy, openness.

ANIMATING FOR LEARNING: Aliveness, grace, harmony, joy, love, presence, wonder.

ENRICHING RELATIONSHIPS: Authenticity, consistency, morality, respect, tolerance, values.

PRIMING FOR LEARNING: Awareness, eagerness, expectancy, openness, presence, sensitivity, unfoldment, willingness.

MOVING TOWARD WISDOM: Caring, connectedness, love, morality, respect, service.

In our model, spiritual capital was considered as both an amount[578] (in terms of subject/object feelings and feeling activities) and an internal state-of-being (in terms of a condition, nature, or essence), or a quality. Considering capital in terms of stock, spiritual capital represents an individual (or organization, country or world) investment in the process of spiritual growth. In its entangled learning role with human and social capital, spiritual capital expands the individual's threshold of awareness, the functioning space within which knowledge and events make sense.[579]

There was a positive correlation between representative spiritual characteristics and human learning. This makes sense, of course, given the overarching connections between the concepts of spirituality and learning that are embedded by virtue of the concepts themselves. For example, Teasdale explains, "Being spiritual suggests a personal commitment to a process of inner development that engages us in our totality ... the spiritual person is committed to growth as an essential ongoing life goal."[580] In other words, learning (growth) is a life goal of spirituality. Therefore, as was affirmed in the research, it follows that human characteristics spiritual in nature would contribute to learning.

Further, we would agree that spiritual learning supports personal growth, helps us understand reality, helps others, and contributes to the greater good. There are indicators that recognition of the relationship of spirituality to these values is spreading. For example, as we entered the new millennium, a World Commission on Spirituality was inaugurated whose commissioners included Nobel Prize winner Elie Wiesel and Bishop Desmond Tutu.[581] And in the field of adult education, English and Gillen concluded that "a more holistic approach of learning, which includes a spiritual dimension, is what is needed in the field of adult education in the years ahead."[582]

Since every complex adaptive system that survives is learning, we now propose *there are innate characteristics within each individual, whether latent or active, which are spiritual in nature*. In other words, it is impossible for the human mind to disengage from the spiritual. For better or worse, *the material and non-material are married in the conscious and unconscious learning of the human mind*. Wherever the human mind exists, so too does spirituality. The implications of these findings may be profound, ushering in new frames of reference to build environments that not only honor diversity of thought and belief, but diversity of learning, capitalizing on new ways of learning and the higher values of the human soul.

Virtues are Spiritual Characteristics in Form

A virtue is considered morally good behavior or character; a good and moral quality; the good result that comes from something.[583] Thus, virtue and morality are synonymous, with other synonyms including: goodness, righteousness, integrity, dignity, rectitude, honor, decency, respectability, nobility, worthiness, purity, principles and ethics.[584] *Virtues are concepts that when thought about expand consciousness, and when acted upon increase intelligent activity*. As we focus on living the future, never has it been more important than to develop the virtue of good character, which directly affects the growth of self and expansion of consciousness.[585]

Early in human history, philosophers such as Socrates, Plato and Aristotle were exploring and reflecting upon those things which are a crucial part of living a good life. For example, in *Phaedrus* (24) Plato talks about the "ability of the soul to soar up to heaven to behold beauty, wisdom, goodness and the like." From their great body of thought and feeling, we as a humanity began to expand our understanding of virtue in terms of beauty, goodness and truth.

Beauty emerges as a transcendent state. Gardner identifies three things that are required to have an experience of beauty: (1) it has to be interesting to the individual, (2) there has to be something about it that will be remembered, and (3) it has to hold our interest. That pleasurable tingle from our emotional system only happens when these three things are present. There's a direct connection between our senses and beauty, that is, our individual senses (and how we make sense of those senses) and beauty. So, everyone's beauty is different. Further, the experience of "beauty" involves all our senses (external and internal) in a "now" experience, such that when fully engaged there's little room for other thought or feelings. For the endurance of the experience, we are in a high-level consciousness event.

Plato talks about the beauties of earth which are to be used as "steppingstones" as we move from fair forms to fair practices to fair notes to fair living, what he calls the *ladder of love*. "Love" is level 500 on Hawkins levels of consciousness (see Chapter 2). In Plato's steppingstones, form represents the human body. From there we move to the beauty of organizations, working together. "Notes" have to do with learning. We are here in this world experiencing (acting and interacting), and through this process learning and expanding, with thought beauty and beauty in thought, and finally recognizing and accelerating to the beauty of life.

In the world of today, Howard Gardner, a Professor of Cognition and Education at the Harvard Graduate School of Education and author of the groundbreaking work on multiple intelligences, embraced these three virtues as crucial to our existence as a humanity.[586] Through a networked humanity, a diversity of information is readily available such that a thinker can more easily discern truth, an explorer can more widely experience beauty, and a seeker can more readily discover goodness. In this context, the choice appears to be at the behest of self. These virtues—identified by our early philosophers as transcendental and reaffirmed by our thinkers of today as crucial to the existence of humanity—are directly linked to the Intelligent Social Change Journey introduced in Chapter 1.

To an extent, morality is developed collectively. In 1995, Sheldrake's formative causation forwarded that our thoughts and attitudes influence other people and we are influenced by other's thoughts and attitudes.[587] The ICALS findings support this. For example, in terms of mirror neurons (Chapter 9), that **cognitive mimicry transfers active behavior and other cultural norms**, and that there is **a rapid transfer of tacit knowledge that bypasses cognition**, which means this transfer is not consciously happening. We also now understand that there is an electromagnetic field generated by the heart which radiates outside the body, carrying information that affects the environment and others in the environment (Chapter 8). And mathematically, transreal numbers show that knowledge through "sentient justification" (or sentient knowledge) gives us the possibility of knowing truth that is non propositional, including the transfer of feelings and intentions along with thought.[588]

This brings in Kant's categorical imperative as a basis of morality.[589] Kant defined an imperative as a proposition that declares a certain action or inaction to be necessary. A categorical imperative is an unconditional moral law, a universal imperative for humanity.[590] When we understand that our thoughts and feelings, our attitudes and beliefs, influence others, they become our responsibility as well as our actions and words, and this suggests the necessity for developing mental discipline and managing self.

In a 2003 research paper on the positive psychology of morality, Haidt forwarded that admiration of virtue is intrinsically motivating, with that admiration inciting people to take actions meaningful for them and beneficial for the larger society.[591] Immordino-Yang, McColl, Damasio and Damasio pushed that further in a research study of the brain and psycho-physiological correlates of experiencing admiration and compassion, which showed that this admiration involved not only cognitive systems but neural "feeling" systems such as the gut and viscera.[592] Specifically, admiration for another's virtue, as a social emotion, resulted in subcortical neural activations in areas of the brain stem that regulate survival mechanisms, which include respiration, cardiac functioning and consciousness. As Immordino-Yang and Sylvan describe, "When people learn of another's incredible accomplishments, moral fortitude, and determination in the face of difficulties and obstacles, often they become inspired to do meaningful work themselves."[593] Thus, virtuous action could perhaps be described as having a multiplier effect, offering a "level of contagion" sorely needed in an uncertain environment.

The Mind Requires Time for Inner Reflection

In Chapter 5 we introduced the ICALS finding that **thinking is mostly unconscious,** and in Chapter 6 that **the unconscious brain is always processing**. Schacter found that the brain uses 80 percent as much energy when sleeping as it does when awake;[594] and Laughlin found that under deep anesthesia electrical activity is abolished and the metabolic rate of the whole brain is reduced by 50 percent.[595] From these findings, it appears that the unconscious mind uses about 30 percent of the brain's energy for maintenance during sleep, and this would presume that when the mind is quieted while in the waking state, there is up to 50 percent of processing capacity available. Of course, these are not absolute numbers, rather "on the order of", with other researchers differing in their percentages, and each individual mind unique.

We also recognize that **meditation quiets the mind** (Chapter 5), reducing the noise that "bedevils the untrained mind, in which an individual's focus darts from one sight or sound or thought to another like a hyperactive dragonfly, and replaces it with attentional stability and clarity"[596] (the idea of monkey chatter). As Schafer states, "Still the noise in the mind: that is the first task—then everything else will follow in time."[597] Meditation can produce a state of passive rest. See TOOL 10, Quieting the Mind.

****

TOOL 10: Quieting the Mind

STEP (1): *Location.* Find a quiet and comfortable place to sit or lie for a half hour (or more), keeping a pen and pad of paper nearby for emerging thoughts. This may be inside or outside, depending on your comfort level.

STEP (2): *Clearing your mind.* Close your eyes and use your imagination to create a mental exercise that will allow you to empty your mind of past and present worries and concerns. As an example, here is an exercise adapted from The Monroe Institute.[598] Imagine a large box with a heavy lead top, which is open.

(a) In your hands is a checklist. On that checklist are all the incidents in your life that have troubled or bothered you in any way, the names of any people with whom you have had an altercation, all the worries that are currently in your mind, and any future commitments that are prominent in your mind. You do not have to bring these things into conscious thought. Just acknowledge that the list is complete, fold it up and put it in your box.

(b) Now, do a quick scan of your body for any aches and pains. Focus on the place where the ache or pain is manifesting, imagine it as clay, and reach in and pull out all the clay, rolling it into a ball and bouncing it into your box. Do this for each area where an ache or pain is manifesting.

(c) Next, do a quick scan for fear, all fears, large or small, wherever fear resides in your body. Focus on each place and—imagining the fear as a stream of yellow, orange or red light—stream it into the box.

(d) Next, focus inside your brain and imagine all the monkey chatter underway as old-fashioned tickertape, the stuff you've seen in the movie *Miracle on 24th Street* being thrown out of windows during the Macy's holiday parade. Grab two or three big handfuls of this tickertape, clearing out the monkey chatter and putting each handful in your box.

(e) Finally, reach inside your chest, pull out your ego, and put it inside the box. Don't worry, you can always retrieve it later (or the part of it you choose to retrieve).

(f) Now, close that heavy lead top and push the box around behind you. Imagine a vacuum cleaner hose coming from that box to your left shoulder. Should any negative image, ache, fear or monkey chatter come up during your quiet time, just send it up that vacuum cleaner hose to your box.

STEP (3): *Float*. For those who regularly meditate, clearing your mind may have already taken you into a place of floating, a quiet mind state. One way to achieve this is to focus on the Fontanelle, the soft spot at the top (center) of your head that served you as an infant. This focus is not accomplished by thought, but rather a feeling of the inner eyes looking upward. This becomes easier with practice.

STEP (4): Once you have achieved this quiet place, relax and enjoy it, letting free-flowing thoughts and visuals play in this space, opening your eyes and briefly jotting down notes when something of meaning to you emerges, then continuing your float. A common visual during this event is the opening and dissolving of colorful energy bubbles. Enjoy the energy and allow your mind and body to relax.

STEP (5): When you feel complete, bring your awareness back to your outer enriched environment, knowing that you can return to this quiet place whenever you choose.

****

Relatively new theories involving the brain's functional architecture show that brain networks support two alternating attention systems, one focused outward on goal-directed tasks,[599] and a second in a state of rest, a time of inner reflection, considered the "default mode" of operation.[600] Immordino-Yang et al. identify this at-rest network as primarily the "regions along the midline of the brain, in both the parietal and the frontal lobes, along with more lateral regions in the inferior part of the parietal lobe and the medial part of the temporal lobe",[601] with the term "network" describing activity in functionally coordinated brain regions. Neurological experiments showed that the activity in this area heightened during passive rest,[602] with participants asked to relax with their eyes open (a non-referential state), or to stare at a plus sign in the mid-range of their view (a referential state).

This research shows that there indeed appear to be two alternating attention systems; with a direct correlation between increasing engagement in one and decreasing engagement in the other.[603] They appear to be codependent systems and to coregulate one another, with some research effectively relating the quality of brain activity during rest to the quality of neural and behavior responses to information coming in from the environment.[604] Similarly, Andreasen forwards that Davidson's work with Buddhist monks suggests that meditation—in this study, non-referential meditation—creates high levels of gamma synchrony, and that this "gamma power" was the greatest "in the association cortices that are the reservoir of creativity—frontal, temporal, and parietal association regions."[605] Andreasen concluded that:

> *People can change their brains by training them in the practice of meditation, so that they improve the quality of their moment-to-moment awareness not only during meditation but also during the routine of everyday life.*[606]

In the "at rest" state of meditation, the mind is not idle, but rather there is the free flow of thought full of memories, imagination, and future planning, non-attentive but awake mental states,[607] as well as self-awareness and reflection, feeling emotions about events and others, and making moral judgments.[608] Building on this research, Immordino-Yang and her colleagues forward that this leads to a new neuroscientific conception of the brain functioning at rest, specifically, that

> *... neural processing during lapses in outward attention may be related to self and social processing and to thought that transcends concrete, semantic representation, and the brain's efficient monitoring and control of task-direct and non-task-directed states (or*

of outwardly and inwardly directed attention) may underlie important dimensions of psycho-logical functioning.[609]

The ramifications of this are meaningful for learning. Whether in children or adults, time to quietly reflect and daydream is *critical for social-emotional well-being—including development of moral emotions and skills*—as well as for the ability to apply attention to tasks at hand and connect decisions and actions to future goals.

There Is a Brain-Heart/Mind-Soul Learning Continuum

That people are holistic beings is foundational to this book, and with that understanding we have recognized the heart and mind as an integrated, biological and complex part of the embodied human system. Within this system, neuroscience findings give us hints of what is possible. There is a brain-heart/mind-soul learning continuum which brings to mind and soul the potential for *an existential state of learning* while recognizing the mind/brain human life is still focused in the physical material reality.

While this bears more discussion, let's first be sure we have a common understanding of our terms as previously introduced. The *brain* is the physical structure that hosts our thoughts. It is a molecular structure that floats inside our skull, with fluids flowing within and through that structure. The *mind* refers to the patterns of neuronal firings and their connections that occur in the brain and throughout the body, encompassing all our thoughts. The pattern of neuron connections, the flow of small electrical impulses through the neuron axons and dendrites, together with the flow of molecules through the synaptic junctions, creates the patterns within the mind/brain.[610]

While historically the brain has been presented as the seat of control—and it certainly plays a continuous role in the process of thought—the body-mind acts as an information network with no fixed hierarchy.[611] *Neurons exist—and learning occurs—throughout the body.* For example, there are neurons in the spinal cord, the heart, the peripheral nervous system and the gut. While these neurons are largely focused on autonomic functions—that is, automatic activity not under voluntary control of the individual such as breathing—all these neurons provide sensory feedback to the brain, which affects emotions. As we addressed in Chapter 8, the heart is very much at the center of this activity, with its powerful magnetic field radiating beyond the body, carrying information that affects the environment and others in the environment, and communicating with the brain and the body in four ways: neurological communication through the nervous system, biochemical communication through hormones, biophysical communication through pulse waves, and energetic communication through electromagnetic fields.[612] Thus, the heart, mind/brain, and body are all part of our experiential learning system, *with continuously changing patterns*. As we interact with life, our neuronal circuitry rewires itself in response to stimulation. Neurons are not bound to each other physically and thus have the flexibility to repeatedly create, break and recreate relationships with other neurons, which is the process of plasticity introduced earlier.

Figure 14 is a conceptual representation of the brain-heart/mind-soul learning continuum. The soul is represented by a conceptual thought form (higher mental thought), which is similar as a visual to the thought form forwarded by Besant and Leadbeater representing vague intellectual pleasure, or "the determination to solve some problem—the intention to know and to understand."[613] Remember: *Energy follows thought*. The material world is an effect, not a cause, with change occurring from the inside out. As the Dalai Lama shares so beautifully, "In order to change conditions outside ourselves, whether they concern the environment or relations with others, we must first change within ourselves."[614] Thus, the thoughts and images generated within have a profound creative and motivating power on human consciousness, with the heart-mind (thoughts and feelings related to those thoughts AND/OR feelings and the thoughts related to those feelings) *controlling energy and building form*. These are not physical forms, but rather "energy complexes on the subtle levels of reality that are analogous to physical things. They are forms made of emotional-mental matter.[615]

Note that while the term "continuum" is used when referring to the brain-heart/mind-soul learning continuum loosely illustrated in Figure 14, this coherent whole does *not* insinuate a *linear progression* but rather a collective working together to achieve growth and expansion, with each element having a different role to play in learning that learning, and with each unable to achieve that goal without the other. While they are interdependent, there is a *focus* in terms of the *origin of learning* and in *attraction based on the frequency of thought*. Experiential learning is focused from the viewpoint of the material physical reality. Existential learning is focused from the viewpoint of the soul, that is, from the domain of the higher mental faculties (called out in the definition of "soul"), which have access to the individual's experiential learning and the larger consciousness field, as well as an understanding of the impact on the emotional and moral development of the individual.

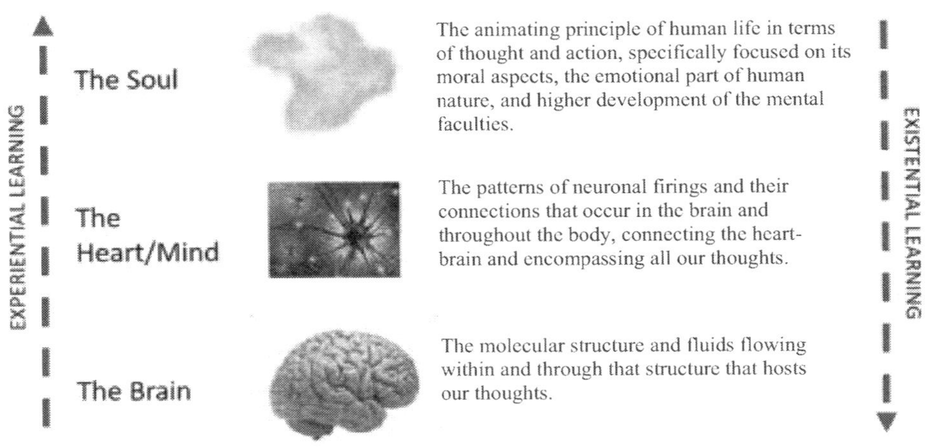

Figure 14. The Brain-Heart/Mind-Soul Learning Continuum.

The human journey of *experiential learning in the physical material body* is one of acting, reacting, then synthesizing, with each new action building on all of the acting-reacting-synthesizing cycles that have come before. We build upon previous learning. Synthesizing is both simplification and explanation, as well as the ability to create a *coherent whole* from various pieces of things. This ability to decide what information to heed, what can be ignored, and how to organize and communicate what we think is important, is a core competency for everyone alive today. See Figure 15.

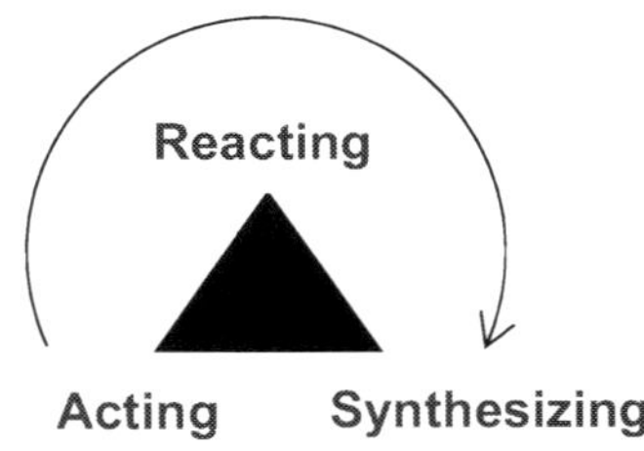

Figure 15. The human experiential state of acting-reacting-synthesizing.

Synthesis Allows Us to Map the Past and the Future

Synthesis—the human ability to knit together information from disparate sources into a coherent whole—is an important human capability, both in the past and as humanity shifts into an unknown future. It could arguably be said that synthesis is a necessary skill for living. It is at play in our everyday life as we tie our life together into *a coherent story*, a concept of self. Moving through a variety of experiences, the individual singles out and accentuates what is significant and connects these events to historic events to create a narrative unity, what—as introduced in Chapter 1— could be described as a *fictionalized history*. As Long forwards "The person makes choices about the importance of persons and events, decides on their meanings … [which are] neither a lie nor 'the truth', but instead a work of imagination, evaluation and memory."[616]

Nobel prize-winning physicist Murray Gell-Mann believes that the mind most at a premium in the 21st century is the mind that can synthesize well.[617] Gardner agrees. As he contends,

> *The ability to knit together information from disparate sources into a coherent whole is vital today. The amount of accumulated knowledge is reportedly doubling every two or three years (wisdom presumably accrues more slowly!), sources of information are vast and disparate, and individuals crave coherence and integration.*[618]

There are literally thousands—and probably more—approaches to synthesis. While certainly some level of synthesis skills are innate, early development and expansion of this skill set comes with doing book reviews, identifying plots, describing characters, linking character traits and actions to the purpose you think the author is trying to convey, and so on, looking for relationships, solving word problems, identifying themes, considering cause and effect, and exploring structure and message. During research studies, to interpret meaning in qualitative approaches, researchers search for themes and descriptions, often winnowing through large amounts of data for what is most important to their research. Experts in many fields "chunk" ideas and concepts, creating understanding through the development of significant patterns useful for solving problems and anticipating future behavior within their area of focus. For example, a study of chess players showed that master players—or experts—examined the chessboard patterns over and over again, studying them, looking at nuances, trying small changes to perturb the outcome (sense and response), generally "playing with" and studying these *patterns.*[619] In other words, they used long-term working memory, pattern recognition and chunking rather than logic as a means of understanding and decision-making.

An important insight is the recognition that *when facing complex problems which do not allow reasoning or cause and effect analysis because of their complexity, the solution will most likely lie in synthesis*, studying patterns and chunking those patterns—organizing at several levels—to enable a tacit capacity to anticipate and develop solutions. This was demonstrated in the movie *A Beautiful Mind* staring Russell Crowe as a brilliant mathematician on the brink of international acclaim who becomes entangled in a mysterious conspiracy.

In today's global world where we have recognized the power of social interaction, synthesis is taking on an expanded meaning to include bringing people together to share and integrate their thoughts, honoring individuation as value to collective collaboration. No doubt several new treatments of this human capability will appear in books over the next few years. TOOL 11 is focused on the ability to recognize patterns.

****

TOOL 11: Thinking Patterns

In the latter part of the 20th century, the authors included the phrase *every decision is a guess about the future* in all of our briefings on decision-making. However, in working with the implementation of knowledge management in the U.S. Department of the Navy at the turn of the century, it became clear that the practices considered as *best* in one part of the organization failed in another part of the organization. This was early in the implementation process, when the understanding was just emerging that knowledge is context sensitive and situation dependent. *Yet there were similarities*; the *types* of things that were similar.

As more implementation examples of similar practices became available, recognizable patterns began to emerge. At this point, we began to purposefully look for patterns. If we found similar patterns occurring, we could more accurately predict the outcome of our implementation process. In other words, it was no longer a random guess about the future. Similarly, Lipton and Bhaerman found, "If a pattern can be recognized, then the accuracy of predicting a future event is relatively high ... if events are found to be random, then all predictions are essentially guesses with an accuracy based on chance."[620] This tool is focused on tapping into the innate human ability to recognize patterns.

STEP (1): Find a place where you will not be disturbed and briefly close your eyes and take several deep breaths. Open your eyes. Then sit comfortably, ready to take notes.

STEP (2): Consider the group of things or series of incidents in which you are searching for connections. If this is a group of things, briefly write down each item's characteristics such as how it looks and what it is made of, how it is created/developed, its purpose, how it is used, etc. If this is a series of incidents, briefly write down for each incident the subject (who or what) and the verb (action occurring) and any descriptive adjectives that come to mind describing the incident, the people involved, the place and timing, the outcome, and the why (if known).

STEP (3): Look across the group and consider the *differences* among the things or incidents. Note these as characteristics.

STEP (4): Look across the group and consider the *similarities* among the things or incidents. Note these as characteristics.

STEP (5): Considering both differences and similarities, identify categories into which these differences and/or similarities could fall. Keep searching until you can bring two or more characteristics of *different things or incidents* together into a category. Repeat until you have discovered all the categories that connect the things or incidents.

STEP (6): Now look at how the categories fit together. *ASK:* How do these categories relate? Are these different categories the same *types* of things? Are there patterns emerging?

STEP (7): Repeat Steps 3-6 until you are satisfied you have discovered all there is to discover.

HINT: This process works well in a facilitated collaborative group looking from multiple frames of reference!

****

Planning Is Core to Existential Learning

The existential learning process from the viewpoint of the soul—relating to the development and expansion of thought and action in terms of morality, emotions, and the higher mental faculties—is at the pinnacle of all that has been learned in the physical experience. As we know, energy conserves itself. There is never any waste, merely transitions. While action is no longer a focal point in the existential state, all the patterns created during the acting-reacting-synthesizing experience remain. Nothing of value is wasted.

The existentialism concept that emerged with 19th and 20th century philosophers put the individual at the center of reality. Everything was subjective, and the meaning of life created by each individual is all that existed. The Universe was unfathomable, and each individual must assume ultimate responsibility for acts of free will without any certain knowledge of what is "right or wrong" or "good or bad". We certainly agree that the Universe IS unfathomable from our individual point of view in this physical material reality, yet we also recognize that there are patterns that *play themselves out in nature at various levels of focus*. For example, all the mineral substances that are part of the earth's crust can be described by the Platonic solids, five shapes that each have equal faces, lines and angles. These are the tetrahedron (4 triangles), the cube (6 squares), the octahedron (8 triangles), the dodecahedron (12 pentagons) and the icosahedron (20 triangles). These five simple shapes are a template for all three-dimensional forms in the Universe.

Recognizing that patterns repeat themselves throughout nature, we can imagine a "being" cycle based on thought versus action, howbeit thought no longer bounded by the limitations imposed by the physical connections of the mind/brain, rather fueled from the viewpoint of the soul. While patterns created during the physically-focused experiential cycle certainly serve as the foundation of the soul's existential state, as we've discovered in our earlier discussion of the mind/brain and experiential learning, these are indeed patterns, which are not exact replicas, that is, blow-by-blow details of events and experiences. See Figure 16.

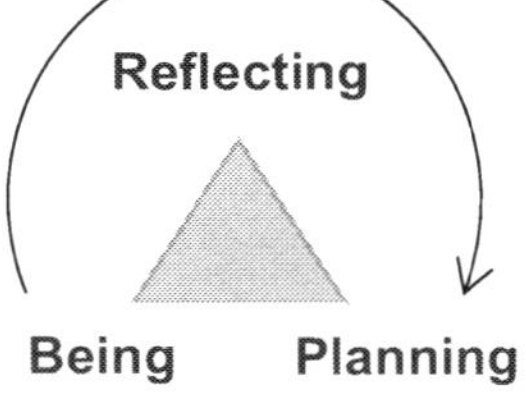

Figure 16. The human soul existential state of being-reflecting-planning.

Planning is forethought—a mental activity focused on achieving a specific goal. From the neuroscience perspective, we now understand that in developing our individual internal model of the external world, the cortical columns in the neocortex develop reference frames and track locations which allow us to plan and create movements. As universal properties of the neocortex,[621] each cortical column has locations and reference frames that enable the association of incoming sensory input— knowing the location of that input relative to the object being sensed— and the planning of future movements through prediction.

When we expand consciousness, we expand our ability to see patterns and to develop good predictive success, and help others do the same. (See Chapter 3 on Contemplating the Future.) Thus planning, related to forecasting and predicting the future, brings the three parts of time together (past-present-future) and is an interactive part of consciousness. As a property of intelligent activity, planning always has a purpose, and includes formulating, evaluating, selecting and sequencing thoughts to move toward a desired goal. From the instant those thoughts regarding what is planned emerge, we are on the path of creation. As Mulford describes:

> *When we form a plan for any business, any invention, any undertaking, we are making something of that unseen element, our thought, as real, though unseen, as any machine of iron or wood. That plan or thought begins, as soon as made, to draw to itself, in more unseen elements, power to carry itself out, power to materialize itself in physical or visible substance.*[622]

In the Industrial Age, plans became the mechanism to create the conditions (resources and events in time and space) to maximize the likelihood of future success. As Alberts and Hayes assert, there was the general faith that a systematic approach consisting of decomposition, specialization and optimization of components would handle even the really challenging problems.[623] Military planning includes five elements: missions, assets, boundaries, schedules and contingencies. In a steady environment this approach worked most of the time. However, it became problematic in more dynamic environments.

While planning was centralized, there was also the acknowledgement that in a changing, uncertain and complex environment, execution must be decentralized, where "the flexibility and innovation necessary for accomplishing the mission typically resided with those implementing the plans much more than with those developing them."[634] In the field of knowledge management this is called decisions made at the point of action. Through cultivating deliberate acts and movement, you increasingly lay a solid foundation of courage, both moral and physical; thus, deliberate planning was the process to achieve force synchronization.

Planning is a learning process which includes two parts: the creation of myths about social realities and the process of emergence.[625] For an expansion of our planning capability, we look to the ICALS model, which adds the fifth mode of social engagement to the earlier experiential learning model, recognizing the global connectivity of people and the need for interoperability in the strategic planning approach, which brings an emergent quality into the process. Built on decentralized execution, an example of this is the *power to the edge* approach adopted by the U.S. Department of Defense in response to accelerated change and increased volatility.[626] This approach moves toward the metaphor of evolution in support of continuing innovation, concluding that organizations *need to experiment* rather than plan. As Alberts and Hayes describe, "various kinds of experimentation activities ... need to be orchestrated as part of a concept-based experimentation campaign to conceive, refine, and fully mature innovation."[627] They contend that the planning process needs to move away from a centralized, top-down, engineered process to,

> *... a process that works bottom-up; one that creates fertility, seeds ideas, nurtures them, selects the most promising, weeds out the losers, and fertilizes the winners. Only an empirically-based experimentation process that employs an appropriate set of measures can accomplish this.*[628]

From the innovation viewpoint, a first step is the redesign of traditional exercises into incubators of innovation as a complement to experimentation, which they see as the fundamental mechanism to cope with change and fuel adaptation. They also recognize the necessity of interoperability and the power of sharing and collaborating addressed in Chapter 9.

While mental thinking can potentially prepare people for future events, it is recognized that there is no linear extrapolation into the future in a changing, uncertain and complex environment (CUCA), the co-evolving environment of Phase 2 of the Intelligent Social Change Journey.[629] This is why Alberts and Hayes feel that specific scenarios of interest, largely scripted, cannot provide sufficient freedom to produce the disruptive innovation that is needed. However, the experimentation and *power to the edge* approach advocated does *not* take into account the power of thought and the power of intention (see Chapter 7).

There is always an element of the unseen when we create a plan, or any business, invention or undertaking. Repeating for emphasis, this element is our thought, which, while it may be unseen, is *as real as any visible material object*. Remember, as soon as it is made, that plan or thought draws to itself other unseen elements which provide the power to execute the plan or the thought. This is the "power to materialize itself in physical or visible substance."[630] Unfortunately, when we have fear or expect negative consequences, we are also creating a construction of unseen elements which draw forces similar to that thought. This is consistent with related ICALS findings that **emotional fear inhibits learning** (Chapter 8), **emotions can increase or decrease neuronal activity** (Chapter 8), and **what we believe leads to what we think leads to our knowledge base, which leads to our actions, which determines outcomes** (Chapter 6). These findings support what is described by Mulford as building in unseen substance, the things we think about, "a construction which will draw to us forces or elements to aid us or hurt us, according to the character of thought we think or put out."[631]

Marrying the mental faculties and the creative imagination, the planning process includes identifying goals and objectives, creating strategies to achieve these goals and objectives, organizing the resources and means to accomplish the strategies, and implementing those strategies. As introduced in Chapter 3, Damasio says that the management of life regulations in a complex environment is dependent on taking right actions, which are "greatly improved by purposeful preview and manipulation of images in mind and optimal planning", which process is allowed by consciousness connecting the "two disparate aspects of the process—inner life regulation and image making".[632] The secret of success is ensuring that the planning process represents intelligent activity, which, as previously defined, is *a state of interaction where intent, purpose, direction, values and expected outcomes are clearly understood and communicated among all parties, reflecting wisdom and achieving a higher truth.*

A tool for the future thinking of planning is scenario building. Recall the ICALS finding that **the neocortex constantly tries to predict the next experience** (Chapter 3), and our more recent understanding that the brain creates a predictive model of the world through experience (movement) which is always changing in response to the environment (input) and the instant-by-instant changes (learning) that occur as we move through life.[633] Planning, related to forecasting and predicting the future, is an interactive part of consciousness. Higher-order consciousness, as seen in humans, includes the ability to build past and future scenarios, which is also related to a synthesis capability.

Scenarios are a form of story that provide a structured process for consciously thinking about—and planning for—the longer-term future. Introduced in the 1950's by Herman Kahn, scenario planning was first used in military war games in the 1960's, with a focus on the "predict and control" approach. Later, the emphasis shifted to analyze "cause and effect" relationships in order to better prepare for the future. As a foresight methodology, it is used to consider possible, plausible, probable and preferable outcomes. Possible outcomes (what might happen) are based on future knowledge; plausible outcomes (what could happen) are based on current knowledge; probable outcomes (what will most likely happen) are based on current trends; and preferable outcomes (what you want to happen) are based on value judgments.[634] See TOOL 12 below.

****

TOOL 12: Scenario Building

Scenario Building is a creative and shared process that involves groups or teams while allowing time for reflection and creative dialogue about the current and future environment and what the organization may become. This specific process was developed for the U.S. Department of the Navy.

STEP (1): Identify a central issue or question. This step involves clarifying the strategic decisions the organization faces that are critical to the future, establishing the inter-relationship among them, and establishing a horizon time.

STEP (2): Identify the key decision factors within the immediate (micro) environment. This is generally done through brainstorming and answering a series of relevant structured questions.

STEP (3): Identify the larger drivers of the key decision factors in the macro environment (social, technological, economic, environmental and political).

STEP (4): Develop the structure for the scenario by grouping high impact/high uncertainty forces and drivers and potential responses. Wild cards, that is, low-probability, high-impact events such as a terrorist attack or disrupted water supply may be considered. Pre-thinking these events offers the opportunity to examine their implications and make better decisions should they occur. This process develops a ranking of possibilities.

STEP (5): Explore the implications of each scenario in relation to current and future strategy, and identify early indicators. This step should result in core strategies and contingency strategies. The aim is to identify strategies that are robust across all scenarios, given what is known and what might occur in the future.

STEP (6): Flesh out the details of the scenarios, creating a logical cause-and-effect flow with a timeline.

STEP (7): Refine and rewrite the scenarios to take them into a narrative form that is easily understood.

HINT: Consider the tenets of Appreciative Inquiry (defined in Chapter 11) as you develop and refine these scenarios.[635]

While these scenarios may prove useful, a significant advantage of this process is the perceptions, awareness and attitudes of the participating strategists relative to their organization, its environment and its evolution into the future. In other words, they will become much more sensitive to changes and potential risks and opportunities that may arise, and potentially more open, and responsive, to unknown and unperceived events. Remember our earlier discussion on Conversations that Matter and our learning related to Relationship Network Management (Chapter 9) that the conversations we have today drive the decisions made tomorrow.

****

The focus on critical uncertainties necessary to move through scenario planning is often quite difficult. Wade argues that scenario planning is all about thinking the unthinkable, although in today's reality the "unthinkable" emerges in our everyday lives.[636] Yet, simultaneously, we are moving closer to understanding that we are co-creators of our reality and the powerful impact of attention and intention (Chapter 7) and thought (Chapter 5) in this process of co-creation. Thus, an intelligent approach is necessary for effective scenario planning, that is, building scenarios that are based on facts while simultaneously ensuring the opportunities offered by every situation are creatively planned into the scenario, always ending with multiple options, multiple choices, and flexibility at the "point of action".

Ultimately, planning is about the future and, as we all have learned along the road of life, living is a continuous journey of change. There is also a moral challenge associated with planning. Planning is intended as a fundamental guideline, with the objective of planning to *maximize the options of future decision*-makers.[637] While we cannot control the future, we can influence and nurture its unfolding, and prepare for the potential of creative leaps. Thus, planning itself becomes a tool for change, and helping others learn how to use this tool fully—a *pass it on* strategy—becomes a powerful way to help the best future emerge.

The Concept of Existentialism Takes on New Meaning

Now we offer a new direction for the concept of existentialism, no longer tethering the concept to the physical body at the lower end of the brain-heart/mind-soul continuum, but rather to the *divine nature of the individual human soul* at the higher end of that continuum. Recognizing that "imperfect" is the state of perfection for a learning/expanding Universe, the term "divine" refers to "perfect"—derived from the Latin *perfectus* (completed, accomplished, exquisite, excellent)—meaning complete capacity, or godlike. Repeating for ease of this conversation, recall that the soul represents the animating principle of human life in terms of thought and action, specifically focused on its *moral aspects*, the *emotional part of human nature*, and *higher development of the mental faculties* and, philosophically, the vital, sensitive, or *rational principle in human beings*. From this viewpoint, you exist, you still exist, you always exist as an individuated expression. Yet this concept of existentialism is in *full relationship with concrete human learning experience*, starting with the authentic thinking, acting, feeling individual who has freedom in terms of agency or choice, and the learning and expansion that result from that agency. **It is an existential "experience"** *orchestrated from the viewpoint of the soul*.

We touch this state during meditation, prayer, during mindfulness, or in selfless service to others—when we achieve an expanded awareness of our connectedness to the larger consciousness field, especially when our energy is focused outward rather than inward toward self (see Chapter 9).

In a recent mindfulness study by Worawichayavongsa and Bennet focused on engagement at the non-managerial level, findings show that "mindfulness practice is experienced as transformative, empowering, and positive introspection; and induces a wide breadth of emotive and behavioral experiences and changes at intra- and interpersonal levels in both a personal and professional context."[638] The changes were driven by "intrinsic motivational factors stemming from conscious and reflective observance of the self."[639] The practice of mindfulness is bringing the inner and outer worlds closer together to actively inform the lived experience.

The inner state can also be achieved through spiritual discipline, psycho-active drugs such as peyote, brain entrainment devices, or various audio recordings of binaural beat technologies such as hemispheric synchronization. The latter will serve as our example. Hemispheric synchronization is the use of sound coupled with a binaural beat to bring both hemispheres of the brain into unison.[640] In the human mind, binaural beats are detected with carrier tones (audio tones of slightly different frequencies, one to each ear) below approximately 1500 Hz.[641] The mind perceives the frequency differences of the sound coming into each ear, mixing the two sounds to produce a fluctuating rhythm and thereby creating a beat or difference frequency. Because each side of the body sends signals to the opposite hemisphere of the brain, both hemispheres must work together to "hear" the difference frequency. This inner-hemispheric communication is the setting for brain-wave coherence, which facilitates whole-brain cognition.[642] What can occur is a physiologically reduced state of arousal while maintaining conscious awareness. This doorway into the unconscious provides greater access to the larger consciousness field which can facilitate creativity and learning.

Following these inner experiences, when we move back into the focus on our physical "reality", the existential experience engaging our creative imagination is similar to—and takes on—the properties of a dream. We may or may not remember parts of this dream. In other words, when we enter an expanded state, we have an existential experience (at a higher vibrational level), and return to our "normal" physical material world focus with the learning from (and specific memories of) this "dream" having very much to do with the level of our consciousness (introduced in Chapter 2). We are there, then we are here—with these two states (the experiential and existential) very separate and distinct "experiences".

However, in today's world of expanding consciousness coupled with technological advancement, the experiential and existential worlds are no longer separate. Co-creating with our creative imagination, humans are moving closer and closer to discovering—and more fully developing and engaging—our higher mental faculties. Through the sciences, we have been given hints of this continuing learning journey through such discoveries as (1) how the brain stores memories as invariant forms so that they can be applied to future situations that are similar but not identical; (2) the understanding that appresentation—what is perceived as an external experience—can stimulate and expand a rich internal experience of the whole; (3) the mindfulness experience; and (4) recognizing the phenomenon of mirror neurons, where the act of observing someone's movement—as associative learning—activates the same brain areas activated by the physical movements themselves.

These serve as BIG hints for the reflective state of existential being, which, in essence is fully "experiencing" … only the experiencing is not in the physical experiential cycle, but rather in the existential state, perhaps with the same neuronal firing sequences occurring as if we were physically experiencing the event. In other words, today's thoughts and actions not only empower the experiential cycle of learning but creatively connect, expand and flavor the **existential learning experience**. "Pieces of experience" of the individual—AND the experiences of others "experienced" through the phenomena of apprehension, mirror neurons, and such or occurring in the physical reality—are connected together in support of a conceptually larger existential "experience", with the experiencing playing out in the mind every bit as "real" as our perceived physical reality. This existential state of creating, of imagining, of experiencing, of learning fully engages the individual creative imagination while simultaneously engaging (or engaged by) the higher mental faculties, the domain of the soul.

Our senses play a large role in physically-focused experiential learning, whether through physical movement, mental thought, emotional feelings or spiritual grounding through beliefs and values. Note that our "senses" are NOT bound to physical expression but are very much connected to "inner" expression related to thoughts, feelings and beliefs. For example, interestingly—very observable in any human—remember that while a little stress may serve as an incentive to learning, when there are strong negative emotions being expressed or felt, little or no learning can occur. And while a feeling of "love" improves the senses, strong emotions reduce them. Further, in our thinking (thought), there is an internal weighting (often unconscious) of the "truth" value of our thought, words, and actions, with truth expanding our consciousness and non-truth reducing our consciousness, whether we are aware of it or not.

In an existential setting, there is high value in the memories of a life well-lived. Life's experiences, and the rich and meaningful moments supported by strong emotional tags they are built upon, complete with inner visuals and feelings, are part of the expanded learning of the existential experience, all contributing to the intelligent complex adaptive learning system that is the human.

The point made here is that the human being exists on three focused levels—the physical, mental, and emotional—while being very much grounded (or not) by spiritual beliefs, or the lack of same. This leads us to ask: What possibilities are offered to each of us with the potential for more fully engaging these personal resources? We've already started down this path with virtual realities that fully engage the creative imagination—offering "sensory" engagement while challenging mental thinking at the conceptual level, encouraging empathic avatars at intuitive levels—which enable players to move upward into imaginal realms while concurrently remaining focused in the physical. What might this existential state of learning look like? How would this work? One of the author's existential experiences is shared in Appendix H.

Exploring existential experiences from the advancing technological frame of reference, three-dimensional virtual reality games such as Occular are expanded to create *psych*ecology educational games (PEGs), that is, virtual games that cross the boundaries between the physical "real" and "virtual", fully engaging the mental, emotional and spiritual. PEGs use symbolic languages and an artificial neural network which correlates recursive input-output, self-organizing map lattices. In this virtual environment, meaning is contextually embedded in the dramatic sequence of scenes, and meaningful insights ensue as the player gradually manipulates his/her plot by making choices that lead to the realization of the premise.[643] As noted above, this virtual reality being played out in the mind is just as "real" as our perceived reality, *with the same neuronal patterns being created in the mind* as if it was being played out in that perceived reality. A deeper description of PEGs is included in Chapter 12.

Our experiential and existential worlds are colliding as we as a humanity move closer and closer to discovering our expanding self, and more fully engaging our higher mental faculties. In these times of a shifting humanity, as intelligent complex adaptive learning systems there is so much more of us to discover … if we dare!

Through Music We Connect the Self and the Soul

The hemispheric-synchronization experience introduced above occurs through the use of sound. Relating the importance of sound to the quantum field, musician and author R. Murray Schafer says, "Today all sounds belong to a continuous field of possibilities lying within the comprehensive dominion of music. Behold the new orchestra: the sonic Universe! And the musicians: anyone and anything that sounds!"[644] While we are both excited and intrigued with this thought, the power of music has been recognized by humans for thousands of years. Plato referred to music as a moral law, giving *soul to the Universe, wings to the mind, flight to the imagination, and charm and gaiety to life and to everything*. We could say that the internal and external worlds collide through sound.

For the past 30 years, and perhaps longer, there have been studies in the mainstream touting the connections between music and mind/brain activity from the viewpoint of psychology, music, education, etc., and another expanding set of studies not as mainstream from the viewpoint of spirituality and consciousness. As our thought and understanding as a species is expanding, these areas of focus are openly acknowledging each other and learning together. It is no longer necessary or desirable to limit our thoughts to one frame of reference, nor to place boundaries on our mental capacity and ability to expand or contract that capacity.

Music and the human mind have a unique relationship that is not fully understood. As Hodges forwards,

> *By studying the effects of music, neuroscientists are able to discover things about the brain that they cannot know through other cognitive processes. Likewise, through music we are able to discover, share, express, and know about aspects of the human experience that we cannot know through other means. Musical insights into the human condition are uniquely powerful experiences that cannot be replaced by any other form of experience.*[645]

The human brain may very well be hardwired for music. This is not surprising since sound is vibrational with different combinations of frequencies inciting reactions across the physical, mental, emotional, and spiritual dimensions. As Weinberger, a neuroscientist at the University of California at Irvine, says: "An increasing number of findings support the theory that the brain is specialized for the building blocks of music."[646] Wilson, a biologist, goes even farther when he states, "...all of us have a biologic guarantee of musicianship, *the capacity to respond to and participate in the music of our environment* [emphasis added]."[647] Sousa provides four proofs[648] that support the biological basis for music: (1) it is universal (past-present, all cultures[649]); (2) it reveals itself early in life (infants three months old can learn and remember to move an overhead crib mobile when a song is played,[650] and within a few months can recognize melodies and tones[651]); (3) it should exist in other animals besides humans (monkeys can form musical abstractions[652]); and (4) we might expect the brain to have specialized areas for music.

Exactly where this hardwiring might be located would be difficult to say. For example, even though there is an area in adults identified as the auditory cortex, visual information goes into the auditory cortex, just as auditory information goes into the visual cortex. That is why certain types of music can stimulate memory recall and visual imagery.[653] Further, the auditory cortex is not inherently different from the visual cortex. Yet in fetus development, the ear is the first organ to develop to its full size (which happens around 18 weeks), during life hearing through the auditory nerve has 90 times greater range than our eyesight, and hearing is the last sense at death. Berendt, who says the ear is the primary sense organ for collecting and processing information, calls it the organ of transcendence, directly linking the ear to compassion and peacefulness.[654]

Thus, "Brain specialization is not a function of anatomy or dictated by genes. It is a result of experience."[655] This process of specialization through experience begins shortly after the time of conception, selecting and connecting. Many of the interconnections remain into adulthood, or perhaps throughout life. While these connections are not exercised in most adults—they are more like back road connections—when the brain is deprived of one sense (for example, hearing or seeing), a radical reorganization occurs in the cortex, and connections that heretofore lay dormant are used to expand the remaining senses.[656]

In the early phases of neuronal growth, during the first few months of life, there is an explosion of synapses in preparation for learning.[657] Yet beginning around the age of eight months through sixteen months, tens of billions of synapses in the audio and visual cortices are lost.[658] Chugani says that this loss is concurrent with synaptic death, with experiences determining which synapses live or die.[659] As Zull explains, before eight months of age synapses are being formed faster than they are being lost. Then things shift, and we begin to lose more synapses than we create.[660] The brain is sculpting itself through interaction with its environment, with the reactions of the brain determining its own architecture.

This process of selection continues as the rest of life is played out. This is the process of learning, selecting, connecting and changing our neuronal patterns.[661] Music plays a core role in this process. Jensen contends that, "music can actually prime the brain's neural pathways."[662] We *do* know that unique combinations of sound frequencies facilitate various states of consciousness, enabling listeners to accomplish goals by achieving a focused, productive, coherent mind/brain state. Brainwaves can be measured with a computerized electroencephalograph, displayed as a map of brainwave activity, and correlated with feelings and actions. Delta waves (0-3 beats per second) are associated with deep sleep; theta waves (3-7 beats per second)—a state of short duration right before and after sleep—are associated with deep relaxation; alpha waves (8-12 beats per second) create a relaxed but awake state of balance and calmness—a good state to process information; beta waves (13-35 beats per second) support peak performance and learning; and gamma waves (35-100 beats per second) create high bursts of creativity and AHA moments. Note that musicians have a high amount of alphas waves, which promote selective attention and visualization. Interestingly, research discovered that while theta waves offer the best learning state, beta waves produce a better problem-solving state. This problem was solved by superimposing a beta signal on the theta wave, which produced a relaxed alertness.[663]

The Mozart Effect emerged in 1993 with a brief paper published in *Nature* by Frances Rauscher, Gordon Shaw and Katherine Ky. To discover whether a brief exposure to certain music increased cognitive ability, the researchers divided 36 college students into three groups and used standard intelligence subtests to measure spatial/temporal reasoning. Spatial/temporal reasoning is considered "the ability to form mental images from physical objects, or to see patterns in time and space."[664] During the subtests one group worked in silence, one group listened to a tape of relaxation instructions, and the third group listened to a Mozart piano sonata (specifically, Mozart's *Sonata for Two Pianos in D*). There were significantly higher results in the Mozart group, although the effect was brief, lasting only 10-15 minutes.[665]

The Mozart Effect quickly became a meme, taking on a life of its own completely out of context of the findings. Perhaps this was because it was the first study relating music and mental spatial reasoning, suggesting that listening to music actually increased brain performance. High media coverage ensued, with the emphasis placed on the most sensational findings. However, the details of the study—specifically, that these findings were limited to spatial reasoning not general intelligence, and that the effect was short-lived (10-15 minutes)—were not part of the meme.

The question of if and how music improves the mind is often couched as a question of transfer effects. This refers to the transfer of learning that occurs when improvement of one cognitive ability or motor skill is facilitated by prior learning or practice in another area[666] For example, riding a bike, often used to represent embodied tacit knowledge, is a motor skill (in descriptive terms, learning to maintain balance while moving forward while peddling) that can facilitate learning to skate or ski.

In cognitive and brain sciences the transfer of learning is a fundamental issue. While it has been argued that simply using a brain region for one activity does not necessarily increase competence in other skills or activities based in the same region,[667] with our recent understanding of the power of thought patterns, one discipline is not completely independent of another.[668] For example, a melody can act as a vehicle for a powerful communication transfer at both the conscious and unconscious levels.[669] Thus, "Music acts as a premium signal carrier, whose rhythms, patterns, contrasts, and varying tonalities encode any new information."[670] By "encode" is meant "to facilitate remembering". An example is the *Alphabet Song* sung to the tune of *Twinkle, Twinkle Little Star*.

Substantiating the long-held "knowing" that music is beneficial to human beings, Hodges outlines five basic premises that establish a link between the human brain and the ability to learn. The first two confirm our earlier discussion of the brain as being hardwired for—or at least having a proclivity for—music. The latter three are pertinent to the impact of musical instruction on the learning mind/brain. As Hodges forwards (with some paraphrasing): (1) the human brain has the ability to respond to and participate in music; (2) the musical brain operates at birth and persists throughout life; (3) early and ongoing musical training affects the organization of the musical brain; (4) the musical brain consists of extensive neural systems involving widely distributed, but locally specialized regions of the brain; and (5) the musical brain is highly resilient.[671]

Why this rather lengthy discussion is included as part of our exploration of the self and the soul is because music is a universal language, felt more than learned, with inner and outer connectivity, and inner and outer effects. Music can tune the heart—affecting an individual's mood, connecting with memories and triggering responses throughout the body. Thus, music, which offers a diversity of choices specific to the listener, provides a means of managing our thoughts and feelings. And, somehow, music is connected to our soul, bringing with it the power to harmonize the mind, body and soul into coherence.

Chapter 11
The Human Gift of Humility

A Contrast of Opposites Can Deepen Our Understanding ... The Meaning of Humility has Shifted Across History ... MQI Research Surfaces Strong Beliefs in the Power of Humility ... Egotism and Arrogance are Barriers to Learning ... Humility is Finding that Balance ... There are Related Positive Characteristics that Lead to an Upward Spiral ... Humility is Part of Human Nature

SUCCESS FACTOR: Humility

FIGURE: (17) Relationship of humility, ego and arrogance.

TABLE: (2) Characteristics compatible with and counter to humility.

TOOLS: (13) Grounding through Nature; (14) The Choice of Humility.

The rapid development of the human intellect has forever changed the focus of work and play. In this context, we have introduced three forces into our world: (1) the necessity of learning for our survival (Chapter 2); (2) the expansion of ego and arrogance due to mental acceleration; and (3) the emergence of technological advancements, which are replacing people in the workplace (Chapter 3). All three of these forces are directly related to the need for, and significance of, choosing humility.

For purposes of this discussion, we use the definitions of humility forwarded by psychologist Joshua Hook, who blogs extensively on healing, growth and learning. Focused from the individual viewpoint, humility is having an accurate view of yourself, neither too high nor too low. This would include knowing your strengths and abilities as well as your weaknesses and limitations, and being honest about them to yourself and others. Focused from the collective viewpoint—the interpersonal level—humility is being other-oriented, focused beyond self. This would include development of empathy, knowing the needs and wants of others, and taking those into consideration in your decisions and actions. Humility is directly linked to the social engagement mode of the experiential learning model and, specifically, to the sub-elements of open mind, a willingness to listen and learn, and creating resonance with people and ideas.

Embracing humility requires a grounding *of* self rather than a grounding *in* self. **Grounding in self** includes all those aspects that come along with self-focus and an expanded view of self, being Anthropocentric. For example, it is quite likely that most of us have special material goods that help ground us, providing feelings of comfort, safety, and continuity. This might include an heirloom ring that's been passed down for generations, or a musical instrument that you periodically play, serving to release tension and bring you into a place of peace. This is materialism that is a grounding of self. However, in advanced materialism where desire has become compulsive and obsessive, perhaps the ONLY thing that is important, this is a grounding in self. And with advanced materialism comes the characteristics of superficiality, exclusion, covetousness, fear, consumption and entitlement.

As a second example of grounding in self consider an individual who is focused on their physical looks, valuing themselves based on their appearance. Looks, of course, are finite in time, and, unless there is a shift of grounding as an individual matures, that individual's survival may be threatened by age! Another example of grounding in self is when an individual identifies with their ideas, often accompanied by expanded ego and movement into arrogance. This is discussed in some depth later in this chapter. Humility is impossible when an individual is grounded *in* self.

In contrast, **grounding of self** is an important and necessary part of life, and there are many ways to achieve this. For example, nature is a structured hierarchical system that can serve as both an external and internal grounding mechanism. We know that we are part of a larger ecosystem that is Earth, whose energies—water, air, light, sound—surge through our physical bodies. When we physically walk through the grass, lean against a tree, or splash in a running stream, we are energetically connecting to these natural energy sources, which promote balance and well-being in our energetic system.

There is a recent movement called Earthing that is rapidly expanding based on the discovery that grounding to the Earth promotes vibrant health. It's the same idea as having an energetic, electrical connection to the Earth and, generally, being comfortable in our physical body. Let's borrow an exercise from the Conscious Look Book on Grounding.[672]

****

TOOL 13 Grounding through Nature

STEP 1: Find a comfortable place outside to stand quietly in the grass with bare feet. Empty your head (using your creative imagination to do so). Closing your eyes, gently wiggle your toes in the grass. Feel the energy of the Earth rising up through your feet, your legs, your torso, then simultaneously flowing down your arms and up your neck into your head, where it mixes with the white light streaming into your head from above, and then moves back down your body and into the Earth. This cycle repeats. Take deep breaths in rhythm with the energy rising and returning, rising and returning. Continue until you are ready to move.

STEP 2: Open your eyes and walk to a nearby tree, feeling a surge of Earth energy into your being with every footstep. When you reach the tree, spread your palms against the outer bark, close your eyes, and ever so slowly use your creative imagination to move your consciousness into the middle of the tree. There is a pinkness there, and a pulsing, rising and falling. Feel this life pulse of the tree and let your heart beat at the same rhythm. Become one with the tree. When you are ready to move on, thank the tree, slowly bringing your awareness back into your body and opening your eyes.

STEP 3: Walk to some nearby bushes. Sit on the ground or in a chair in front of them, facing them. Feel the solidity of the ground beneath your feet. You are connected. Pick a specific leaf or small clump of leaves on the bush and focus on those. Note that they are alive. Reflect on their size, color, how they are attached to the stem and one another, their beauty. If there is a breeze, watch their movement. Reach out your hand and gently touch the leaves, closing your eyes and pulling their energy into your hand. There is a warmth in their feel, almost a kiss. When you are ready, send loving thoughts to this plant, open your eyes, and gently bring yourself back into a standing position.

STEP 4: Turning your face upwards, feel the light of the sun (day) or the gentle echo of the moon (night) against your face. Invite the light to enter through the pores of your skin. Feel it caressing you, filling you and gently rolling over your skin, and, like the gentle dripping of a soft shower, moving into the earth below you. When you are complete, thank the light and slowly bring yourself back into your reality.

REFLECT: Which energy of nature did you identify with closely? Why? The more you repeat this exercise, the more you will "feel" at one with the connectedness of all things.

****

Consider the words: **Strong. Bold. Competent. Courageous. Adventurous. Humble.** A large body of work in the Old Testament of the Bible relates to *humility* and its anthesis *pride*. For example, Proverbs 16:5 says, "Everyone who is proud in heart is an abomination ..." and Proverbs 26:12 says, "Do you see a man wise in his own eyes? There is more hope for a fool than for him." In the Greek language there are two different words for pride, one that infers being haughty or magnifying one's self; the other referring to being blind. Pride is a mindset of self, self-exaltation, with service to self in the context of a burning desire—and actions driven by that desire—to control and use all things for self.[673] Yet, *there is no conflict among the string of words in the first line of this paragraph*, with "humble" complementary to the other descriptive terms representing qualities we admire in our leaders, friends and avatars.

A Contrast of Opposites Can Deepen Our Understanding

There are multiple descriptive terms, states, or characteristics for humility, and various other states or characteristics that are considered as counter to humility. To get a grasp on these perceived differences, major texts (in terms of leading authors in their field) in the areas of philosophy, religion, psychology and leadership were researched to look at the contrasts of opposites in terms of humility. This resulted in Table 2 below.[674]

	Characteristics compatible with humility	Characteristics counter to humility
1	Willing to listen; Good listener; Honor and seek truth; Unnecessary to receive rewards for right actions	On broadcast; Talking too much; Voicing/pushing preferences or opinions when not asked; Bragging; Boasting; Using attention-getting tactics; Ostentatious
2	Receptive to difference and new ways of thinking; Having a teachable spirit	Arrogant; "What I have to say is more important"; Inflated view of importance, gifts and abilities; "I'm better than others"; Unteachable

3	Honor others; Serve others; Focus on others in service; Others over self	Selfishly ambitious; Greedy; Lack of service; Serve me; Meet my needs
4	Seek input and perspectives of others; Seek and follow good counsel; Thankful to others for criticism and reproof; Quickness in admitting you are wrong; Repenting wrong actions	Defensive of criticism; Devastated or angered by criticism; Dismiss instruction or correction
5	Honest/open about who you are and areas you need growth; Awareness of faults; Openly address faults; No need to elevate self; Seeing yourself and others as equals; Seeking to build others up; Minimizing other's wrong doings/ shortcomings	Perfectionism; Hide faults; Minimizing own short-comings; Lack of admitting when you are wrong; Defensive; Blame-shifting; Being deceitful by covering up faults and mistakes
6	Gladly submissive and obedient to those in authority	Resisting authority or being disrespectful; Leveling of those in authority; Demeaning; Being sarcastic, hurtful or degrading
7	Gentle and patient; Gratitude; Thankful and grateful to Life; Genuinely glad for others	Scornful; Angry; Contemptuous; Impatient or irritable; Jealous or envious; Lack of compassion
8	Accurate view of your gifts and abilities	Victim complex; Poor me; Focus on lack of gifts and abilities; Complaining; Consumed by what others think of you
9	Possessing close relationships; Recognize value in others; Willingness to ask forgiveness; Talking about others only good or for their good	Not having close relationships; Passing judgment; Using others; Ignoring others; Talking negative about others; Gossiping; Lack of forgiveness
10	Strong, yet flexible	Willful; Stubborn
11	Theocentric; Recognition of being part of larger ecosystem; Realizing higher power	Anthropocentric; Exalts self; or "He is here for me"

Table 2. Characteristics compatible with and counter to humility.

An important approach that is supportive of the humanness of humility is Appreciative Inquiry (with the "AI" acronym in this conversation not to be confused with Artificial Intelligence). AI is an approach that discovers and promotes (appreciates) the best in people and those things around us, emerging positive emotions. To "appreciate" is to value; "inquiry" is the act of exploration and discovery. asking questions, *learning*. Since we now recognize that there is no behavior or thought that is not impacted by emotions in some way, and that positive emotions open the door to learning, it is clear that emotions are the undercurrent of cognitive thought. Appreciative Inquiry engages this energy in a shared learning experience.

AI is based on the simple premise that individuals, organizations, teams, and communities grow in the direction of what they are repeatedly asked questions about and therefore focus their attention on.[675] (Recall the discussion on attention in Chapter 7.) The principles of AI as translated by Hammond and Hall—consistent with what has been discovered through neuroscience research—have a great deal to do with learning. They are as follows: (1) In every society, organization or group, something works. (2) What we focus on becomes our reality. (3) Reality is created in the moment and there are multiple realities. (4) The act of asking questions of an organization or group influences the group in some way. (5) People have more confidence and comfort to journey to the future (the unknown) when they carry forward parts of the past (the known). (6) If we carry parts of the past forward, they should be what is best about the past. (7) It is important to value differences. (8) The language we use creates our reality.[676]

As can be seen, AI is an approach to engage others and ourselves in a positive learning experience as we co-create reality. However, whether we are in the role of researcher, teacher, leader or learner, a coupling with humility is necessary to reap the full benefits of what we can learn from others.

Before exploring humility further through the results and insights from a Mountain Quest Institute research study, let's first take a deeper look into historical meanings—particularly in terms of religion and philosophy—of the term "humility", which offer a contrast to *the powerful meaning it is assuming in today's wicked environment.*

The Meaning of Humility has Shifted Across History

As introduced at the beginning of this chapter, in Biblical terms pride itself is viewed as a sin, and thus humility is "often identified with repentance and remorse."[677] This is the context directly voiced or alluded to throughout the Old Testament, where the lessons that are forwarded are related to power, and sometimes fear itself. For example, at the end of Isiah 66:1-2, which begins with, "Thus says the Lord," and then it is written, "but this is the one to whom I will look: he who is humble and contrite in spirit and trembles at my word." Note the direct link to repentance and remorse, and the relation to "trembling," which connotes an element of power whether perceived as fear or as profound respect.[678] Yet Scott reminds us that there are several terms that were translated as "humility" or "humble" in the Old Testament, which primarily allude to "bowing low" or "crouching down", such that the actual intent can be lost in the translation. Yet this translation has weighed heavily on the meaning associated with this term.

Similarly, in the early Greco-Roman culture, humility was a negative concept which was considered a sign of weakness, and some of this has crept into our organizational, institutional and political mindsets. Yet, the works of Socrates and Plato, along with the Biblical scriptures that were the foundation of the Old Testament, are very much a part of our human history, and these ancient texts and the emergent books that were to become the New Testament are filled with the *virtue of humility*. Note that, as with the Old Testament, there are several meanings of "humility" emerging from New Testament translations. Two words for humility were used, with one meaning "*servile, base* or *groveling* and the other meaning *gentle, meek* or *yielding*."[679]

There are quite literally hundreds of texts, and most likely thousands of articles, that frame humility as a virtue. St. Francis de Sales considered humility as *the highest of all human virtues*. Philosopher Dietrich von Hildebrand agreed, seeing humility as "the genuineness, the beauty, and the truth of all virtue."[680] Augustine of Hipopo, a philosopher and theologian that lived from 354-430, recognized the strength in humility. He felt that a life of greatness could be compared to constructing a great building. As he wrote,

> *Thou wishest to be great, begin from the least. Through art thinking to construct some mighty fabric in height; first think of the foundation of humility. And how great soever a mass of building one may wish and design to place above it, the greater the building is to be, the deeper does he dig the foundation.*[681]

An example of this combination of strength and humility is Mother Teresa, in terms of physical size a very small woman, but with a huge heart, dedicating herself to serving the suffering children, lepers and destitute ill in the midst of India, her adopted country. There are all sorts of stories that demonstrate that strength in terms of persistence, faith and assertiveness. For example, on one occasion she went into a grocery store with no money, gathered up several shopping carts full of food, waiting in the long check-out lines, and when she reached the front of the line said loudly, "I have no money to pay for these groceries. But I am buying food for poor, starving people—and I am not moving out of this line until the other patrons in this store come up with the donations to pay for this food."[682] While no doubt this produced an amount of grumbling in the line behind her, Mother Teresa left with the food fully paid. One of the authors had the opportunity to spend a day with Mother Teresa, capturing her words, "Hunger is not for bread alone ... hunger is for love."[683] This woman well understood the needs of living, discovering a balance between strength and humility, between selflessness and service, achieving a greatness in the midst of compassion.

In the New Testament, the teachings in the Epistles are largely taught without reference to sin. Indeed, humility is related to the concept of "holiness", a filling with God that displaces self, and as a concept is second only to the emerging lessons of love throughout the New Testament. For example, in a parable related in Luke 14:7-11, Jesus talks about the dangers of exalting oneself when invited to a wedding feast, ending with "for everyone who exalts himself will be humbled, and he who humbles himself will be exalted." Similarly, we are reminded in Proverbs 27:2, "Let another praise you, and not your own mouth; a stranger, and not your own lips."

But it is not in words alone that humility was exalted; but in the actions that demonstrated this behavior. For example, a well-known biblical story is that of Jesus washing his disciples' feet at the Last supper (John 13:1-11). Similarly, Bridges, a well-known Christian writer and speaker, says that the behavior characteristics related in the Beatitudes are all expressions of what he calls *humility in action*.[684] It is in the New Testament, then, that the focus is on humility, this softer trait of one's character.

Humility is considered a virtue in both the Islam and Buddhist faiths. In the Quran there are several Arabic words, *tada'a* and *khasha'a*, which carry the meaning of humility. Totally submitting to the will of Allah is considered a state of humbleness, which is the companion of Paradise. *In the state of humility, you accept the truth, regardless of who is speaking that truth.* In Buddhism, humility is one of the ten sacred qualities which have been attributed to Avalokite Bodhisattva, the Buddha of

Compassion, and leads to nirvana. Mahayana Buddhism specifically advocates humility as a moral precept.[685]

From an intellectual viewpoint, Socrates is credited with saying, "I know nothing except the fact of my ignorance." This is similar to the quote credited to Confucius, "Real knowledge is to know the extent of one's ignorance." Both of these quotes exhibit *intellectual humility*. In a 2016 review of two current research studies, Davis and his colleagues found evidence for distinguishing general humility and intellectual humility. According to Davis, general humility involves "(a) an accurate view of self, and (b) the ability to regulate egotism and cultivate an other-oriented stance." This is consistent with our definitions of humility from the individual viewpoint and the collective viewpoint. Conversely, Davis saw intellectual humility (a subset of general humility) as involving "(a) having an accurate view of one's intellectual strengths and limitations, and (b) the ability to *negotiate ideas* in a fair and inoffensive manner."[686] This distinction takes on a level of increased importance in the age of today and as we move into an uncertain future.

MQI Research Surfaces Strong Beliefs in the Power of Humility

Through the latter half of 2018 and continuing into 2019, as visitors moved through the Inn at Mountain Quest, we collected survey data on some pretty deep questions having to do with the current state of the world. Several of these had to do with humility. Moving into 2020—and before the pandemic stopped the flow of international visitors through the Inn—there were 176 responders, with 12.5% of those born outside of the United States.

Of these 176 responders, 58% were female and 42% male, and the 12.5% born outside the country came from Brazil, Canada, China, Ecuador, Germany, India, Ireland, Mexico, Poland, Romania, Thailand, the UK and Wales. Further, when asked about their heritage, responders offered combinations of the following: African, American Indian, Arabic, Asian, Austrian, Belgian, Black Hawk, British Isles, Cherokee, Czechoslovakian, Dutch, English, Finish, French, German, Greek, Hungarian, Irish, Italian, Japanese, Jewish, Latvian, Nordic, Norwegian, Polish, Portuguese, Scandinavian, Scottish, Sicilian, Slovakian, Spanish, Swedish, Ukrainian, and Welsh. Only two answered "American" and one of those put "mixed" in parentheses as part of their response. Three answered "white"; and one each answered European, Latino and Hispanic.

Specific age groups of the responders were: 17% over 65; 21.2% age 51-65; 25.9% age 36-50; 15.9% age 22-35; 8.2% age 17-21; and 11.8% under 17. Professions ran the gamut from blue collar workers (such as carpenter, school bus driver, electrician, pipeliner, waiter, truck driver, logger, hair dresser) to white collar (such as doctor, dentist, manager, professor, lawyer, realtor, investor, tax advisor, astronomer), with lots of wonderful diversity (such as artist, coach, speaker, stage manager, and author); multiple technology roles, various other service roles (such as nurse, psychologist, counselor, physical trainer, security, and elderly care); and more generic categories (such as student, business owner and retired).

A large number of the human insights from this research—which we will refer to as the SOW2018 research study—are pertinent to learning. As one responder stated, *being humble equals being teachable*. Similarly, another said *humility is the key to learning and growth*, and another noted that each of us walks a unique path in what we call life; hence, "we should exercise humility and accept the fact that we always have something to learn from one another." The quite impressive response from a 13-year-old was, "I think that humility is amazing to have. If you didn't have humility, then you would never learn or know how to correct what you just did."

In the SOW2018 study, 27% of responders agreed that humility is letting things be new in each moment, opening the self to others' thoughts and ideas, and providing an opportunity for listening, reflecting, learning and expanding. One said quite simply, "I believe humility is an essential characteristic for communication and idea-sharing." Specifically focusing on listening, another responder stated that, "One has to set aside what one is thinking in order to truly listen and engage with others." Another noted that, "If one is to truly engage with another, being humble is crucial to forming a true connection." Nine (5%) of the responders focused on this connection in various ways, many tied directly to open and free communication. This was spoken about in terms of humility helping others feel more comfortable; allowing others to open up; providing a non-threatening environment; engaging as equals, leaving social ranks and status aside; and eliminating judgment. And yet another said, "Humility allows for a deeper connection and understanding between individuals for it allows only selfless conversation to follow." These thoughts reflect the idea of moving beyond self, and that idea formed the basis for development of the critical human learning tool simply called *The Choice of Humility*. See TOOL 14 below.

****

TOOL 14: The Choice of Humility

Simple, yet profound, the conscious choice of humility is a powerful tool for the discovery of truth. The greatest barriers to learning and change are egotism and arrogance, which are fundamental difficulties in a rapidly-developing, mentally-focused business environment. Egotism says, "I am right." When egotism advances to arrogance, it says, "I am right. You are wrong. And I don't care what you think or say." As can be seen, egotism shuts the door to learning, and arrogance ceases to listen to or consider others at all, which is necessary for the discovery of truth as well as growth and expansion of an individual or organization. Since others are irrelevant and largely non-existent, an arrogant individual often does not care what harm is inflicted on others.

Both egotism and arrogance increase the forces being produced. Conversely, humility opens the self to others' thoughts and ideas, providing an opportunity for listening, reflecting, learning and expanding, and the discovery of new truths. Humility is the choice of letting things be new in each moment.

STEP (1): To develop humility, first *open your mind* to accept that, by nature, at this point of development human beings have egos and desires, both of which can have strong emotional tags connected to them. It can be quite difficult for an individual to recognize egotism and arrogance in themselves. Remember, the personality, not the self, is often in control, so the individual may or may not be aware of their projection or position. This is potentially true of the individual with whom you are interacting, as well as yourself.

STEP (2): Second, *assume the other is right*. Set aside personal opinions and beliefs for the moment, accept what is being said, this idea or concept, and reflect on this new perspective in the search for truth. While this may prove quite difficult for an individual who is highly dependent on ego and arrogance to survive in what can be a challenging world, almost every individual has someone or something they love more than themselves. Try imagining that this new idea is coming from that someone whom you love, or emerging from something that you love. This simple trick will help increase your ability to engage humility.

STEP (3): Adopting this new idea or concept, *try to prove it is right*, pulling up as many examples as you can and testing the logic of it. If all the examples you can pull up fit this new perspective, then you have discovered a new level of truth. If the examples contradict the concept, then bring in your ideas and test the logic of those. Again, if the examples do not all fit, continue your search for a bigger concept that conveys a higher level of truth. The critical element in this learning approach is giving up your way of thinking so that you can understand thoughts different than your own, discovering new truths. You can compare the various concepts, asking which is more complete.

One issue that may emerge is the inclination for people to think how they feel first, then think about the logical part to determine truth. The "feeling" has already colored their higher conceptual thinking, which may result in it being untrue. It is necessary for us to develop a new sense of self that does not require us to be right in order to feel good about our self.

STEP (4): Once we come to a conclusion, we need to take action. It is time to affirm our incorrectness as appropriate to those with whom we have potentially lacked humility, and to *show gratitude for them sharing their thoughts with us*. Note that the expression of appreciation and gratitude reduces forces. It is not enough to say that you were wrong, *nor is that an important issue*. What *is* important is to acknowledge that someone else is right, and that you are appreciative of learning from them.

STEP (5): Finally, *ensure that your motive for adopting humility is your search for truth*. Motive eventually comes out, and the wrong motive will defeat the purpose at hand. In this search for truth, you are using mental discipline to develop greater wisdom. It is difficult to overcome the urge to "look good" and to be "righter" than others. But remember, when we are "full" there is no room for new thought. When choosing humility as part of our learning journey, we discover that it is not about being right, rather it is about the continuous search for a higher truth.

* * * * * * * * * * * * * *

Egotism and Arrogance are Barriers to Learning

As introduced in Chapter 2, accelerated mental development driven by the innate human hunger for more—the need to survive and desire to succeed—coupled with a focus on hard competition and economic wealth as an indicator of success has led to expansion of the ego into arrogance, plaguing both individuals and the organizations of which we are a part. And somewhere along this road we seem to have forgotten the nobler parts of humanism, individually and collectively seeking potential value and goodness of humanity through rational thought. Egotism closes the door to learning; arrogance builds forces that can lock that door. The relationship of humility, ego and arrogance is represented in Figure 17.

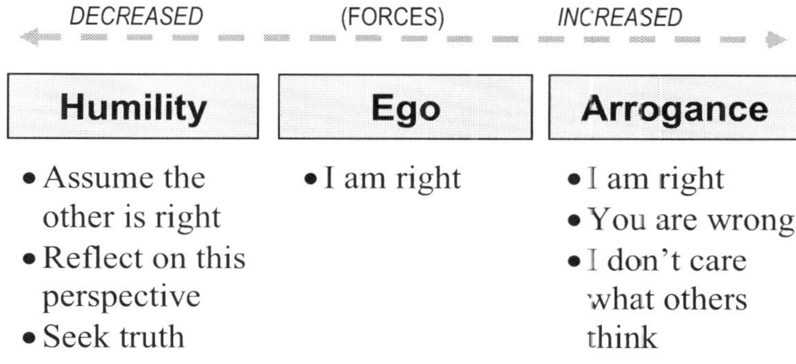

Figure 17: Relationship of humility, ego and arrogance.

When you remedy egotism, the self grows. Since the self is now listening to and considering others' ideas, there is greater opportunity for the bisociation of ideas, and creativity expands. See Chapter 5. Even a small amount of change can have a large impact on an individual, or humanity at large!

In the MQI Research Study, nearly 10% of the responders specifically pointed out ego and arrogance as barriers to communication. For example, as one responder said, "Ego can be self-limiting and potentially offensive to others, thereby building barriers to human connectedness." Another said, "Arrogance only drives people away." And a third quite bluntly said, "I am tired of dealing with and watching egotistical maniacs!"

One responder said that humility comes with respect. This person believes that "the human being is equipped with a very powerful organ, the brain, which has the capability of creating wonder. The least one can do is respect it!" Twelve (6.8%) of the responders agreed with this strong connection between humility and respect. Specifically, this respect was for the individual's brain, experiences, lifestyles, nationality, culture, beliefs, and knowledge. One responder went beyond respect to embracing other's thoughts, saying, "Respect and embracing others' beliefs and valued knowledge expands our own contextual understanding and can better the whole of humanity. Love and kindness are more easily embraced by others if we walk on the same level." Humility was also looked at as "honoring others."

Humility is Finding That Balance

In the Conscious Look Book entitled *All Things in Balance*, we begin the first Idea with, "Every structure we see in the Universe is a result of the balancing between opposing forces of Nature." And as part of Nature, so it is with us. Yet, as Arthur Shelley, an Australian educator and businessman, says, "The natural balance that exists in Nature is something rarely achieved in human systems."[687] Arthur then asks if we could learn from nature how to better balance ourselves. Can we?

In the SOW2018 study, six (3.4%) of the "yes" responders—and nearly all of the responders in the "no" category (5%)—shared a concern with unhealthy extremes. One responder said "meekness" was unhealthy; another noted that being able to put others ahead of yourself is important "but not to the point of being a doormat." A third said, "Humility is okay to a certain extent, but you have to be willing to defend what and who you love. And also, be willing to speak your mind openly and honestly. If not, you are betraying your life." As a fourth agreed, "Humility demonstrates an openness to others and willingness to listen and grow your understanding of others," then added, "It is also a double-edged sword which others may take advantage of if you are not careful!"

Surprisingly, seven (3.9%) of yes responders made remarks that were concerned with what others would think of them. For example, one responder said, "Yes, because arrogance can give people the wrong idea or cause them to draw bad conclusions [about me]." Another said, "I try to be humble most of the time so I am likable." A third said, "This quality makes your character more appealing to others." And a fourth noted, "It's easier to get things under control when you are calm." This last has levels of meaning. Clearly, the responder feels the need to "control" interactions with others. What do you think the need to control by one party does to a conversation? Here is another negative aspect that could do with a bit of humility!

Finally, another "yes" responder clearly summed up this idea of balance. "I think balance is important. We all possess unique abilities and can do so much for the world. But we should never consider ourselves better than others."

Several of the responders in the "no" group (5%) were quite clear that the idea of humility tilted the scale too far. For example, as one said, "Pride in work and personal accomplishment contribute a lot to conversations and allow people to get to know another and share interests. It also demonstrates passion, which gives human connection. Bragging and hubris are different than a well-deserved feeling of accomplishment, and by toning that down you belittle yourself as well." Another shared, "I regret having been too humble. Most people might not be able to relate to that, but if you never let people know your abilities/knowledge/ accomplishments, you miss opportunities not only to use those, but to build relationships based on respect." Taking a very negative stance on the meaning of humility, another "no" responder said, "I do think being humiliated every once-in-a-while is good for you because it allows you to learn from previous mistakes." How do *you* feel about these responses? What is your opinion regarding this balance?

There are Related Positive Characteristics that Lead to an Upward Spiral

Earlier in this chapter we explored the relationship of pride and humility historically in the context of philosophy and religion, and took a look at various characteristics that were compatible with and counter to humility. However, there are also positive characteristics that are not part of humility but often related to humility. In life the combining of positive characteristics can lead to an upward spiral, with the emergence of benefits that are beyond the sum of the two characteristics themselves in terms of learning and the expansion of consciousness. These characteristics—which could be considered in terms of feelings, and largely as positive cognitive conveyors (introduced in Chapter 1)—would include confidence, curiosity, empathy, gratitude, optimism, and even the positive aspect of perfection. We will briefly explore these concepts and their relationship with humility.

Confidence. We've known for years that self-confidence is a large contributing factor to success. Self-confidence has to do with your thoughts about yourself; which can be differentiated from self-esteem, which is what you think others think about you. Confidence was introduced in Chapter 7 in the discussion of an enriched environment; in Chapter 9 in the discussion of the First Knowing; and in Chapter 9 as a necessary attribute of interdependency.

While we might perceive self-confidence and humility as opposites, they actually work quite well together. Jack Zenger, who writes about leadership development, sees six things as elements of self-confidence: mindset, dress and grooming, posture, overall manner, speech and communication practices.[688] For example, he relates mindset to optimism and general happiness, and having a warm interaction with others. Overall manner would relate to behaviors such as walking briskly to laughing with others to interacting with many people in large gatherings. Communication practices would include asking questions, showing intense interest in what others have to say, expressing ideas respectfully—that is, avoiding confrontation—and using metaphors and stories.

Last year Gary Vaynerchuk, who calls himself an investor and serial entrepreneur, wrote a fascinating blog on confidence and humility.[689] Vaynerchuk is the CEO of a full-service advertising agency that services Fortune 100 clients and the chairman of a modern-day media and communications holding company. He says he tries to get himself into a place where he *simultaneously knows he's great and knows he's insignificant.* This, of course, is the actual case for each of us. From a neuroscience viewpoint, our mind/brain is magnificent and where all the action in the world begins, yet when we perceive ourselves as a separate entity in a world population of nearly 8 billion people, it is easy to feel insignificant. While each of us can be considered one neuronal firing in a Universal brain, if there were no neurons firing there would be no action in the brain, no patterns forming, no learning or expansion happening. As small as we are, as important we are!

Similarly, a respondent in the SOW2018 survey acknowledged that while humility is essential in positive socializing, it should be balanced with self-confidence. "Too much of one or the other leads to pride and arrogance or low self-esteem, lack of direction, or to unquestioningly following others." In this regard, Vaynerchuk says "it's incredibly important to know when to turn on your confidence, especially when people push against you, and when people are razzing you, trolling you, or doubting you and I think it's equally important to know when to deploy your humility when people say that 'you're the best' or you are a marketing genius or the best business person or anything of that nature."[690]

The idea that humility is context-sensitive and situation-dependent also emerged in the SOW2018 survey. Similar to Vaynerchuk's approach, one responder says bluntly that his choice of humility depends on the people and the topic. A second says humility is a good idea when others act accordingly and, if that fails, then adopt "whatever attitude is needed." A third responder said it is good to adapt to each situation. "If someone is very shy or has little self-confidence, it wouldn't be kind or respectful to overwhelm them with one's fantastic accomplishments. If one meets someone with similar experiences, I'm not hesitant to share, and I like to learn from others. I don't like bragging, but I like sharing." Unfortunately, sometimes it may be difficult to tell the difference between the two!

What's really impressive about Vaynerchuk's blog is that he's writing this blog because he feels he has a lot to give, and a responsibility to give it, while simultaneously admitting that he's just figuring out how to do this. He notes that there are billions of people out there who are "unbelievably talented, incredible, noble, special people," many of whom have praised him, then responds, "But I'm unable to really accept it."

Curiosity. Curiosity was introduced in Chapter 2 as a quality that serves as an attractor of learning, and in Chapter 5 as an attribute of highly creative people. Psychologist Cynthia Hardwick says that humble curiosity is a pathway to meaningful connection.[691] Humble curiosity means not only being nonjudgmental and letting go of preconceived notions, but also letting things unfold naturally. Thus, the impact of humility is expanded by adding curiosity, and there is a greater potential for deeper connections when humility is combined with curiosity. In short, humility and curiosity, when used together, strengthen each other.

Hardwick forwards that humble curiosity comes from a place of love, which is a place where connections are built among people. As she says, "Curiosity creates space for an innocent journey of meaningful connection and possibility. Whether you are asking powerful questions, encouraging new learning, or soothing frayed tensions, an attitude of humble curiosity lays the groundwork for creating deeper connections and more willing engagement."[692]

Empathy. In the working definition of humility forwarded by psychologist Joshua Hook,[693] which was introduced in Chapter 3, the primary characteristic of humility at the interpersonal level was being other-oriented. By definition, then, the other-oriented part of humility includes development of empathy, knowing the needs and wants of others and taking those into consideration in our decisions and actions. Empathy was introduced in Chapter 1 as an element of the co-evolving phase of the Intelligent Social Change Journey; in Chapter 3 as a fundamental value for the Millennials; in Chapter 3 in the discussion of the future of work; in Chapter 8 in the introduction of the heart-mind connection; in Chapter 9 in the discussion of mirror neurons; extensively in Chapter 9 as a heart-mind tool of connection; and in Chapter 10 as a characteristic of both learning and spirituality.

Much like sympathy, empathy has its Greek roots in the term *pathos*, which means feeling or suffering. Empathy is the entanglement of intuition, resonance, and sympathy simultaneously processed through physical, mental and emotional channels. This combination is an objective attempt to try and live the inner life of another person.

A good starting point to developing empathy is to develop a good understanding of your *self*, including your physical and mental capabilities and an understanding of your emotional guidance system. Without this understanding, the boundaries between self and others can easily blur, such that the inner states of others are assumed to be identical with your inner states. This fusion prevents an objective perspective and leads to confusion in relationships and difficulties in co-evolving. People are not the same.

Nine of the responders in the SOW2018 study used some form of the word empathy in agreeing that humility is a valuable characteristic when engaging others. One responder said that humility is a throughway to empathy. Another was quite explicit, "Humility is essential to empathizing with others. Communication without empathy may be useful for transmitting data, but *not reaching minds*."

A key component of empathy is feelings, and there are findings from neuroscience that suggest that through feelings there is an active link between *our own* bodies and minds, and the bodies and minds of *those around us* (see Chapter 8 and 9). It has been discovered that the insula cortex and anterior cingulate in the brain actually becomes activated either *when we experience pain* **or** *when a loved one experiences pain*! And the degree of activation of these two structures has been shown to correlate with measures of empathy.[693]

Gratitude. Gratitude was introduced as a trigger in the Self-Belief Assessment tool in Chapter 6, and as part of the Releasing Emotions Technique tool in Chapter 8. Humility and gratitude mutually reinforce each other. Similar to humility, gratitude is an "opening" concept, which impacts our thoughts, feelings, and our physical body! As Hess and Ludwig forward, "Studies of gratitude have discovered wide-ranging physical and psychological benefits associated with it, including immune system improvement, lower blood pressure, increased and longer-lasting positivity, and decreased stress, anxiety, and depression."[695] If we reflect on this, it is not surprising. By being grateful to others, you are focusing positively on them and the value they are bringing into the conversation, or into your life.

Gratitude plays a strong role in learning. As an example, Athlos Academies, a charter school network dedicated to helping schools succeed, recognizes the relationship between humility and gratitude. They feel that humility is crucial to the success of students in terms of their social-emotional development. The effects they see from humility include "the cultivation of meaningful relationships, a willingness to be vulnerable, and the ability to practice gratitude. Humility makes each of these possible."[696]

Three studies performed by Kruse and his colleagues, show that gratitude, which inhibits internal focus while promoting external focus, can increase humility. Further, these researchers discovered that the state of humility actually "facilitates greater sensitivity to gratitude." As they explain, "Because humility involves less self-orientation and more other-orientation, humble individuals have an enhanced capacity to notice others' needs and offer assistance."[697]

Optimism. Optimism is a balancing concept with humility, ensuring that the concept of being humble does not fall into a low valuing of self or "poor me" attitude. In a Yale psychology course, educator Paul Bloom ended the course with humility. He pointed out that humanity doesn't yet fully understanding what sorts of experiences make us what and who we are, and then he fully moved into optimism, which focused on the joy of discovery of the unknown elements that the future offers.[698]

Optimism links the power of positive thinking to the future. Anticipation of the future is introduced in Chapter 2; as a sub-element of the adult experiential learning model in Chapters 4 and 5; as part of the discussion in Chapter 6 on bringing the three parts of time together; and in the discussion on focusing attention in Chapter 7. The benefits of positive thought are immeasurable. "The more good-feeling thoughts you focus upon, the more you allow the cells of your body to thrive. You will notice a marked improvement in clarity, agility, stamina, and vigor, for you are literally breathing your way to well-Being."[699]

There are several important organizational processes where successful collaboration requires a combination of humility and optimism. For example, a group visioning process is most successful when participants are outward focused and optimistic about the possible future.

The Appreciative Inquiry approach combines humility and appreciation with optimism regarding the future. AI locates and tries to understand that which is working, learning from it and amplifying it to improve the future.[700] As introduced earlier in this chapter, AI is based on the simple premise that organizations (teams, communities, countries) grow in the direction or what they are repeatedly asked questions about and therefore focus their attention upon.

Perfection. On the surface it would appear that perfection and humility are polar opposites. And, indeed, that does occur, depending on the definition you give to perfection and the personality attitudes that can come along with the idea of perfection. For example, someone who perceives themselves as "perfect" would have no need to listen to anything anyone else has to say, sliding fully into the mode of arrogance. This is where a good balance with humility comes into play.

We also know in our world of mental acceleration that striving for perfection in specific things or areas of our lives—which is an impossible goal to achieve in a changing, uncertain and complex world—can lead to mental, emotional and/or physical breakdown! Simultaneously, striving for perfection can be a motivator to become the best we can be in the areas where we focus our attention.

From its Greek origins, the idea of being perfect is "complete" or "whole." This does not mean being "best" at something, but rather "whole," a full system. Similarly, the Christian concept of perfection is not as a state of being, but rather as a capacity, trying to be better with negative issues rather than an ability to achieve great things. Mathew 5:48 reads, "You are to be perfect, as your heavenly Father is perfect." This is achieving "wholeness", the perfection of the soul. Humility is also connected to "perfection" in Islam and Buddhism. As introduced above, in Islam humility is the companion of Paradise, and in Buddhism humility leads to nirvana.

Knowing your own limits in terms of the physical, mental and emotional is part of this perfection. There are always limits. For example, if you abide by ethics, you are choosing to be limited. So, in a large sense, limits are a way of exercising humility. Used intelligently, limits are powerful tools for growth and change. When referring to ideas, each of us lives in a field of possibilities. The limits imposed by defining ideas within a framework encourages deeper understanding and spurs the emergence of new ideas, with the potential for those ideas to go far beyond the defining framework of their birth. In other words, setting limits (humility) provides focus that can lead to new thought, which offers a possible new way of defining "perfection." Perfection was introduced in Chapter 10 as the descriptive term for the "imperfect" learning/expanding Universe.

As a closing thought, while humility is clearly situated in the group of positive states which bring value to individuals and organizations (in terms of confidence, curiosity, empathy, gratitude, optimism and the "wholeness" of perfection), it has also been found that humility may actually "build social resources through externally focused emotions such as gratitude"[701] which provides long-term benefits.

Humility is Part of Human Nature

In the culture of today, one which promotes self-esteem, at least in the Western world, it is not surprising that we often perceive ourselves as better than we actually are. Noting that few people would describe themselves as humble, Julian Stodd, a researcher and Captain at Sea Salt Learning, a global partner for change, blogs that, "The property of humility may be one that is bestowed."[702] This perspective would mean that the state or condition of humility is not one to be sought after or attained, but rather it is perhaps "a light that is shown upon our actions."[703] This new frame of reference may help us make sense of the conundrum of humility, that is, while humility is perceived as a highly desirable state—whether culturally expected or desirable in terms of leadership, virtue or faith—the act of professing humility of self is contradictory to what humility represents! *The condition of humility is demonstrated, not professed.*

From this frame of reference, it is also clear that humility is a perceived social phenomenon, something that is observed in a social setting. Observing someone demonstrating humility, there is a compliment in the saying. Yet saying that you are humble does not convey humility, even though an honest (not over-valued or devalued) understanding of your self is necessary to successful living. Here we have a conundrum.

That which is observed as humility may also be a conditioned response in context. We each wear many hats, and have lived through many experiences, which means we have conditioned responses—often in the context of sub-personalities—in certain situations. For example, if we are continuously belittled when we are young, perhaps being told over and over again that we are stupid, we might have developed a defense mechanism such that anything said by another perceived as in that same context would produce an immediate forceful defensive response. This, of course, results in poor communication, with a reduction of listening capability and perhaps even an escalation of forces. There are Conscious Look Books on *The Emoting Guidance System* and *Engaging Forces*.

As noted earlier, humility can also be context-sensitive and situation-dependent. For example, the will of self may be more open and receptive in a work situation than in a home environment, or vice versa. In the first instance, I've learned that if I listen closely to a colleague's ideas there's a greater opportunity to solve a problem, improve a process, or expand into new markets! Conversely, the pressures of work may be such while at home that I've not yet realized how important it is to stop and listen to my five-year-old's thoughts.

The really good news is that a majority of people from around the world *do* believe that humility is a valuable characteristic when engaging others. Recall that 12.5% of the responders in the SOW2018 study are people who were born outside the United States, specifically, in Brazil, Canada, China, Ecuador, Germany, India, Ireland, Mexico, Poland, Romania, Thailand, the UK and Wales. Further, responders largely described their heritage as follows: African, American Indian, Arabic, Asian, Austrian, Belgian, Black Hawk, British Isles, Cherokee, Czechoslovakian, Dutch, English, Finish, French, German, Greek, Hungarian, Irish, Italian, Japanese, Jewish, Latvian, Nordic, Norwegian, Polish, Portuguese, Scandinavian, Scottish, Sicilian, Slovakian, Spanish, Swedish, Ukrainian, and Welsh.

In this participant group, as described in the Introduction, **MORE THAN 95% said they believed that humility is valuable when engaging others**. Yet, this is clearly not an easy thing to accomplish. When asked to rate how often they engaged humility when interacting with others on a scale from 1 (not at all) to 10 (all the time), these same responders averaged 7.85, with a median of 8.C. Many of them wrote comments with their responses such as: Sadly, only about ...; Aspiring to 10; I wish it were 10; And should be higher; Want to be a 10 all the time; etc.

These results bode well for humanity, demonstrating a willingness and desire to listen and learn from others in our day-to-day lives, even in this real and perceived mix of ethnicities that still is America and is becoming the world. As one responder in the SOW2018 study concluded, "Being humble is an awareness of the greatness in which we are existing and that Life is One."

Chapter 12
Looking from the Inside Out

Engage in Paradoxical Thinking ... Design *Psyche*cology Learning Games (PEGs) ... Reset Your Emotional Preferences ... Value Story and Embrace Storytelling ... Connect Social Influencers ... Create Your Personal Reality "Television" ... Choose Intelligent Intercourse ... Closing Thoughts

FIGURE: (18) The power of setting intention to co-create the future.

There are so many pieces to unleashing the mind, to becoming the fullness of human possibility. Ah! But then, you have seen that as you moved through this book, and there was so much left out! Undoubtedly, we are complex adaptive systems; indisputably, we are physical, mental, emotional and spiritual; undeniably, we have entangled inner and outer consciousness systems. We think, feel and act. And in support of all that we are, learning is traveling down a new and different path—unfolding with the expansion of consciousness—which is often occurring outside our conscious awareness, although perhaps not beyond our ability to identify, and perhaps mitigate, negative learning spirals. *Reflect:* How can we engage the power of the human mind in positive learning experiences? This is a question we've asked before. But now, through neuroscience findings—looking from the inside out—we can more fully understand the social nature of learning, which widens the focus of our question to include: How can we more fully embrace the positive learning experiences of others?

The pandemic has brought with it opportunities for our educational systems. We are no longer tethered to historical educational rituals difficult to break through in an economically situated setting. There are so many learning situations that we as humans have recognized in the past, and which now we can more fully understand the value associated with those experiences. For example, in terms of expanding the mind, we've always recognized the value of traveling in different countries, exposure to different work environments, and immersion into different cultures and languages. There is no question that the very act of living and experiencing these differences continuously changes the models created in the neocortex that create the perceptions of reality, which in turn surface higher truths and expand our consciousness.

There are so many everyday experiences that we can plan into our lives which can help unleash the mind. We invite you to join us in exploring some familiar and a few unique options to expand learning opportunities—suggested by our new understanding of the mind/brain— that are or could be available to all.

Engage in Paradoxical Thinking

The 2008 MQI research study introduced in Chapter 8 found that through exposure to diverse, and specifically opposing, concepts that are well-grounded, it is possible to create a resonance within the listener's mind that amplifies the meaning of the incoming information, increasing its emotional content and receptivity. As previously noted, while it is words that trigger this resonance, it is the current of the search for truth flowing under that linguistically centered thought that amplifies feelings, bringing about the emergence of deeper perceptions and validating the recreation of externally-triggered knowledge in the listener.

The concept of truth has emerged throughout this book. Humans operate from a place of *yearning to know truth*. Truth is a living, dynamic awareness that expands our consciousness. Recall that intelligence is defined as *an aptitude for grasping truths*. It is the neocortex, as the organ of intelligence, that is very much concerned with updating its models of the world based on a continuous stream of incoming sensory input, ever creating new models of the world. Connections are not fixed. As we learn, connections are strengthened; as we forget, they are weakened.

When we experience a positive reaction to someone's spoken words or written thoughts, we are experiencing a *resonance of thought*, with those ideas consistent with the rhythm of our natural frequency, that is, with our truth, beliefs and values. The example provided in Chapter 8 is that of a debate, which provides both sides of a question such that an active listener who has an interest in the knowledge area of the debate is pulled toward one side or the other, which is based on the triggering of previous thought, experiences and feelings.

An even stronger example of this is paradoxical thinking. As we expand our understanding of the way the mind/brain works, we can now recognize that the mind is *primed to support paradoxical thinking*. Let's briefly explore that understanding. A "paradox" is considered sets of inconsistent propositions, with a "set" including an explicit contradiction or entailing one.[704] Alternatively, from other viewpoints, a paradox is considered an apparently unacceptable conclusion derived by apparently acceptable reasoning from apparently acceptable premises[705] or could be thought of as two contradictory propositions to which we are led by apparently sound arguments.[706] This means that two "arguments" or "positions" can both be right when looked at individually, yet when looked at as a set they do not agree, that is, they can't both be right. For example, a paradox which was popular when the authors were young—and still pops up now and again—is posed by the question: Which came first the chicken or the egg? Clearly, a chicken comes from an egg, and eggs come from a chicken. Both are truths. However, when you consider the third element of "which came first?" there is clearly a paradox. Something missing beyond our understanding.

In the duality of today's world, paradox provides the rare opportunity to *end a conceptual argument in a win/win state*, that is, with each theoretical stand "right" when separated from the set and considered in its individuality. As forwarded above, it appears that the human mind is *primed to support paradoxical thinking*. This is because a paradox does not trigger the negative emotional "push" from the old brain (most often happening in the unconscious), that "need" to "take a position and defend it" or ensure that an individual is not seen as "wrong", both related to ego issues. The lack or softening of this triggering eliminates a potential conflict between the old brain (emotions, ego, etc.) and the new brain (the neocortex, the organ of intelligence), enabling the intellect to be open to learning as it navigates the challenge of this thought. As explained by Bennet,

> *The excitement of a paradox simultaneously evokes emotion and triggers conscious thought inducing resonance, engaging the old brain while stimulating the new brain, expediating consciousness expansion, and opening the mind to the exploration of ideas.*[707]

While purposeful events focusing on paradoxical thinking can be quite successful, they are generally limited to those who have a pre-interest in a paradox. One approach to bring paradoxical thinking into a larger context is to identify a paradox related to the domain of focus—for example, when addressing an issue at hand or during a training session—and introduce that paradox in a warm-up session prior to addressing the larger issue. Since paradoxes have been identified dating back to the early days of our Western Philosophers, and probably earlier, there are numerous texts that can serve as guides to identify paradoxes which have "taxed thinkers from Zeno to Galileo, and Lewis Carroll to Bertrand Russell."[708] Two excellent references are *Paradoxes: Their Roots, Range, and Resolution* by Nicholas Rescher (which also includes a history of paradox) and *Paradoxes from A to Z* by Michael Clark.

A second approach would be to introduce a paradox just before a meeting, then break for a few minutes, and spend 15-20 minutes at the start of the meeting soliciting thoughts from participants. A third approach is to make up a short paper including 3-5 paradoxes as a take-away from a more formal interaction, whether face-to-face or virtual.

One of the potential issues with the use of paradox is similar to that occurring in a debate, the emergence of a dissonance or discomfort with one side, thus a "locking in" or polarization that closes off potential integrating thought. However, what is powerful about paradox is that *no one is wrong*, and that is what has to be emphasized and understood "emotionally", not just mentally. One way to do this is the repeated affirmation to both sides of a paradox that "you are right", then focusing on, "Just as ancient philosophers did, what can we learn from this paradox?" In other words, you are bringing participants into the experience as "philosophers", with the recognized wisdom of their lived experience. In a group, you might have people raise their hands regarding one side and you tell them why you believe *they* are right. Then, have the other side believers raise their hands and you tell them why *they* are right.

At this point, you introduce the tool of humility (see Chapter 11). Humility asks us to look at whatever we have a strong belief about from the other's viewpoint, literally trying to PROVE the other viewpoint is right. This is a powerful exercise. Again, the beauty of paradox is that doing this does not mean that YOU are wrong. You are proving it right from the other side of the paradox, not from the viewpoint of the set, that is, the contradictory statement. If an individual is able to engage in both these "arguments" and see the "rightness" of both sides, after a few successful experiences this pattern will begin to expand to other aspects of life, and this ability is sorely needed to navigate the divisiveness of today's world.

Further, you will discover that shared paradoxical thinking has a viral nature. How many times have you repeated, or heard, our earlier example, "Which comes first, the chicken or the egg?" For a developed mind, paradox poses a challenge, and a curious mind asks, "How can this be?" In the quest for a solution—whether ego-driven or knowledge-seeking— insights are shared freely. Your participants will begin looking for paradoxes in their everyday life, and they will find them, and in the finding become more tolerant of difference and the challenging excitement of contradiction. As Quine, a mid-20th century thinker, posed: "More than once in history the discovery of paradox has been the occasion for major reconstruction at the foundation of thought."[709]

Design *Psych*ecology Learning Games (PEGs)

Through 2021, one of the authors had the opportunity to interact as a co-editor with a group of domain experts focused on building the foundational understanding for development of PEGs (*Psych*ecology Video Games). PEGs serve as analogs based on source code that depicts all known electromagnetic energy dynamics or algorithms that contribute to conscious and unconscious humanly embodied dynamics. Embedding symbolic languages containing a narrative-metaphorical common denominator, the PEG template can be thought of as an artificial neural network, somewhat like a Kohenen map, correlating recursive input-output, self-organizing map lattices.

The single-layer self-organizing map is a "feedforward" network, with the output syntaxes arranged on a low dimensional grid, usually in two or three dimensions. Each input is connected to all output neurons, and attached to every neuron is a weight vector with the same dimensionality as the input vectors. The number of input dimensions is usually a lot higher than the output grid dimensions.

Self-organizing maps (SOMs) are popular neural network models used for unsupervised learning, which means that *no human intervention is needed during the learning* and that little needs to be known about the characteristics of the input data. SOMs can be used for clustering data such as contextual personality and story premise, and to detect features inherent to the story problem or premise. The goal of the learning in the self-organizing map is to associate different parts of the SOM lattice to respond similarly to certain input patterns. This is partly motivated by how visual, auditory, or other sensory information is handled in separate parts of the cerebral cortex in the human brain.

What this means is that conscious player responses to the symbolic lattices of game images may be recursively correlated with unconscious affordances in the program-syntax at computational levels of the game.[710] The narrative-metaphorical common denominator can be used to correlate data as well as to generate greater and more meaningful narrative interactivity in the game story real time, depending on the response of the player during the play sequence. In PEGs, the premise is clearly expressed in the beginning (*exposition*), meaning is contextually embedded in the dramatic sequence of scenes, and meaningful insight emerges as the player gradually manipulates his/her plot by making choices that lead to the realization of the premise (*lysis*).[711]

Luca—an independent chartered accountant with an unbridled passion for psychology, social sciences and software—spent years developing an experimental prototype based on a study of graphical interfaces in order to provide the efficient management of knowledge artifacts. From this focused research, a technology and relational multimedia platform emerged—what he calls *Ysarmute*—that offers the potential to revolutionize knowledge flows, linking conceptual models and patterns to logic-based concrete examples while engaging in a new type of multimedia intellectual capital building and consciousness expanding through the use of PEGs. Thus, PEGs are already being integrated into systems that can help develop the learning and coherence necessary for successful and sustainable business in the current and future volatile gaming (and business) environments.[712]

In the discussion of truth in Chapter 8, the mathematical discovery of "sentient justification" was introduced. Research by Gomide using transreal numbers has shown that knowledge by sentience gives us the possibility of knowing everything that is not propositional.[713] In other words, through the narratives ("poetical discourse") involving sensations or feelings that are part of PEGs, there is an apprehension beyond that which is propositional, which would include direct intuition, states of consciousness, and perceptions regarding emotions. The mathematical research using transreal numbers has proven that narratives involving sensations or feelings transfer the full range of sentient knowledge, offering the strong potential to change our belief states.

While these games are not yet available on the market, the theory itself is gaining support widely around the world. For example, in a 2022 paper developed by Schafer on Evolutionary Learning, he says:

> *Psychecology video games (PEGs) have the technological potential for fostering Evolutionary Learning on a large scale in the process of gameplay. This process is based on monitoring frequencies in the electromagnetic spectrum that are coherently induced in gamers during gameplay. Monitoring the vicissitudes of a gamer's state-of-coherent entrainment and increasing the coherent-resonant state of gamers with PEG algorithms is our foremost objective.*[714]

He goes on to explain that the PEG "source code" includes numerous dynamics that are perceived to contribute to this objective, specifically: electromagnetic frequency monitoring (Braintap technology), coherent entrainment (Institute of HeartMath), asymmetrical mathematics (quantum field vectors and quantum collapse), and Jungian compensation/individuation (Myers-Briggs personality types). As Schafer forwards,

> *The more people that function in coherent states as a result of being harmonized with gameplay (leading to game-wins that reflect coherent entrainment & individuation as evolutionary learning objectives), the more problems (disharmonies) will be "healed".*[715]

Global gameplay includes millions of players at any given time, which could be sufficient to alter the electromagnetic geosphere toward a state of harmony. The final step in theoretical development of PEGs is to integrate these algorithms into an executable product, which is exactly what Luca has in mind with the development of Ysarmute.[716]

Reset Your Emotional Preferences

With a larger understanding of how emotions work in the brain, we now recognize that *emotions influence all incoming information and that emotions can increase or decrease neuronal activity.* We also recognize that "molecules of emotion" clan reinforce what is learned and aid in memory recall. Quite literally, and as stated earlier, emotions are a building block of consciousness.[717]

Recognizing that everything that is perceived in our reality was a thought first—and that our perception is based on our subjective model of the world, which is continuously being adjusted based on sensory input strongly affected by emotions and feelings—it just makes sense to master our emotions to facilitate greater learning and expansion. What emotions do you choose for your baseline? We all have preferences of how we address life; for example, as a glass half full or half empty, and it's easy to tell which approach harnesses negative or positive energy! We can now understand from the brain's viewpoint why *the power of positive thinking* became such a strong personal development tool and management movement! And yes, from the brain's viewpoint, we also recognize that *emotional fear inhibits learning.*

But how do you reset your emotions? Of course, you can't change anything until you are aware of it, so we must first be aware of those emotions. And through the process of patterning, recalling similar responses in the past, we can become more aware of the emotional presets we have developed in our unconscious over years of living. Remember, the first step to change of awareness requires *conscious attention* (see Chapter 7).

A second important necessity is the *desire* to release or shift negative emotions (see Chapter 8 for a discussion of desire). Recognizing that desire when emotional surges surrounding various events—or specific types of thought, or a particular set of people—occur, we can choose to either remove those things from our life or change our perception of, and thus our feelings about, those things. Recall that what we perceive is our personal model of the world, not the world itself, and that our perception is being continuously updated as we experience life, and remember, we are "verbs" not "nouns". Thus, when emotions do not serve us, that is, when we've learned all we can learn from them, and we have the desire to do so, we can *choose* to diminish or release emotions (see TOOL 8, Releasing Emotions Technique, in Chapter 8).

When negative emotions have been diminished or released, there is an opportunity to reset your emotions. Activities of the mind and body are the vehicles of a reset or, if you choose, a preset. As a behavioral model, it has historically been recognized that vigorous physical activity requiring a focus of energy can trump, at least temporarily, negative thoughts and feelings. Rigorous positive thought can serve this same purpose, that is, by choosing our thoughts and refocusing our attention, we can determine or adjust our emotional experience. One approach to do this is mood shifting. In this exercise, you prepare a card in your wallet and carry it around with you. On one side of the card write five things you are or have been a part of that you feel good about. On the other side write down five things you have done in your life to serve others.

Small and large perturbations or changes have a way of working themselves into our lives, interfering with the flow. When a life perturbation occurs and your emotions tank, first recognize your emotional response (awareness) and honor it (attention), being sure to note any lessons learned. To do this honoring, one of the authors wraps her arms around herself in a self-hug, closes her eyes, gently rocking left to right while saying out loud: "I'm having a human moment." However, you honor your emotion, when you are ready, pull out the card and read whichever side is most appropriate for the situation at hand. Spend several minutes—or as much time as needed—remembering and reflecting on the events or situations represented on your card, "feeling" them. When you sense your emotions positively shift, you are ready to work with or address the issue at hand from a different viewpoint.

A similar approach is that used by actors. When a specific emotion is desired, they recall a life event that surfaced those emotions and stay focused on that event for several minutes. Amazingly, *it can take as little as 17 seconds of focused feeling to shift from one emotional state to another!* So, polish up your acting skills and become the character in life you choose to be, portraying the emotion you choose to either shift into in the present or to preset your system for the future. And have fun!

As one last thought, we now know that there is an existential learning state beyond the physical/mental/emotional experiential learning cycle (see Chapter 10). In this use of existentialism, the term is no longer tethered to the physical body, yet is in full relationship with concrete human experience, starting with the authentic thinking, acting, feeling individual who has freedom in terms of agency or choice, and, as previously noted, the learning and expansion that result from that agency.

In short, we are describing an existential "experience". For example, we touch this state during meditation, mindfulness, prayer, or in selfless service to others—whenever we achieve an expanded awareness of our connectedness to a larger field, especially when our energy is focused outward rather than inward toward self (see Figure 12 in Chapter 9). This means you can expand, build on, or imagine experiences and emotions BEYOND those in your "memory banks". Being "human" is becoming ever more exciting. Use your creative imagination to co-create your feelings … and your reality.

Value Story and Embrace Storytelling

In Chapter 11 we introduced Appreciative Inquiry and its relationship to humility and learning. Recall that AI is based on the premise that what people are repeatedly asked questions about is *what they focus their attention on.*[718] Four of the principles of AI, as translated by Hammond and Hall, are pertinent to valuing stories and storytelling. These are:

- What we focus on becomes our reality.
- People have more confidence and comfort to journey to the future (the unknown) when they carry forward parts of the past (the known).
- If we carry parts of the past forward, they should be what is best about the past.
- The language we use creates our reality.[719]

Briefly consider the appropriateness of these points when reflected through story.

Throughout the process of living, we are walking through a temporal sequence of how things are done, with one pattern evoking the next, with the recall of our memories following a pathway of association. Thus, we can say that human memories are story-based (see Chapter 6). Effective memories also have memory traces, which are labels that attach to previously stored memories. For example, these could be decisions, conclusions, places, attitudes, feelings, questions, etc. And what we have come to understand is that memory recall is improved through temporal sequences of associated patterns, that is, stories and songs.

Stories convey knowledge, and our knowledge of the world is more or less equivalent to our experiences. They are a living, breathing example of the "what" and "why" of our experiences. In Chapter 1, we introduced the idea that the story of "you", a fictionalized history, lives in the mind. In Chapter 6, we addressed memories as stories, and in Chapter 9, while exploring our social interactions, we are reminded that a main job of consciousness is keeping our life tied together in a coherent story, a concept of self very much related to the current environment and social situation as well as our expectations of the future.[719]

People do not just "hear" stories, they "relate" to them, often in unconscious ways. Looking from the viewpoint of social learning while simultaneously taking advantage of the power of ownership, it appears that there is the opportunity here to engage learners through the early capture of their individuated stories relative to the subject at hand, as well as a continuing process of story capture as part of the learning experience, and the sharing of those stories with the larger learning community, perhaps accompanied by discussion and questions. For example, in one intervention using story to convey organizational values, these guidelines were followed: (1) tell stories in a dynamic and entertaining manner and facilitate, don't lecture; (2) use coaching to get listeners thinking for themselves; (3) let listeners create their own meaning; (4) use small-group dialogues to share and enhance understanding; (5) build core value ownership through reflection and journaling and guiding; and (6) let the attraction of stories spread the core values.

An interesting approach in one organization was creation of an archetypical storyteller who represented the "older and wiser" leaders in the organization who had "moved through" critical organizational learning experiences. This archetype storyteller was assigned characteristics related to previous "highly regarded" leaders no longer with the company. A number of stories were prepared for this storyteller to convey to different groups, portraying situations and solutions consistent with the organization's values and direction. Further, each group was challenged to build a story for the storyteller to convey to future employees. As the repository of story situations increased—all directly related to organizational learning—these were placed into an "Ask [subject]" system such that all employees had virtual access to these stories through the designed representation of the archetypal storyteller.

A value of developing archetypes in both of these ways is that they take on a life of their own as nurtured through their continuous exploits and resolves focused around "real" organizational issues. While contemplating these real stories, these archetypes provide a closer source of identity for new employees operating beyond their threshold and level of knowledge comprehension. In addition, archetypes (in the first example) and the characters in the archetypal storyteller's stories (in the second example) can make mistakes, perhaps helping new employees avoid those same mistakes.

There are so many ways to embed story and storytelling more fully into your everyday lives. We use the term "more fully" because you are already doing this! Reflect on how you answer "What did you do today?" or "What did you do on your vacation?" or "How did your daughter become so good playing the piano?" and so forth. There is no list of "facts" that goes along with your responses, which are pulled from your memories and perceptive viewpoint and conveyed as a story.

A simple story tool is keeping a daily diary—not so much about your everyday mundane actions, but about the thoughts prompting your actions, allowing those thoughts to flow freely consistent with your desires, purposely bringing into conscious awareness the FUTURE YOU, living the life you choose to live, and WRITING THAT DOWN. Then, periodically go back and scan your notes, making any changes you choose to make to that future. Finally, FEEL yourself embedded in that future, and think and act accordingly in your everyday life. You are consciously co-creating your reality and designing the storyline of your life.

As story has woven itself throughout the evolution of the human, all sorts of behaviors and traditions have emerged. Native American myths and legends could only be told in a certain way at particular times; for example, during winter months and in many cases only at night.[721] The Iroquois say that if you tell stories in summertime a bee will fly into your dwelling lodge and sting you. The Abenaki believe that if you tell stories during the growing season, snakes will come into your house.

Pourquoi stories, part of the American Indian oral tradition, are transformational stories. Named after the French word meaning *Why,* and focused on how and why, these stories center around the creation, naming, and characterization of almost everything in nature. They are instructive, teaching lessons in a gentle, non-threatening way. Animals and unspecified settings are used to provide a comfortable distance for the listener. Specific characteristics of *pourquoi* stories include: (1) settings that are unclear as far as time and place; (2) characters defined by only those elements that are essential to the story; and (3) a dilemma that is common to the listener. The transformational aspects of these stories often involves supernatural or magical elements, which satisfy a need to make sense of that which is unknown or not understood.

Many of these tales provide moral instruction on how we should and should not behave. For example, one Algonquin story tells about a very hard-working man who has a lazy wife who never cleans their home. One day, Wakonda (one of the Great Spirits) is going to visit. The man begs his wife to clean. When she doesn't, the man is told to take some dirt and throw it at his wife. These specks of dirt became the first mosquitoes. As can be seen, these stories make lessons easier to remember, communicating simple truths without singling anyone out. *Pourquoi* stories are also used as a means of conveying survival information.

The Anishinabe medicine woman/storyteller begins every story with an offering of tobacco to the ancestors, and begins each story in song. Also, to mention the names of certain characters in the stories outside of the story itself is considered bad luck. In California Indian tribes if you mention the word "Coyote" outside the story he will come visit you and cause mischief.

All that being said—given none of these storytelling limitations are part of your personal belief set—we're going to encourage the use of story—as has been historically proven and now is understood from the inside out—as a powerful approach to unleashing the mind. There are many books available full of the effective use of story in personal and professional situations. One of these published at the turn of the century— yet still echoing a leading voice in storytelling as a business tool—is Denning's *The Springboard*. In this text, insights are offered on how to use stories for personal and professional growth in an elegant, simple, and effective manner.

The springboard story is a transformational story enabling the listener to take a personal leap in understanding how an organization or community or complex system can change. Denning purports that the intent of this type of story (told in oral tradition) is not to transfer information, but to serve as a catalyst for creating understanding within the listener. As Denning says,

> *It can enable listeners to visualize from a story in one context that is involved in a large-scale transformation in an analogous context. It can enable them to grasp the ideas as a whole not only very simply and quickly, but also in a non-threatening way. In effect, it invites*

them to see analogies from their own backgrounds, their own contexts, their own fields of expertise.[722]

The stories that Denning feels are successful in this transformation have specific characteristics:

(1) They are told from the perspective of a single protagonist who is known to the audience and actually in the predicament being told.

(2) There is an element of strangeness or incongruity to the listeners which can capture their attention and imagination.

(3) The story has a degree of plausibility and a premonition of what the future might be like.

(4) There is a happy ending, which makes "it easy for the listeners to make the imaginative leap from the explicit story that I was telling, to the implicit story that I was trying to elicit in their minds."[722]

This has been a very light treatment of story, a wonderfully large field of study that has played—and continues to play—an important role in human evolution. The difference now is that we understand the power of story in relationship to how the brain learns as we create the story of us, and perhaps we realize that the gift of story and storytelling is something all of us can engage.

In understanding the significance of the sense-making structure of story, we can question why we have primarily focused its benefits within the field of communications, that is, focusing on storytelling which uses story as a communications strategy instead of focusing on *story thinking* which uses story as an operational strategy. What if the power of story was applied to the models that drive other topics, such as change, learning, and leadership? This was the question explored by John Lewis in his book *Story Thinking*, which compares story structure to 30 of the most common models related to change and found that our institutional models break down because they do not align with our expectations of story. If the models that currently drive the institutions of education and policy-making do not make sense to you, there is a good reason—they do not align with story. Imagine the transparency and growth that will come when story is adopted beyond the field of communications! We are really just beginning to understand the significance—and power—of story.

Connect Social Influencers

This has to do with individuals buying into the decisions of others *with discernment*. Influencers are individuals who are recognized experts in a particular niche or domain of knowledge—and/or inspirational leaders who use positive or negative emotions, producing a resonance to attract followers—who actively and regularly engage their followers mentally and emotionally. Whether earned or not, influencers develop trust within their followers. There is a common focus of interest, an openness in this context, and a continuous flow of information, and, over time, an interdependency develops. When the relationship between an influencer and their followers is based on truth and openness, this supports growth and learning. When the relationship is based on propaganda, misinformation and disinformation—usually inclusive of fear—a deception occurs which precludes growth and learning, which results in decay.

In the summer of 2016, Ray-Ban, a global leader in the premium eyewear market, decided that Act Six of their *#It Takes Courage* campaign was "the courage to ditch the screen and truly connect in the real world."[724] Ray-Ban asked seven people from around the world who were online influencers and digital content creators, with millions of followers on various social media, to spend 48 unplugged hours in the vicinity of Greenbank, West Virginia, a cell phone tower-free location. These influencers included a YouTuber-Twitterer from Brazil; an Instagrammer from Holland; a Fashionista/Blogger from China; an Electronic Musician from the UK; a Digital Artist from Italy; an Instagrammer from the US; and a Photographer from South Africa.

They stayed at the Mountain Quest Inn and Retreat Center in Frost, West Virginia, which provided the opportunity for the Mountain Quest Institute to do questionnaires before and after their experience as well as observe their behaviors *during* the experience. One insightful observation by an influencer was, "There's no 'social' in social media." And, as he added, "I learned that with the right attitude new and strange experiences can yield interesting reflections on self and society."

What was fascinating is that for this influencer, the "new and strange experience" was actual face-to-face interaction. They all admitted that there was an interdependency between them and their virtual followers. They find joy in being globally connected, chatting with people 24/7, and getting messages out to their audience instantaneously, with an immediate response from followers. In short, they truly enjoy influencing and being an inspiration to others. They also acknowledged learning from their followers, and the excitement of completing idea "sparks" and linking disparate ideas across cultures and countries.

In the knowledge management field, communities of practice and interest emerged as a web of individuals connected through a common language and set of goals, the pursuit of knowledge and understanding in a specific domain of knowledge, and a focus on issues and opportunities in that field.[725] There was an evolving agenda defined by knowledge, not tasks, with the focus on value added, mutual exchange and continuous learning. In this community of learning, an important facilitative role was that of a *champion and expert—an influencer*—someone who could keep the flow of information interesting and informative, and who could challenge others' ideas as well as provide ideas to challenge others. This role was so important that in the U.S. Department of the Navy retiring "experts" were asked to participate in community dialogues, and, even after retirement, could be reengaged in these interactions when issues arose around their area of expertise. This moves a step beyond shadowing and mentoring.

Building on these models and understanding the importance of a trusted other in social learning, communities of learners can benefit from identification of and interaction with an influencer tied to that domain of knowledge. This expert would participate fully as an influencer, taking full advantage of social media over a bounded period of time consistent with the classroom or virtual experience related to the focused domain of knowledge.

There is value in developing and embracing the role of influencer, whether you are the expert or you identify experts in your organization's primary domains of knowledge. It used to be that people would attend conferences and build their network through that venue. Today, Twitter has taken hold, Zoom is a norm, and, in a domain-focused recognized network, sharing is both open and immediate.

Experts in an individual's knowledge domain of focus (as influencers) can be built into an individual's relationship network. We now recognize that the unconscious is receiving incoming information continuously as we move through situations and among other sentient beings. This places a huge responsibility on each of us to ensure that the activities and people in our lives are those who we WANT to learn from and who can weigh in positively—physically, mentally, emotionally and spiritually—as part of our learning experience.

In Chapter 9 we introduced Relationship Network Management (RNM), which was developed by the U.S. Department of the Navy at the turn of the century concurrent with the creation and sophistication of instruments to measure brain activity such as functional magnetic resonance imaging (fMRI), the electroencephalograph (EEG, and transcranial magnetic stimulation (TMS). RNM emphasizes the importance of social interactions, and the dramatic impact—largely unconscious—of these interactions on future decision-making. As supported by the ICALS study, **physical mechanisms have developed in our brain to enable us to learn through social interactions**. With this understanding, it just makes sense that we look closely at the people who are part of our network. RNM provides an approach to do just that; and Appendix F provides a Relationship Network Management Assessment Chart in support of that process.

Create Your Personal Reality "Television"

Back when, soap operas provided emotional excitement for the suburban housewife. In the 1990's, reality television took the spotlight, offering the opportunity to break away from the everyday doldrum of a known life and step into the unscripted and emotional "real world" of another. Examples of these shows are Survivor, Idols, the Bachelor, and the Kardashians, which all became global franchises. While "reality" television often reflects deceptive situations filled with disinformation and misinformation, often actually pre-structured and sometimes honoring the negative attributes of humanity which promote conflict, they create a resonance with watchers, who can imagine themselves in these situations and reflect on how they would respond. The perception is that these events are happening to "ordinary people", and television watchers consider themselves "ordinary people". In addition to personal identification with stars, the appeal of various reality shows also includes aspects of entertainment, competition, distant (safe) participation, a break from the boredom of daily life, excitement of the different/unknown/unexpected, and a desire for expanded experiences and human insights.

So, how to bring what is good about reality television and what we are learning from neuroscience about the mind/brain together to address the needs of a shifting and unsteady learning environment? We now recognize that the experience of others can become the experience of the learner. Acknowledging the caveat voiced in Chapter 9, by creating the same neuronal patterns in your mind that are in the mind of another person, the need for cognitive thinking is bypassed and tacit knowledge can be immediately transferred.

Consider a classroom of people—whether virtual or in-person, whether young or old—as a team of learners, sharing the same context and focus of the learning experience. In the past, in-person field trips punctuated learning, and in today's environment virtual field trips offer a vast array of learning experiences. Taking these experiences one step further, what would be the advantages of having a known member of the learning team (a member of the class, perhaps along with the instructor) participate live in what becomes an interactive virtual learning experience for the rest of the team? Reflect on the "connected" individuals who have a network of supporters who pay for the privilege of "voting" on their every action! In our scenario, the class members would be fully participative in determining the focus of the learning experience AS IT IS OCCURRING.

And, of course, participants are selected through a competitive process such as writing an essay on why they would like to participate in a specific learning experience. If the learning experience involves a specific skill set, participants would be selected for their specific ability in that context. An example would be a behind-the-scenes visit at a jazz concert, where the participant(s) selected would have the opportunity to briefly play with a jazz combo, which would require the ability to play a musical instrument. The participation of one or two "reality" participants could provide "reality" learning for hundreds of virtual classmates as well as contributing to a resource library for future use.

While much of this has been done spottily and somewhat incidentally, by using our creative imagination we can perceive powerful learning opportunities in a highly reactive, context-sensitive learning environment.

Choose Intelligent Intercourse

Let's make sure we understand the terms "intelligent" and "intercourse"! While the term "intercourse" is largely used to represent sexual interaction, it means so much more, representing communication among individuals, groups, organizations and countries, *an interchange of thoughts and feelings.* Intelligent in this usage refers to "intelligent activity", which, consistent with our earlier definition, is a state of interaction where intent, purpose, direction, values and expected outcomes are clearly understood and communicated among all parties, reflecting wisdom and achieving a higher truth.[726]

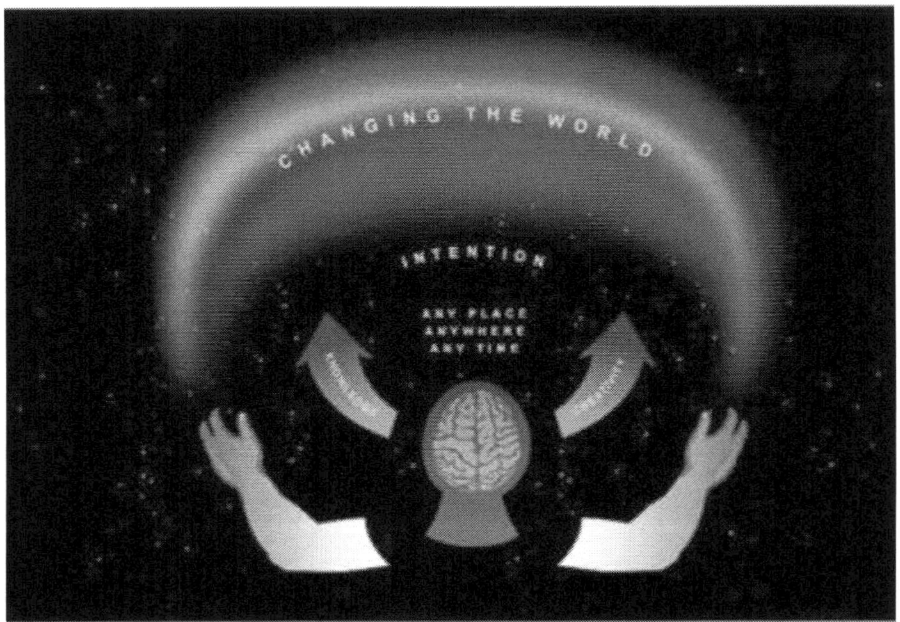

Figure 18. The power of setting intention to co-create the future.

A strong example of intelligent intercourse introduced in Chapter 9 is Shelley's model for Co-Creating Conversations that Matter.[727] The process begins with an image. For example, Figure 18 (howbeit used in color) is one of many images designed to stimulate visioning the future, not to define the path, but to engage people in exploring their perspectives, emphasizing the potential impact each individual has in this co-created journey into the future. This process occurs with the tenants of Appreciative Inquiry in mind (see Chapter 11).

People in a group, team, or organization can perceive their current position very differently. Experiences using this image (and others with a different focus) have demonstrated the power of exploring individually—and then in pairs or small groups—before engaging in a whole-group wider discussion, which maximizes the range of perspectives available through divergent thinking before convergent dialogue begins. Cycling between divergent (focusing on openly exploring and option generating) conversation and convergent interactions (focusing on merging, prioritizing and reducing ideas and concepts) optimizes the outcomes. This iterative approach has been successfully adopted in several design and problem-solving approaches such as Design Thinking.[728]

The wider the scope (diversity) of participants in the conversations, the greater the outcomes can be. For example, an early success with stimulating open conversations includes the public gamification of the design of biochemicals. Researchers have found that novice voices can add new perspectives. Some novel RNA (Ribonucleic Acid) molecules have been designed and folding patterns solved by providing a gamified environment asking online gamers to solve the "puzzle" (faster than supercomputers were able to achieve). Another example is mapping of astronomical bodies by enabling thousands of amateur astronomers to add data to the collective library. The learning happens much faster, and the extensive data generated enables astronomers to shift their focus from data gathering to data analysis, from which emerges the deep understanding and discovery.

The simple process is to first show the image or object in question, and then ASK a question. The question can be changed depending on the desired outcomes. It can be open-ended such as "Tell me what you think about this image?" or it can be somewhat leading to get a different focus such as "Where do you think our organization fits into this image and why?" The best results come when you ask each person to write down a few quick bullet points. You want them to capture their initial FEELINGS about the question (coming from the old brain) before the thinking mind (the neocortex) begins to over analyze, which can wait for the wider conversation.

Engaging intelligent rules of etiquette for dialogue, start that wider conversation. There is an amazing set of themes that come out of such conversations. As introduced earlier, some people inherently see the pessimistic side of their situation and highlight barriers to progress. And some do the opposite and talk of the positives, perhaps even over-estimating the quality of what is being done. Some see the component parts of the organization, while others take a more holistic or systems point of view. The key is to engage participants in exploring the *reasons behind the differences* to share why there are multiple perspectives. This is where the insights come from as ideas shared stimulate others to respond and new knowledge is co-created.

Combining a creative and out-of-context stimulant with a provocative question and open and inclusive facilitation generates optimal outcomes, leveraging the diversity of views of engaged participants to create new knowledge and insights, which form new options. Synergies emerge from connections between thoughts and ideas and emotions, with each component critical to the richness and success of the interaction. The key to remember about such interactions is that, as the facilitator, your aim is not to lead the participants to a predetermined outcome, but to co-create a set of options that did not previously exist, and then intermix these to generate a range of options to co-create a future that does not yet exist. This is an example of intelligent activity.

In a follow-on book titled *Becoming Adaptable*, Shelley introduces the Collaborative Conversations Spiral, which shows how participants can amplify the social aspects of being into their interactions. There are eight steps moving participants toward higher levels of understanding, with Shelley pointing out that each new conversation builds on earlier conversation and in turn leads to another conversation. These steps for successful conversations are (1) acknowledging differences; (2) respecting other's perspectives; (3) conversing to explore; (4) inquiring to learn; (5) adapting to build capabilities; (6) becoming more capable; (7) engaging in collaboration; and (8) co-creating social value.

Closing Thoughts

Humankind has engaged in so much rich learning from the outside-in, watching behaviors, recording patterns, and observing results. And now we have some of the tools to understand the *why* and *how* of those results. While to a large extent physical libraries are becoming a thing of the past, digitally there are massive amounts of data and information available at the stroke of a key, and our upcoming generations are already masters at stroking those keys!

So, we challenge you to have some fun with this, to reflect on those old favorite learning approaches through new "eyes", looking at them from the inside-out, perhaps using your imagination to create new settings and details for their expanded use. Simultaneously, with humility, open your "self" to other's ideas, expanding the learning beyond your individuated frame of reference, unleashing your mind to reach toward new beginnings.

UNLEASHING Litmus Test #4*

*After reading **Chapters 10-12**, and considering earlier chapters, reflect on each question for one minute (Reflective Observation) prior to answering silently, verbally or in writing.*

1. **AWARENESS** means something has come to your attention, it is perceived, it has been mentally engaged.
   Ask: How does reflection affect my capacity for change?

2. **UNDERSTANDING** includes your perception of the situation—the who, what, where, when, and why, and the anticipated results. As the situation becomes more complex, you need to re-create your understanding.
   Ask: In what ways does an existential learning experience impact my SELF?

3. **BELIEF** means you accept what you are aware of as true and understand it exists. Beliefs which dominate other patterns are prominent. Strong patterns are created by experiences and are closely related to emotions.
   Ask: How does humility enrich my learning opportunities with others?

4. **FEELINGS** are the foundation of learning—positive feelings make actions important to you and worthy of your efforts. Reason cannot operate without emotions.
   Ask: What is a story I tell about my past that I could tell in a more positive way?

5. **OWNERSHIP** implies a personal commitment for you to take responsibility and act.
   Ask: How can I encourage others to think about beneficial experiences they had during the Covid pandemic?

6. **EMPOWERMENT** refers to self-empowerment, that is, having the *knowledge* to make the necessary change and the *courage to act* on what you have learned.
   Ask: Based on what I've learned so far, what can I expect to take away from the final chapter and what changes will I make in my life?

*These questions are based on the Individual Change Model.[729] CHANGE comes from within, that is, unleashing your mind is YOUR choice.

Chapter 13

New Beginnings

A Nexus of Choice ... Moving Toward AGI ... Human Identify ... The Intelligent Complex Adaptive Learning System

While the neocortex is continuously engaged in predicting the future based on our personal models of the external reality, the future is largely unknown. Indeed, with ego and arrogance-driven extremists creating forces, the freelancing of biological threats, the self-aggrandizing pursuit of power and control, the disregard for morality and virtue, the blatant manipulation of others, the failures of governing systems, the devastating crushing of life, and the very real threat of nuclear weapons, what kind of a future will there be? In the midst of the mayhem of today, this question rears its ugly head again and again.

Humankind seems to wait until the last instant of breath to surge forward, coming from behind to squarely face survival, and then somehow—at least historically—discovering solutions. As the doomsday clock ticks closer and closer to midnight, will this happen now?

A Nexus of Choice

Detail does not need to be provided regarding the current state of the learning environment within which we find ourselves. We are living through it together. What we can all agree on is that our technological advances are—and have the potential to be—both demons and angels, destroyers and builders. For example, along with the expansion and opportunity offered through global connectivity comes "fake" media and cognitive hacking coupled with out-of-control social networking, all interwoven with ever-increasing fear, confusion and dysfunctional thinking as our grounding—whether connected to economics, health, politics, or the Earth's ecosystem—gives way to an uncertain future for humanity. All of this is part of a larger consciousness shift occurring in response to a troubled and unbalanced world, with each individual choosing to either open or close their learning door, to either grow or decline. Remember, we are complex adaptive systems, which cannot remain in stasis and survive.

The challenges to learning and educational institutions are immense. Yes, all the innovative ideas that have emerged relative to virtual learning and new educational approaches in our advanced technological age—necessitated by a persistent pandemic—are in play, still in many cases necessitating learning in place. Throughout this book, we've addressed skill sets necessary to resiliently navigate an unknown future, which include managing self from the inside out, knowledge capacities, presencing, managing relationships, synthesizing, planning, and otherness.

Self is the ground of learning.[730] **Knowledge capacities** are sets of ideas and ways of acting that change our reference points of perception.[731] **Presencing** is the art and practice of entering a creative space consciously and focusing that creativity such that innovation can take place.[732] **Managing relationships** in both physical and virtual settings is driven through relationship network management, seeking attuned others, and engaging heart, mind and soul in personal and professional relationships.[733] **Synthesizing**—part of the experiential learning cycle—is both simplification and explanation, as well as the ability to create a coherent whole, tying together the concept of self.[734] **Planning**—part of the existential learning cycle—is forethought, and as we expand consciousness, we expand our ability to see patterns and access future thinking.[735] Remember, prediction is a natural state of mind. And while **otherness** certainly includes a respect of and empathy for others, and the value of cooperation and collaboration, it also includes recognition that we are part of a larger whole. *There is a multiplier effect of ideas as they are shared, with all benefiting.*[736]

Only, even these skills are not enough. We now understand the humanness of learning and *the need for and power of humility*; the necessity for trusted others in the learning process; the unconscious learning (both positive and negative) continuously underway through social networking such that we are motivated to consciously manage our relationship network; the powerful opportunities offered by our mind/brain through apprehension, mirror neurons and plasticity; *the need for a spiritual counterbalance to accelerated mental development*; the continued importance of critical thinking, creativity, and innovative thinking; the need for humans to focus on quality versus quantity as AI replaces former human jobs; the existential learning opportunity bestowed through advanced consciousness; the potential offered with the development of *psych*ecological educational games, particularly with our larger understanding of knowledge through sentience, which is both sensory and mental; and—of such great importance as we honor individualism and diversity as the power of our greater whole, humanity— how to think, how to listen, how to learn and experiment through inquiry, how to emotionally engage and manage emotions, how to collaborate, and how to embrace mistakes as learning opportunities.[736]

It is clear that we are at a point of new beginnings, a nexus of choice.

Moving Toward AGI

The idea behind Artificial General Intelligence, introduced in Chapter 2, is the creation of machines that mirror certain human intelligence capabilities, that is, having the ability to rapidly learn new tasks, see the relationships among tasks, and the *flexibility* of connecting ideas and patterns from one domain of knowledge to another domain of knowledge in a CUCA environment. This is not a new idea, but has so far proven too difficult and has given way to artificial neural networks—referred to as deep learning networks—which are able to out-perform humans in specific tasks or narrow domains of knowledge. They do not, however, have the cognitive flexibility possessed by a five-year-old child who has already acquired a wide range and large amount of everyday knowledge.

Knowledge infers action, whether we reflect on the philosophical definition of "justified true belief" or a more modern rendition used by the authors as the capacity (potential or actual) to take effective action.[738] Stating a simple fact, or surface knowledge, can be easily replicated, but, as Hawkins notes, "The difficult part of knowledge is not stating a fact, but representing that fact in a useful way."[739] This problem of knowledge representation, organizing knowledge, cannot be solved using various software structures and schemas. The world is too complex. A more recent approach involves training large artificial neural networks on lots of text, what are described as language networks. However, relying on statistics and lots of data, they still fall short of the flexibility offered by the human brain.

Hawkins proposes that his "Thousand Brains Theory" solves this problem, developing universal machines with the focus placed on learning models of the world using map-like reference frames such as those formed by the neocortex.[740] This theory forwards that all the cortical columns (some 150,000 of them distributed across the neocortex), including low-level sensory regions, have the capability of learning and distinguishing complete objects. As Hawkins says, "A column that senses only a small part of an object can learn a model of the entire object by integrating its inputs over time."[741] This infers that there is minimal hierarchical structure of cortical regions, which is counter to previous understanding regarding hierarchy in the brain. This also means that a specific object would be represented by multiple models, complementary but created from different sensory input. Thus intelligence, like knowledge, is distributed with nothing stored in one place.

In 2014, the Bennets published a journal article in a special issue of JEMI focused on connecting theory and practice.[742] Theories reflect higher-order patterns, that is, not the facts themselves but rather the basic source of recognition and meaning of the broader patterns, models for acting. This is consistent with the work of Bohm,[742] who sees theories as a form of insight, a way of looking at the world, with a continuous shifting as new insights emerge through experience. We concluded that a balance needs to be found between the conscious awareness/understanding of higher-order patterns (theories) and the actions we take; between overarching theory to guide us and the experiential freedom necessary to address context-rich situations in a CUCA world. Effective action relies on a good model of how things work and how things *work together*. This symbiotic relationship between models and effective action cannot be overemphasized.[744]

From Hawkins' research, it is clear that the models formed across the cortical columns in the neocortex are self-organizing, providing the flexibility to understand and respond to the various changing contexts of the environment. Only—even if we are able to mirror this approach and achieve AGI—from whence comes the direction, the goals and values? In the human system, it is the emotions that largely direct our behaviors, often framed as desire and passion—and sometimes driven by fear—which in the human system are determined from the "old" brain through our emotional system. Which brings our focus back to the human mind/brain. Our search for artificial intelligence has prodded us to understand more fully the *amazing reality and potential of human intelligence.* Remember, individual change begins with awareness and understanding, which needs to be coupled with believing in and feeling good about the change. With those in place, our intelligence can take the lead, exploring the why, the who and the how, as we build the confidence and courage to act.

Human Identity

Kissinger, Schmidt and Huttenlocher dive into the concept of "replacement theory" in the larger context of AI. Since a large part of human identity is connected with the work people do—and AI is increasingly capable of replacing formerly human tasks and can exceed our performance in some—then what constitutes our identity as human beings? While we explored the future of work briefly in Chapter 3, this now moves us beyond our personal relationship with technology to consider our *relationship with our self* as an individuated human in an increasingly technologically advanced society. As Kissinger and his colleagues say,

> *Human identity may continue to rest on the pinnacle of animate intelligence, but human reason will cease to describe the full sweep of the intelligence that works to comprehend reality. To make sense of our place in this world, our emphasis may need to shift from the centrality of human reason to the centrality of human dignity and autonomy.*[745]

Sustaining human autonomy would mean determining the permissible and impermissible uses of AI, carving out decision-making domains that are limited to human administration and oversight. This all may happen faster than we expect, so it is time to prepare to navigate this future. "We must draw on our deepest resources—reason, faith, tradition, and technology—to adapt our relationship with reality so it remains human."[746]

Human dignity conjures up the need for a return to morality and virtue, two concepts that have declined over the years of technological advancement, or at least lost their focus in the highly competitive world of technology, which has brought with it economic dividends along with the magnetism of power and growth of arrogance.

The daily news says it all, no matter which networks you follow, or the level of truth you personally assign to those networks. Acts of destruction and death are prevalent all around us; Putin's War on Ukraine, the flood of mass shootings, the rise of hate crimes, the contagious threats of politics, the corruption of democracy. When one author was a teenager, her history class subscribed to *Pravda*, at that time a weekly publication of the Communist Party of the Soviet Union. What was remarkable was the contrast of headlines and content with American papers. *Pravda* highlighted stories about reunions, or discovered treasures, or rescue of animals, and such. Happy stories. While the local Washington, D.C. paper, while certainly including some happy events, was primarily headlined with events related to murder, theft, economic shifts, etc. Indeed, *Prava* was a propaganda vehicle for the CPSU, both *internally and externally*. Yet after observation and reflection during the year-long following by a high school class, it was agreed that "news" was *that which was unusual*, different from everyday happenings. In the U.S. at that time the tensions of the 50's and 60's seemed to be on the mend; at least it appeared that way in the suburbs. So, for the U.S. negative events were "news"— conjuring up the same excitement and curiosity as a neighbor who hopped in the car to chase fire engines. Scanning through the headlines of the internet MSNBC news shorts, and noting the many stories about animals and rescues, what does this say about the U.S. of today?

This shift has occurred over time, creeping into economic, political, governing and educational systems around the world. The warnings were there, signs of a higher valuing of wealth and power accompanied by a waning focus on people. And then it seemed better, only there was a segment of humanity that continued to listen to the old brain, that continued to be driven by the desire for power, and by fear and need. How do we regain the morality and virtue which largely lay dormant within the human?

Through continuing research in neuroscience, we are expanding our understanding of the potential power of the new brain, the neocortex, as the organ of intelligence, capable of breaking through the old brain's intuitions and automatic responses when attended by the conscious mind. Much resides in the unconscious, with the brain seeking to conserve energy, pursuing the path of least resistance. Thus, humans develop subpersonalities (subsystems) that automatically repeat historic preferences or successes to recurring life situations. The only way to shift these responses is to recognize them intellectually, attending to them and choosing to think and act differently. Thus, *it is the growth of intelligence coupled with the expansion of consciousness that offers the key to the enlightenment of humanity*, the potential to create a world beyond that which we currently perceive as surrounding us.

Regardless of historic patterns culturally, educationally, and environmentally embedded in the life path, and the emotions tied to these patterns, awareness and attention start the journey of intellectual development. We must attend to our thoughts and actions to make conscious change. And here is where we need to ask the difficult questions. Who are we? What future do we choose? What are my highest personal goals? What are the values I purport, and are my thoughts and actions consistent with those values? What contribution can I personally make to the larger consciousness field, to a new Golden Age that rises up for all humankind? Am I ready to fulfill my full potential as an Intelligent Complex Adaptive Learning System for such an unprecedented age of human endeavor?

The Intelligent Complex Adaptive Learning System

Neuroscience research continues to move forward at an amazing pace. When we originally published the ICALS theory in 2018, we noted the billion-dollar HBP Human Brain Project,[747] collaborative research efforts based in Europe across 26 countries with 135 partner institutes. As HBP is currently organized, it is one of the world's largest research projects with over 500 scientists and engineers. It was launched in 2013 as a ten-year project and during its final phase it is concentrating research on brain networks, their role in consciousness, and artificial neural networks, providing the strategy and infrastructure design for future brain research. In the United States, the Brain Research through Advancing Innovative Neurotechnologies (BRAIN)[748] at the National Institute of Health (NIH) began as a Presidential initiative, part of the early global research engagement focused on understanding the human brain as a model for machine learning and artificial intelligence. Research emerging from that project, while still underway, is revolutionizing our understanding of brain circuits as well as forcing us to consider the ethical implications of ongoing research.

It is increasingly challenging to stay abreast of the volume and diversity of neuroscience research as it has expanded to broader brain analysis and the inclusion of additional sciences and disciplines. *The more we know, the more there is to know, and the more we want to know*. For example, currently the U.S. National Library of Medicine lists over 600 neurological diseases[749] affecting more than 10 percent of our U.S. population. We anticipate that future treatments will include advanced medical procedures for conditions that will involve new specialized learning processes, as well as new psychotherapeutic techniques based on innovative learning interventions. We can also expect that the research will yield pharmaceutical discoveries enabling rehabilitative and enhanced learning results for specific disabilities. A rapidly growing indicator of the progress being made in researching the human brain is the 10th edition of *Neuroanatomy Atlas in Clinical Context: Structures, Sections, Systems, and Syndromes*[750] with over 300 pages of maps, illustrations and descriptions. Concurrent mapping documentation for the cerebral cortex has expanded in this past decade from 160 regions in the left and right hemispheres to a new total of 360 regions. Our hope is that research in the coming decade will yield supplemental maps, even an atlas, of the learning processes in the human brain.

In this book, we've attempted to coalesce ICALS with a sprinkling of new theory that is just beginning to spread across the neuroscience-driven learning community. A large contribution to the field of experiential learning, interjected throughout this work, is the theory evolving from Numenta. This theory forwards a new understanding of intelligence and is clearly voiced by Jeff Hawkins in his book *A Thousand Brains: A New Theory of Intelligence*. We also deeply honor the insights emerging from the research of Mary Helen Immordino-Yang which can be found in her book *Emotions, Learning, and the Brain: Exploring the Educational Implications of Affective Neuroscience*.

Recall that the "old brain" is the seat of emotion, that emotion influences all incoming information, and that emotions can increase or decrease neuronal activity. To a large extent, with all our new understanding, we are still a slave to the emotions and feelings driven by the old brain, still responding to the innate drive for survival, pursuit of pleasure, and avoidance of pain. Much of this is unconscious, with emotions such as fear dominating our thoughts and actions. As advocated throughout this text, it is only when we choose to unleash our minds, attending to these drives through conscious awareness and an openness to intellectual exploration, that we can release patterns of the past—our own experiential past as well as the human past that sponsored our survival. Through conscious awareness and attention, we have choice.

ICALS presents a new and expansive theory and model of human experiential learning based on what we have learned about our mind/brain, which is vastly more systemic and comprehensive than previously known. This is especially true in terms of the influence of our environment, including the environment of the natural world, and the developmental potential of all in the human family. Because of ICALS's completeness of theory—built on the minds of John Dewey, Kurt Lewin, Jean Piaget, David Kolb and J. E. Zull—biologist James Grier Miller,[751] author of the landmark work *Living Systems,* if he were alive, would be excited about ICALS. Not only does this theory place the human species in a central and emergent role in living systems, it presents the human self as having the unique capacity for highly purposeful learning, which will increase stability and evolution across the full range of living systems.

Since there is comprehensiveness in ICALS fundamental constructs, it poses critical questions at the core: Why is there functionality in our human brain, like memory and neuron capacities, that exceeds our current human needs? For example, the calculated memory storage capacity may be as high as one petabyte, ten times larger than previously measured. Moreover, where does the phenomenon of accelerated learning ability and gifted thinking that some humans demonstrate come from? What is the human evolutionary process explanation for these capacities that are beyond normal adaptation explanations?

A related area of questioning concerns the natural world around us. What is our capacity to learn from the natural world? More precisely, is there intelligence in nature that may significantly support knowledge and intelligence in humans? Using transreal numbers, the idea of sentient knowledge would support this. And artificial intelligence is seeking linkages with our brain: What about increased augmentation with the natural world?

These are just a few questions that come to mind; and as we continue to learn about and observe the power of social learning—reflecting on its role in our emerging reality—no doubt there are amazing opportunities to unfold. However, even in this new setting, Kolb's specific wording descriptive of experiential learning is still relevant, that is, the "transformation of experience" is very much what is occurring; howbeit with our new understanding, it is expanded to include *the transformation of others' experience* as well as our own.

When we step back and weigh our experience with the ICALS Quest, three things rise to the top. Before we share those here, we would invite you to be sure to read the Afterword. There, we have a closer view of our experience. Back to the three. First, the breakthrough scientific findings that change the horizon of human experience. Second, the quantum leap in the creative and innovative capacity of the ICALS Consilience Framework—this is not merely measurable progress into knowing the human brain/mind-heart/soul—this is a state change in scientific procedure and exploration. Third, there is new profound hope in what is possible for our species.

We are humans. We are complex adaptive systems with the potential of expanding our intelligence beyond experience and, if we choose, unleashing our minds from the drivers of the old brain and our self-imposed limitations, individually and collectively developing ever changing models of a new reality. Being human need no longer be seen as a limitation; we are not destined by the perceived boundaries imposed through our environment. As we explore our human potential, there just appear to be more and more possibilities, with human capabilities and capacities beyond our perceived needs such that the human can now be expressed in infinite terms! This perception of human capacity is echoed in the new and profound discoveries of neuroscientists, in the vigilance and brilliance of those who seek the finest pedagogy, and in the lives of millions who yearn to expand their self through learning, knowing and acting.

For us, we take solace in the affirmation of William Faulkner proclaimed in his 1950 Nobel Banquet speech after receiving the Nobel Prize for Literature. "I believe that man will not merely endure: he will prevail. He is immortal, not because he alone among creatures has an inexhaustible voice, but because he has a soul, a spirit capable of compassion and sacrifice and endurance."[752]

To that end we endeavor through unleashing the human mind.

AFTERWORD
The ICALS Consilience Framework
A Life, A Journey, A Quest

As we marvel at the breadth and depth of this work, we simultaneously recognize that it only breaks the surface of what it is to be human. And we look back and wonder how this book came into being.

David Bennet: *I stand in the middle of the Mountain Quest library, surveying the shelves and shelves of books, many thousands, knowing that each one represents at least one person's life focus, and a large number have many more authors participating. Each person is so unique, and through individuation has so much to add to the whole. And I can't help it. Some books I pull down to read for a third time, underlining and notating, each time learning something new.*

Was this an ICALS moment? Was this the moment when David reached deep and asked: *What is the nature of our Intelligent Complex Adaptive Learning System?* How does the human mind/brain learn and think to enable all this to be created? What David saw that day he had seen and felt a myriad of times over decades of his quest for knowledge, consciousness, and meaning, a questing that had taken him from nuclear physics to organizational systems and back again endless times—only now delving into the quantum realms. The library sitting at the foot of the mountain raised centuries of knowledge and wisdom to a higher accessible level, enabling inquiry of the human ability to conceive and create in all reaches of human experience.

David knew—as he had always known—that the human brain was not only *central* to the human experience, it was the most complex organism in our existence. He wanted to know how it worked, how we learn and how we think. And at some point it came to him, not loudly, but clearly. Then it began to raise its voice. The answer was *not* buried here or there, partly hidden over in this corner and partly hidden over there. It was resident in a kaleidoscope of insights, represented in dozens of shelves and a multitude of disciplines, sciences, and fields of inquiry. Alas, *the vantage point was consilience*, the aggregation point of synergy across seemingly disparate, but causal relationships.

Soon emerged the consilience fit between learning and neuroscience. Close behind came *epistemology*, the study of information and knowledge, and then *constructivism*, the search for meaning through experience, and a more current discipline, *complex adaptive systems*, pointing to the nature of the mind-heart/brain-soul. With these four constructs the framework for the ICALS research was conceived. This was a breakthrough! While the path for exploration was now broad enough to be inclusive, it was also focused enough to facilitate discovery of advanced levels of the Intelligent Complex Adaptive Learning System.

Looking at the human learning process as a complex adaptive learning system and using specific definitions of information and knowledge, neuroscience findings emerging after the turn of the century—when technology enabled exploring the workings of the brain from the inside out—were conceptually related to the accepted modes of the behavioral experiential learning model emerging in the 1980's. And then the discovery of learning from the inside out began. The four foundational constructs essential to this work—consilience, epistemology, constructivism and complex adaptive systems—are briefly shared below.

Consilience

Consilience means a "jumping together" of knowledge through the dynamic linking of facts and fact-based theories across disciplines to create a common groundwork of explanation.[753] The ICALS research was concerned with any disciplines that played significant roles in adult experiential learning. These include education, adult learning, cognitive psychology, sociology, biology, systems and complexity, and neuroscience. Each of these looks at experiential learning from a different perspective, each has a contribution to make, and each has its own way of solving problems and validating findings. In addition, each discipline may hold a specific, primary frame of reference. Although individuals in these fields may be studying the same general phenomena, their frame of reference will influence what is considered valid data or acceptable conclusions. Further, as systems become more complex, their behavior and characteristics change, requiring different (or multiple) approaches to understanding. [754] Thus, the emergence of a number of hybrid disciplines over the past several decades, such as evolutionary psychology, cognitive psychology, and neurobiology.[755] this trend is important to the ICALS study since neuroscience findings fall out of a number of different disciplines.

When applying these learnings to managing self, it is critical to address the human condition, meaning, and human values along with other fields such as knowledge, change, physics, decision-making, leadership and management, strategy and planning, social sciences, human an organizational development, linguistics, language arts, sustainability, consciousness and spirituality—basically, the humanities (and more) come into play. While clearly necessary, integrating these disciplines can be challenging, particularly when one discipline leans towards the "hard" sciences (neuroscience) and others lean toward the "soft" sciences (psychology, education), and other fields fall outside the parameters of science as we understand it (history, spirituality). In addition, each field of study may have a different ontology as well as epistemology while simultaneously likely to have an impact on the other.

Today we are finally discovering the reality of what C. P. Snow wrote about the two cultures of science and humanism as early as 1959, that is, the importance of their interdependence. As he believed, "Nothing fundamental separates the course of human history from the course of physical history, whether in the stars or in organic diversity."[756] The issues we face today demand that we use all of our resources—all of our understanding across a myriad of fields of learning—to produce a clearer view of reality.

Epistemology

This field of philosophy is often referred to as the study of knowledge, which is understandable since the word is a combination of the Greek words *epistēmē*, which means knowledge, and *logos*, which means reason. Knowledge requires information, so it is important to explore the nature of information in order to pursue the nature of knowledge. To understand **information**, we offer the wisdom of theoretical biologist Tom Stonier:

> Specifically, a system may be said to contain information if such a system exhibits organization. That is, just as mass is a reflection of a system containing matter, and heat is a reflection of a system containing energy, so is organization a reflection of a system containing information.[757]

By organization, Stonier means *the existence of non-random patterns*. From this perspective, information is a fundamental property of the universe and takes equal status with energy and matter.[758]

The interpretation of information offers three advantages. First, it recognizes the foundational nature of the concept and, as such, information becomes a part of all life and the physical world as we understand it at this time. Second, information is precisely defined such that the definition circumvents the confusion of multiple interpretations. Third, this definition should be applicable to both the natural sciences and the humanities, at least to the extent that where interpretations contradict each other they can be recognized and acknowledged as personal opinion.

Knowledge has been defined earlier in this book as the capacity (potential or actual) to take effective action, which is consistent with both the philosophers of old ("justified true belief") and leaders in the field of Knowledge Management. More specifically, deep knowledge can be considered as having the following attributes: understanding, meaning, insight, creativity, judgment, and the ability to anticipate the outcomes of actions. Putting this together, this means that deep knowledge is the capacity to understand situations, recognize their meaning and implications, identify underlying problems (versus symptoms), create solutions, make decisions, and implement effective actions, all of which is necessary in managing self.

An advantage of these definitions of information and knowledge is that their meaning is the same within the mind/brain and in social communication. While knowledge has an information component called Knowledge (Informing), it also has a process component called Knowledge (Proceeding) that consists of the capacity to put information together in different ways by recalling, selecting, combining, and integrating internal and external information to create and implement effective actions.[759] Thus, the necessary skills of Synthesizing and Planning for navigating the internal and external landscapes.

Constructivism

Constructivism is very much about the individual search for meaning. As Edelman and Tononi pointed out at the turn of the century,

> A biological observation that is also connected to the evolutionary assumption is that during learning and in many matters of human comprehension, doing generally precedes understanding.[760]

As we have learned in this book, our bodies sense incoming signals from the external environment and then process, associate, assimilate, and mix them with internal neuronal patterns to make "sense" of our external reality (Associative Patterning). Neurologically we create internal patterns that "re-present" our external world (in Cortical Columns across the Neocortex). This process of making sense of the world is Constructivism or the Constructivist orientation. It is how people build understanding and construct meaning from their experiences whether individual or social.[761]

Constructivists understand that knowledge is constructed by the mind and not by procedures.[762] "They see the knower as an integral part of the known." (p. 90). Thus, constructivists—constructing meaning from experience—learn from a whole range of approaches that externally may include listening, concrete experience, dialogue, and social networking, and internally from their own intuition, feelings, and insights.

Note that as with any field, constructionists differ as to the nature of reality, the role of experience, what knowledge is of interest, and whether the process of meaning-making is primarily individual or social.[763]

However, a large number of researchers agree—as do we—that the learning process is one of constructing meaning from experience.[764] This would include both the individual and the social construction of knowledge. Its manifestation in adult learning would include experiential learning, transformational learning, communities of practice, situated learning, and reflective practice.[765] However, that said, we also recognize that sometimes experiences produce learning and sometimes thoughts (from life experiences) produce learning. Thus, the newly recognized role of existential learning discussed in Chapter 10 and Appendix G.

Complex Adaptive Systems

Systems thinking is a conceptual framework, or body of knowledge and tools, that has been developed over the past 50 years to explore the structure of systems and their patterns of change to enable a better understanding of their behavior and to solve problems more effectively. What may not be recognized is that Miller's early work in the late 1970s was based on an *extensive analysis of living systems* in terms of the overall systems perspective, hierarchies, interfaces, and structures.[766] Better known is Senge's seminal book The Fifth Discipline that introduced Systems Thinking to visually and qualitatively understand how system elements interact and affect each other.[767]

In 1984 the Santa Fe Institute was created to better understand complex systems (or complexity) and, specifically, complex adaptive systems (CAS). They took a consilience approach to their study of CAS, developing a consortium of leading researchers in such diverse fields as biology, physics, economics, and management. The Institute defined complexity as,

> The condition of the universe which is integrated and yet too rich and varied for us to understand in simple, common mechanistic or linear ways. We can understand many parts of the universe in these ways but the larger and more intricately related phenomena can only be understood by principles and patterns—not in detail. Complexity deals with emergence, innovation, learning and adaptation.[768]

The human as a complex adaptive system is the first section in the first chapter of this book. Of particular import to the focus of this Addendum are the characteristics of self-organization, non-linearity and operating at some level of perpetual disequilibrium (which denote unpredictable behavior), and emergence (properties which cannot be easily traced back to their origins). Other potential characteristics include feedback loops, time delays, tipping points, power laws, correlations, and butterfly effects.[769] Simultaneously, the system has multiple connections, relationships, and is often surprise prone. The adult experiential learning process is a self-organizing complex adaptive learning process that may carry within its behavior many of the characteristics of complexity. In other words, many of these characteristics apply to you. The mind/brain/body complex is a highly interconnected, complex adaptive biological, living system. Thus, the adult experiential learning process is viewed both internally and externally as a self-organizing adaptive learning system embedded within a complex adaptive environment.

The Model

The simple icon above represents both the reach and the dynamics of the ICALS research methodology. While the four quadrants of Consilience, Epistemology, Constructivism, and Complex Adaptive Systems represent specific choices and parameters for a deeply coalesced process, the dominance and direction of the arrows symbolize the concerted outward reach for new inquiry and discovery.

The Intelligent Complex Adaptive Learning System Consilience Framework mapped in Figure 19 is a representation of the power of this approach in development of the ICALS experiential learning model. The current listings under each quadrant are a synthesis of original ICALS focus, research sources for this updated publication, and identified break throughs that warrant serious attention.

Consilience

The Intersection of adult learning and other fields and disciplines

❏ **Biological and Behavioral Sciences**
Affective Neuroscience, Behavioral Neuroscience, Chemical Neuroanatomy, Cognitive Neuroscience, Computational Neuroscience, Molecular Genetics, Neurobiology, Neuroimaging, Genetics Physics, Sensory Physiology and Biochemistry

❏ **Social Sciences and Humanities**
Cognitive Psychology, Cybernetics, Education and Learning, Evolutionary Psychology, Neuropsychology, Sociology

❏ **Engineering and Mathematics**

❏ **Theory**
ICALS, Quantum Field Theory, Thousand Brain Theory

Complex Adaptive Systems

The nature of the human

❏ **Living Systems**
Systems Theory, Complexity Theory

❏ **Holistic Human**
Body, Mind, Emotions, Soul; Heart-Mind Coherence; the Brain-Heart/Mind-Soul Continuum

❏ **Inner and Outer Self**
Consciousness, Higher Order Resonance

❏ **Social Engagement**
Affective Attunement, Human Social Constructs, Living Social Networks

❏ **Connections**
Brainscapes, Mapping of the Brain, Global Workspace Theory, *Psych*ecology Learning Games

Constructivism

The search for meaning through experience

❏ **Learning**
Experiential Learning; Existential Learning; Spiritual Learning; Theories of Dewey, Lewin, Piaget, Kolb, Zull, & Bennet

❏ **Self**
Development of Self, Change Theory, Humanism, Perception, the Intelligent Social Change Journey (ISCJ)

Epistemology

Information and knowledge

❏ **Theory**
Knowledge Theory, Beliefs, Information Theory, Networking Theory

❏ **Disciplines**
Data Analytics, Information Management, Knowledge Management

Figure 19. The Intelligent Complex Adaptive Learning System (ICALS) Consilience Framework.

What will ICALS Consilience yield? As the disciplines, sciences, and fields of research shift their focus, results, and discoveries, attention and participation will fluctuate and evolve. Nevertheless, it is hoped and anticipated that the ICALS Consilience Framework will serve well in advancing mind-heart/brain-soul exploration in the future, foremost concerning learning. Furthermore, we would offer that Dr. David's Bennet's breakthrough research and the follow-on research at MQI have established the basis for Human Learning to be recognized nationally as an Interdisciplinary Science. Since the ICALS Consilience genesis has broad and deep reaches, identification of a series of viable related scientific branches could be organized with core sets of principles and initial hypothesis.

It is envisioned that an ICALS foundation could be developed as a national program with a diversified advisory board to flesh out an expanded framework with evolving questions for research interests, a web-based forum identifying research projects and achievements, and grants in selected areas. While this program would continue to focus on advances in adult learning, it would share results with other scientific research such as analyzing the human brain for purposes of computer technology advancements and for advancements in the domain of brain/mind and body related health issues.

Questing

At a more pragmatic level, below is a simple exercise—well, simple to understand but perhaps challenging to devote the time and focus to—that will enable you to explore the relationships of cross disciplines to your area of focus, your area of research. The exercise is based on the mnemonic IDENTIFY. Once you get the idea of "Questing", then change parameters to meet your specific requirements in terms of timing, opportunity and desire. This is the "identify" model.

Identify the fields that touch your area of focus, assuming these fields have a contribution to make to your research. Surprisingly, you will discover that nearly every field has something to contribute. Use your "feelings" to help with your selection, and choose the optimum number of fields correlating to the strength of your desire to learn—no more, no less.

Develop field baselines in terms of core definitions and well-referenced models, which requires scanning, simplifying, and selecting *what* is important as well as developing an understanding of *why* it is important and organizing each set of selected materials into a coherent whole representing the field (synthesizing). HINT: The literature review of more current publications will have done much of this organizing for you in terms of historic learning. NOTE: Capture what you think and feel is important and no more, remembering that detail for its own sake is NOT a virtue. Do not get discouraged. Deciding what is important and ignoring what is not is an art that you will cultivate through practice.

Examine the model(s) that "feel" right to your logic sensing. When a model is "right" for you, a resonance will occur. *We are able to access that which we have mentally prepared ourselves to access.* HINT: The contents and indexes are your friends, high-level conceptual titling by the originator of the written text. Noting that no decision is made without emotional input, howbeit often unconsciously, *ask yourself:* What is it about this model that feels right? Have I discovered an important concept or relationship? How can I connect this way of thinking to my field of focus? Continue this process until you have identified several "thoughts" that appear (somewhat) relevant.

Note your thought process and anything that you have learned from your questing, making sure to capture the context of the thought in terms of location (both the specific model and the reference source). Later it can be quite difficult to go back and find a reference.

Tie it all together, looking for patterns of conceptual thought across your notes, across your learning. You may have to pause and take a step back for a short while, such that you can re-engage from a fresh point of view. Take a close look at the Knowledge Capacities in Chapter 5 and Appendix E. These may give you a few ideas in terms of shifting your frame of reference. Be committed to this. Know that there ARE connections; otherwise, *you* wouldn't have proceeded to this step. Your job is to find those connections. Another approach is to Sleep on It (see the tool of that name in Chapter 6).

Integrate, integrate, integrate! Allow your imagination to soar and your thoughts to visualize. SEE the connections, FEEL the connections, then DRAW the connections, writing descriptions and the context in support of what you draw. Keep returning to your original field of research to ground your thoughts, and in your mind's eye, SEE the neuronal connections being created in your mind. These are yours, and yours alone, and perchance something is emerging that can take you one step closer to a larger understanding.

Find the hotspots in current literature, associating these with that which is already known (associative patterning) from your baseline study. In this way you continue to refresh your knowledge, keep the human mind from becoming bored, and even begin the tickling of passion around these new fields of mental exploration.

You're there, able to integrate the entangled knowledge of multiple fields. Only, this is just the beginning. YOU are a verb, not a noun as you were taught in grammar school. And now that we understand systems, we also know that a system cannot exist in stasis, that is, it must be either growing or declining. You are a CAS. *Choose wisely.*

Admittedly, having a 40,000+ volume library aids the learning and research process. This has enabled 20 years of continuous research and writing. Inspired by Robert Turner, see: *The QUEST Where the Mountains meet the Library* by the Drs. David and Alex Bennet. The Mountain Quest library is open to—and has been visited by—researchers from around the world.

Through the years as the books in the library moved in and out of the shelves, scanned and explored, as mirrored in the open quote by David, we have often taken a few moments to reflect on the intellectuals and researchers whose thought resides in those books, sharing their passions across millions (billions?) of pages. How often has the wish been voiced that their thoughts could be discovered by the mere touch of the bindings. And, perhaps, we are moving closer to that possibility? All things are possible for the human!

In the meantime, we have a new viewpoint, and we could add that *Unleashing the Human Brain* doesn't stand alone; it stands with *Expanding the Self* and David's research. Moreover, we have consilience as a cornerstone of the brilliance of ICALS. It's not just that coalescing a range of information and knowledge was a stroke of genius, but, moreover, that David was able to illustrate and analyze the synthesis in the environment of MQI. As he solidified the ICALS consilience framework, revealing the theory and model, he created a new lens through which to explore experiential learning. What consilience showed David was a fresh perspective. Somehow, he needed to make sense out of it and that new comprehension was rooted in understanding the contributions of the disparate sources.

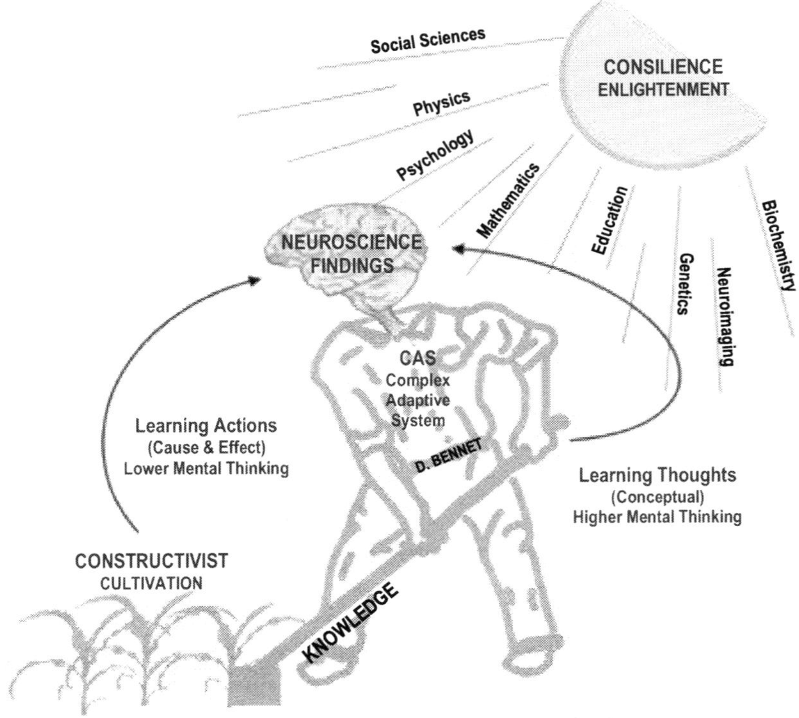

Figure 20. Consilience Enlightenment.

At the point of inception there had to be enough depth of knowledge in a wide range of areas to envision what would produce results. That range of background provided the capacity to choose what was essential for each stage of research. The early choices created enough synergy to point to breakthrough discoveries. Interestingly enough, David was so careful about the brilliance of what he saw that he reconfirmed fundamental issues such as the nature of knowledge itself.

He was clear about this consilience enlightenment approach: "Consilience is said to broaden one's perspective—and so it may. But it also confuses, challenges, and creates more questions than answers. Meaning and knowledge are uphill, icy roads that strain every step yet lead to a beautiful vision—if we can only get there. I welcome those to come, and thank all past and present researchers who travel these slippery and exciting roads. Wherever we end up, we are better for the effort."

We his co-authors are so honored to have worked with David through these experiences, and to bring this work forward for wider sharing. David is now in his older ages, recently quietly celebrating 88 years of continuous learning—reading, reflecting, and contemplating. And while a slow dementia has set in, he still shares momentary glimpses of deep thought, always triggering the learning of those who are listening. As his neurologist describes, although dementia is a form of neuronal decay, David's mental patterns are so full and so entangled from living consilience that she has no doubt death will catch up with his body long before emptying his mind. We know this to be true. And with that thought still fresh, in a moment of "now", David asks, "What would be possible if we elected a leader who could manage his Self and translate this learning into leading the country?"

Now, *that* wish has been voiced.

Appendix A

The Intelligent Social Change Journey
(ISCJ)

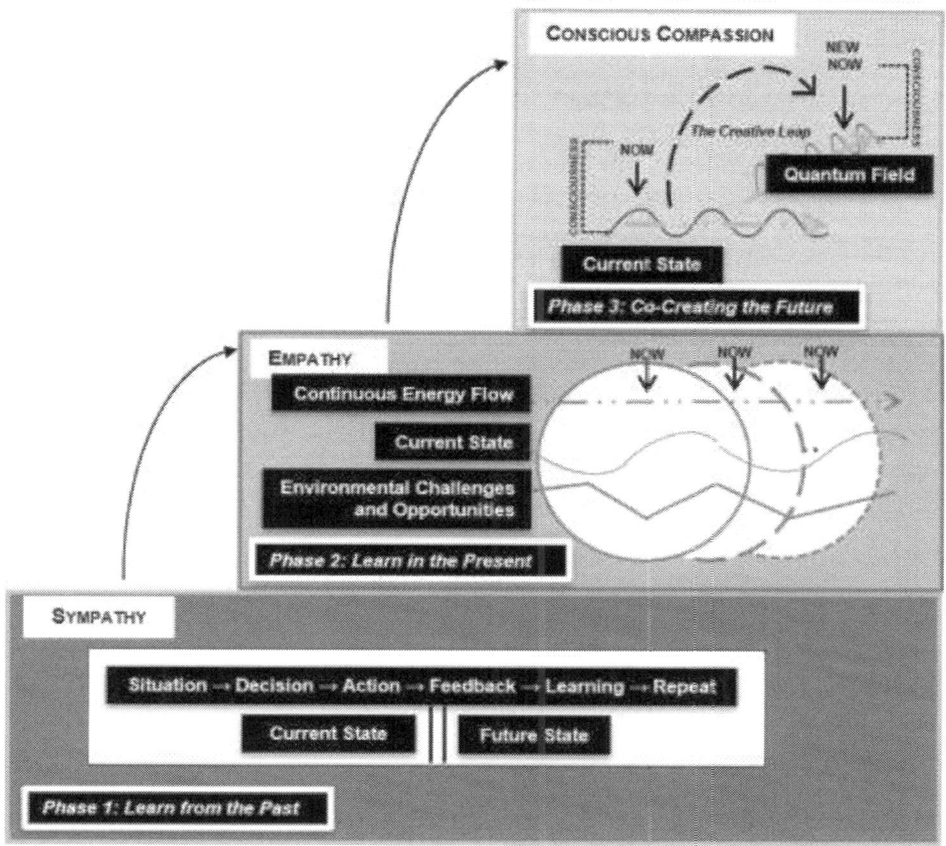

Figure 21. The ISCJ is a journey toward intelligent activity, which is a state of interaction where intent, purpose, direction, values and expected outcomes are clearly understood and communicated among all parties, reflecting wisdom and achieving a higher truth.

In Phase 1, *Learning from the Past*, the nature of knowledge is characterized as a product of the past and, as we now know, knowledge is context sensitive and situation dependent, and partial and incomplete. Reflection during this phase of change is on reviewing the interactions and feedback, and determining cause-and-effect relationships. There is an inward focus, and a questioning of decisions and actions as reflected in the questions: What did I intend? What really happened? Why were there differences? What would I do the same? What would I do differently? The cognitive shifts that are underway during this phase include: (1) recognition of the importance of feedback; (2) the ability to recognize systems and the impact of external forces; (3) recognition and location of "me" in the larger picture (building conscious awareness); and (4) pattern recognition and concept development. These reflections are critical to enabling the phase change to *co-evolving*. (See Table 3.)

Phases of the Intelligent Social Change Journey	ISCJ: Nature of Knowledge	ISCJ: Points of Reflection	ISCJ: Cognitive Shifts
PHASE 1: Cause and Effect (Requires Sympathy) • Linear, and Sequential • Repeatable • Engaging past learning • Starting from current state • Causal relationships	• A product of the past • Knowledge is context-sensitive and situation-dependent • Knowledge is partial and incomplete	• Reviewing the interactions and feedback • Determining cause-and-effect relationships; logic • Inward focus • Questioning of decisions and actions: What did I intend? What really happened? why were there differences? What would I do the same? What would I do differently?	• Recognition of the importance of feedback • Ability to recognize systems and the impact of external forces • Recognition and location of "me" in the larger picture (building conscious awareness) • Beginning pattern recognition and early concept development
PHASE 2: Co-Evolving (Requires Empathy) • Recognition of patterns • Social interaction • Co-evolving with environment through continuous learning, quick	• Engaging knowledge sharing and social learning • Engaging cooperation and collaboration • Questioning of why?	• Deeper development of conceptual thinking (higher mental thought) • Through cooperation and collaboration ability to connect the power of diversity and individuation to the larger whole • Outward focus	• The ability to recognize and apply patterns at all levels within a domain of knowledge to predict outcomes • A growing understanding of complexity • Increased connectedness of choices

response, robustness, flexibility, adaptability, alignment.	• Pursuit of truth	• Recognition of different world views and exploration of information from different perspectives • Expanded knowledge capacities	• Recognition of direction you are heading • Expanded meaning-making • Expanded ability to bisociate ideas resulting in increased creativity
PHASE 3: Creative Leap (Requires Compassion) • Creative imagination • Recognition of global Oneness • Mental in service to the intuitive • Balancing senses • Bringing together past, present and future • Knowing; Beauty; Wisdom	• Recognition that with knowledge comes responsibility • Conscious pursuit of larger truth • Knowledge selectively used as a measure of effectiveness	• Valuing of creative ideas • Asking the larger questions: How does this idea serve humanity? Are there any negative consequences? • Openness to other's ideas; questioning with humility: What if this idea is right? Are my beliefs or other mental models limiting my thought? Are hidden assumptions or feelings interfering with intelligent activity?	• A sense and knowing of Oneness • Development of both the lower (logic) and upper (conceptual) mental faculties, which work in concert with the emotional guidance system • Applies patterns across domains of knowledge for greater good • Recognition of self as a co-creator of reality • The ability to engage in intelligent activity • Developing the ability to tap into the intuitional plane at will

Table 3. The three phases of the ISCJ from the viewpoints of the nature of knowledge, points of reflection and cognitive shifts.

In Phase 2, *Learning in the Present*, the nature of knowledge is characterized in terms of expanded cooperation and collaboration, and knowledge sharing and social learning. There is also the conscious *questioning of why*, and the *pursuit of truth*. Reflection includes a deepening of conceptual thinking and, through cooperation and collaboration, the ability to connect the power of diversity and individuation to the larger whole. There is an increasing outward focus, with the recognition of different world views and the exploration of information from different perspectives, and expanded Knowledge Capacities. Cognitive shifts that are underway include: (1) the ability to recognize and apply patterns at all levels within a domain of knowledge to predict outcomes; (2) a growing understanding of complexity; (3) increased connectedness of choices, recognition of direction you are heading, and expanded meaning-making; and (4) an expanded ability to bisociate ideas resulting in increased creativity and innovation.

In Phase 3, *Co-Creating the Future*, the nature of knowledge is characterized as a recognition that with knowledge comes responsibility. There is a conscious pursuit of larger truth, and knowledge is selectively used as a measure of effectiveness. Reflection includes the valuing of creative ideas, asking the larger questions: How does this idea serve humanity? Are there any negative consequences? There is an openness to other's ideas, a questioning with humility: What if this idea is right? Are my beliefs or other mental models limiting my thought? Are hidden assumptions or feelings interfering with intelligent activity?

Cognitive shifts that are underway include: (1) a sense and knowing of Oneness; (2) development of both the lower (logic) and upper (conceptual) mental faculties, which work in concert with the emotional guidance system; (3) recognition of self as a co-creator of reality; (4) the ability to engage in intelligent activity; and (5) a developing ability to tap into the intuitional plane at will.

Appendix B
The Wisdom Model

[See Chapter 1.] The highest part of mental thought is wisdom, yet it is *more* than mental thought. Representing *completeness and wholeness of thought*, wisdom is universally a lofty consideration, and too often we sense that it eludes us. The more we seek it, the more we understand that it comes through experiencing and learning, and brings with it the desire to learn more.[753]

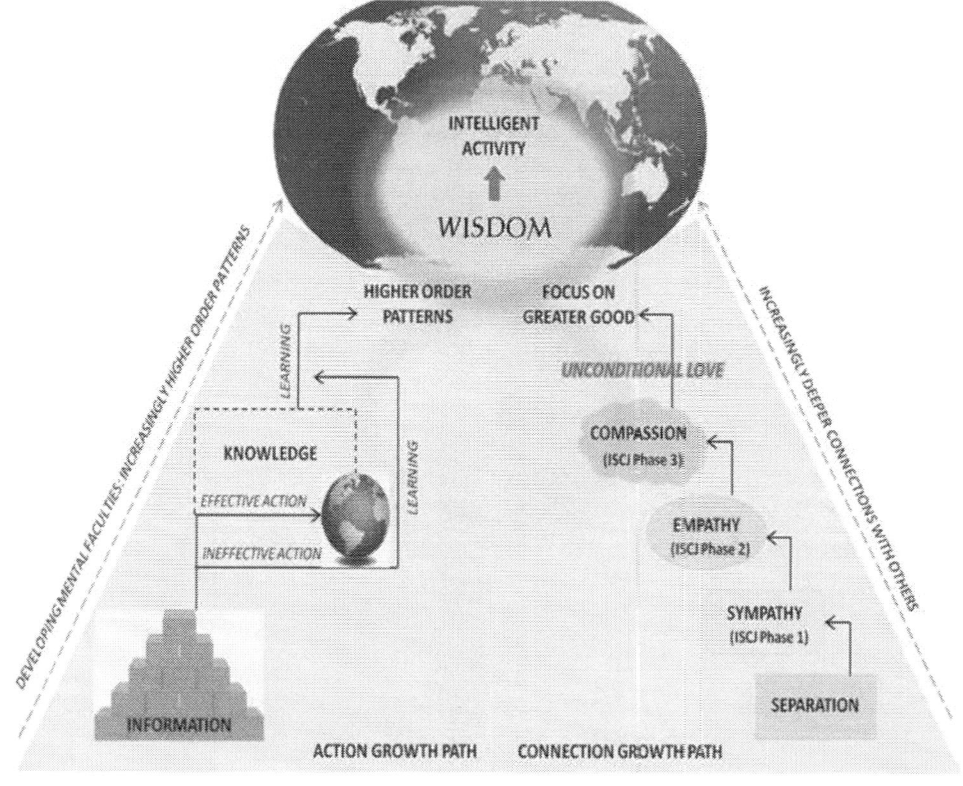

Figure 22. Wisdom reflects the completeness and wholeness of thought.

Intelligent activity is a state of interaction where intent, purpose, direction, values, and expected outcomes are clearly understood and communicated among all parties, reflecting wisdom and achieving a higher truth.

Appendix C
Assumptions the ICALS Theory is Built Upon

Assumption 1: The quantum field—which has many other names such as Noosphere, consciousness field, or God field—refers to an unlimited field of possibilities. When things within this field are heading in the same direction, they group together and create a subfield; uniquely different from the infinite field, pursuing a probability, yet pulling along related elements outside that probability. So it is with consciousness; and in this subfield, uniquely different, we become the co-creators of our Self and our life.

Assumption 2: Learning is the creation and application of knowledge, with knowledge considered as *the capacity (potential or actual) to take effective action in varied and uncertain situations.*[754] Knowledge consists of understanding, insights, meaning, intuition, creativity, judgment, and the ability to anticipate the outcome of our actions.

Assumption 3: Experience is the primary realm for human learning. Learning and knowledge are two aspects of a continuous cycle as we move through life, with learning enabling the creation of knowledge and applied knowledge (effective action) creating the feedback for continuous learning.

Assumption 4: As technology and the complex systems it produces increase at an exponential rate, the magnitude and transfer rate of information is exploding. This information explosion sets the stage for the following hypothesis: World complexity is increasing and, because of this, the continuous creation and application of knowledge and learning is essential for the future welfare of the planet. Complex systems can rarely be understood by analytical thinking or deductive reasoning alone. Therefore, deep knowledge created from effortful practice, the development of intuition and tacit knowledge through experience and continuous learning, and the recognition of and sensitivity to our inner knowing is required.

Assumption 5: Human beings and the organizations we create are complex adaptive systems. A complex adaptive system (CAS) contains many parts that interact with each other. Complex adaptive systems are partially ordered systems that unfold and evolve over time. They are largely self-organizing, learning and adaptive—thus their name. To survive and thrive, they foster and create new ideas, scan the environment, try new approaches, observe the outcomes, and change the way they operate. To continuously adapt, a CAS must operate in perpetual disequilibrium, which can result in unpredictable behavior. Having nonlinear relationships, the CAS creates global properties that are called emergent because they seem to emerge from the multitude of elements within the system and the relationships among these elements. Examples are life, ecosystems, economies, organizations and cultures.

Assumption 6: The human mind is an associative patterner that is continuously re-creating knowledge for the situation at hand. Knowledge exists in the human brain in the form of stored or expressed neural patterns that may be selected, activated, mixed and/or reflected upon through thought. Incoming information is associated with stored information. From this mixing process, new patterns are created that may represent understanding, meaning and the capacity to anticipate, to various degrees, the outcomes of potential actions. Thus, knowledge is context sensitive and situation dependent, with the mind continuously growing, restructuring, and creating increased organization (information) and knowledge for the moment at hand.

Assumption 7: We are social creatures who live in an entangled world; our brains are linked together. We are in continuous interaction with those around us, and the brain is continuously changing in response. Over the course of evolution, mechanisms have developed in our brains to enable us to learn through social interactions.

Assumption 8: The unconscious mind is multidimensional and has a vast store of tacit knowledge available to us. It has only been in the past few decades that cognitive psychology and neuroscience have begun to seriously explore unconscious mental life. Polanyi felt that tacit knowledge consists of *a range* of conceptual and sensory information and images that could be used to make sense of a situation or event.[54] We agree. The unconscious mind is incredibly powerful, on the order of more than half a million times more processing speed than the conscious stream of thought. The challenge is to make better use of our tacit knowledge through creating greater connections with the unconscious, building and expanding the resources stored in the unconscious, deepening areas of resonance, and sharing tacit resources.

Assumption 9: There are still vast workings of the human mind and its connections to higher-order energies that we do not understand. The limitations we as humans place on our capacities and capabilities are created from past reference points that have been developed primarily through the rational and logical workings of the mechanical functioning of our mind/brain, an understanding that has come through extensive intellectual effort. Yet, we now recognize that knowledge is a living form of information, tailored by our minds specifically for situations at hand (see *Assumption 4* above). The totality of knowledge can no easier be codified and stored than can our feelings, nor would it be highly beneficial to do so in a changing and uncertain environment. Thus, in this book, given the limitations of our own perceptions and understanding, we consider and explore areas and phenomena that move beyond our paradigms and beliefs regarding learning and knowledge to the larger area of knowing beyond the basic activities of our cognitive functions to consider the energy patterns within which humanity is immersed.

Appendix D
ICALS: Guidelines for Learning

1. The Infinite Potential of the Human Mind

This research identifies a number of neuroscientific findings that signify the almost unbounded limits of the human mind. While, to our understanding, no one knows the number of neurons and synapses that may be involved in any specific thought or idea, given some reasonable assumptions it is easy to show that the number of possible thoughts in the human mind becomes extremely large, larger than 10^{79}—the estimated number of particles in the Universe. The plasticity of the mind/brain, the ability of thoughts to change the physiological architecture of the brain and to impact the entire body, its potential efficacy throughout our lives, and its penchant for growth and development are clear evidence of the potential and magnificence of this mind/brain we carry with us throughout life.

The effects of a positive environment, good nutrition and health, and regular exercise also provide evidence of the high potential of all individuals. What this indicates is that parents have more influence, and therefore more responsibility, for the support and guidance of their children and their children's environment. It also indicates that adults, no matter where they start, have the tremendous possibility and opportunity to learn, grow, and develop as individual human beings. From a survival—and perhaps sustainability—perspective, taking advantage of our potential as individuals, families, organizations, and nations may be our most important contribution to the existence and advancement of the human race, and perhaps the survival of all of Earth's life forms.

The Republic of Singapore, named the most admired knowledge city in the world in both 2007 and 2008, recognized the need to restructure their complete educational system. As Wong and his colleagues describe, "The educational system was restructured in the last few years in order to foster greater creativity and instill higher-order (i.e., analytical, creative, and systems) thinking skills amongst its school children. There is now a substantial reduction in curriculum content and student assessment in favor of team learning, problem solving and process skills acquisition."[755] These authors also note that the objective is to create a future workforce capable of advanced, continuous learning, un-learning, and relearning. As neuroscience and its related sciences continue to contribute to our knowledge of human learning, we will be able to take better and better advantage of that with which we have been gifted.

2. The Powerful Role of Beliefs

As described earlier in this study, the mind/brain learns through mixing incoming information patterns with internally existing patterns of information and knowledge, thereby creating new patterns that represent understanding and meaning (the process of associative patterning). These internal patterns are significantly influenced by the beliefs and theories created and held by the Self. Lipton makes the point that both positive and negative beliefs impact every aspect of our life, including our health.[756] Because of their power to influence our decisions and actions, beliefs have influence over our learning. A good representation of this idea is that *if an individual believes that they cannot learn, then in fact, they will not be able to learn.* For example, when social relationships and attitudes cause an individual to feel incapable of learning, they experience a self-fulfilling prophecy. On the other hand, if an individual believes they can learn, then in fact, they can and will be able to learn, moving through environmental influence to personal decisions, to beliefs, to frames of reference, to understanding, to knowledge, to action.

Thus, an individual's attitudes toward their own capability and efficacy of learning is critical to their capacity to learn. It is for this reason that we should seriously consider and question our own beliefs to see if they are aligned with our personal objectives and values. If not, then we may want to reinvestigate the basis and assumptions that drive these beliefs and expand our awareness to the possibility of beliefs more in alignment with our learning goals. See the Self-Belief Assessment tool in Chapter 6.

Another role of beliefs is influencing health. The biologist Bruce Lipton forwards that, "If you choose to see a world full of love, your body will respond by growing in health. If you choose to live in a dark world full of fear, your body's health will be compromised as you physiologically close yourself down in a protection response."[757] Referring to his own book, *The Biology of Belief*, Lipton says that the secret of life is all about learning to harness our own minds to promote and accelerate growth.

3. The Influence of the Environment

Influence can be thought of as a power to sway or affect. An individual's relationship with the environment may significantly influence the learner's stress level, self-confidence, and tentativeness or interest in any specific learning. Here is where the learner must recognize what is happening and seek to modify or change the environment to better enable successful learning. At the same time, an individual having a dialogue with another learner needs to recognize the importance of the environment relative to the efficacy of either party's learning process. Often the environmental influence is through embodied experiences without conscious awareness. Other paths to the unconscious include affective, intuitive and spiritual—all contributing to learning. [758] Placing the potential effects of the learning environment, both good and bad, next to the potential power of the human mind places the responsibility for learning and personal development directly in the hands of the adult learner, not the environment.

   Another aspect that came up in this study is the local environment within which cells exist throughout the body. Through epigenetics it becomes clear that the local environment of the cell could have significant influence on whether or not the DNA of the cell was expressed. This means that genes do not have to be destiny and that an individual should in some cases be able to influence the potential effect of their genetic heritage, and some of the influences may be passed on to their offspring. As Church says, "There is mounting evidence that invisible factors of consciousness and intention—such as our beliefs, feelings, prayers, and attitudes—play an important role in the epigenetic control of genes."[759]

4. The Responsibility of Knowledge

This guideline comes from the recognition that most societies hold individuals responsible for their actions. Knowledge has been defined as the capacity to take effective action and learning is the creation and application of knowledge. In other words, those individuals who learn and create knowledge have a duty to implement that knowledge responsibly. As Meacham posits, "The essence of wisdom…lies not in what is known but rather in the manner in which that knowledge is held and in how that knowledge is put to use".[760] Whether that knowledge relates to a small situation or to a world crisis, whoever holds that knowledge has the responsibility to act appropriately and wisely. This is where wisdom and fairness come into play. Knowledge alone has no constraints; but knowledge is not alone when it is applied in the real world. Here humans, as a part of all life, have strong responsibilities towards other life forms and the planet we live on. Thus, since learning creates knowledge, both learning and knowledge include responsibility for application, both pragmatically and ethically.

If we consider deep knowledge, that is, individuals with high expertise in a specific area, one would expect them to have an even higher responsibility in implementing such knowledge. However, with deep knowledge—or perceived deep knowledge—also comes a caution. We are not our knowledge; indeed, the ICALS is a continuous learning function of the human. Since knowledge is context sensitive and situation dependent, it is incomplete, that is, what worked in one situation may not work in another and, as the environment shifts and changes, new knowledge is continuously emerging. Thus, as a learner and part of a global network of learners we carry the additional responsibility of social engagement, to enable learning through sharing and taking action for the broader and higher good of our families, friends and civilization and all other life on the planet.

5. The Power of the Unconscious

Power means the ability or capacity to perform or act effectively, as well as the ability or official capacity to exercise control.[761] As we have seen in this study, the unconscious mind plays a significant role in experiential learning. The brain is always processing information, mostly through the unconscious. When we sleep, we often reflect on the previous day's information. Rock suggests that at night the unconscious evaluates the information that has come in that day for its relative importance and discards the unimportant information while storing what is important.[762] Kandel notes that the unconscious never lies, but it can make mistakes.[763] This is not surprising since our unconscious is a part of who we are and exists to aid in our survival. The unconscious influences our thoughts and emotions without our being fully aware of it. This may be a good thing if we watch what we take in, but as Marshall so eloquently warns,

> ***Beware of the stories you read or tell; subtly, at night. Beneath the waters of consciousness, they are altering your world.***[764]

The unconscious processing of incoming information is significantly greater than conscious processing and includes the memory and autonomous systems with the exception of part of working memory. All tacit knowledge is created and stored in the unconscious. It is the source of dreams, intuition, judgment, knowing, and much creativity. By recognizing, respecting, and working with our unconscious we can improve our capacity to learn, think, make decisions, and take effective actions. However, much more research needs to be done in understanding and explaining the operation and influence of the unconscious in learning.

There is considerable interest and dialogue with respect to the existence of the reality of free will. The relations between the conscious Self, the unconscious, and memory and emotion are likely to play a large role in our learning capacity and intuitively it makes sense that these four phenomena will play a central role in any strong theory of adult learning. The connecting link would likely be through the internal relations within a complex adaptive, self-organizing learning system.

In our book, *The Course of Knowledge: A 21st Century Theory.*[765] we present a deep treatment of tacit knowledge with a chapter dedicated to engaging tacit knowledge. For those interested in exploring the inner realms of tacit, this is a good starting point.

6. The Wisdom of Age

Expanding on our earlier treatment of wisdom, Meacham says that wisdom is not beliefs, attitudes, or sets of facts, rather wisdom is *the attitude taken by individuals toward their beliefs, values and knowledge*. Similarly, Sternberg posits that **"The essence of wisdom is not in what is known but in how that knowledge is held and put to use."**[766] According to Goldberg, age can bring along with it wisdom if the individual so chooses. For the mind/brain to maintain its functional capacity, *it must continuously be used throughout life*. Ideally, all parts of the brain should be exercised both mentally and physically. While the total neuronal population of the brain may decrease with age, some parts of the brain will continue to create new neurons. Goldberg claims that the patterns that represent significant meaning and value, those patterns referred to as wisdom, tend to remain independent of age.[767] This, of course, depends on the nature of the mental and physical life lived by the individual. Thus, we may conclude that wise individuals are not just lucky individuals, they are the people who *continue to learn and work to develop and apply knowledge for higher-level purposes.*

7. The Drive for Certainty

Drive means to push, propel, or press forward; to supply a motive force or power to cause to function. From an evolutionary perspective, survival can be seen as lifelong learning and a search for certainty in a changing, uncertain, and increasingly complex world. One solution to this challenge is the capacity to understand the environment and develop the capability to anticipate its immediate future. As discussed above, this is one function of learning, namely, to be able to take effective actions for survival. Theories, beliefs, and assumptions are used to build mental models or frames of reference to understand the environment. These then form a framework for new learning as well as for guiding actions. If the beliefs are consistent with the external environment, the believers are rewarded through effective actions and empowerment.

Beliefs may also serve to narrow the field of perceived possibilities that could occur in the external world and, if they are strong, absolute beliefs, they can create the potential illusion of certainty. Unfortunately, such adverse drives for certainty serve to limit learning capacity and seek to maintain the status quo—an almost impossible task in today's world. There is no risk aversion, there is only risk management. *Risk requires learning, learning creates knowledge, knowledge leads to action, and action needs wisdom. They are all connected in circular spirals that determine the quality of our lives and the existence of our planet.*

8. The Sacredness of Values

One interpretation of sacred means dedicated or devoted exclusively to a single use, purpose, or person.[768] Although this research does not deal directly with values, they are inherent in all learning and knowledge.[769] Two of the above guidelines relate responsibility and wisdom to learning. Recall the set of relationships: beliefs influence thinking, which influences understanding, which creates knowledge, which directs actions, which impacts the environment. Actions are related to the environment and two other guidelines immediately call up responsibility and wisdom, which in turn are closely related to values. Wisdom suggests what the objective should be, knowledge says what needs to be done to achieve the objective, and values provide guidance on how it should be achieved and what should not be done, which in a sense leads back to wisdom.

One surprise during this research was the central role that learning and the learning process play in a great many aspects of our lives. We cannot escape the role of—and benefits of—learning in our lives, yet we often ignore it. The questions are: What shall we learn? How should we learn? What can we do with our learning? And what are our responsibilities if we have knowledge? Learning is too important to the future of the world to be left to teachers, schools, industry or governments. It must become the responsibility and the activity of every individual!

9. "The Paradox"

Einstein reminds us that everything should be as simple as possible and no simpler. Mountain Quest Institute, the research location for the authors, posits that *before you can simplify something, you had better understand its complexity.*[770] Yet the world continues to change and increase in complexity at a faster and faster rate. So how do we understand the world and its complexity well enough to be able to improve our knowledge and effectiveness of learning? And how can we learn faster and better, thereby keeping up with our world and its increasing complexity? This is the paradox and the challenge of the future! The simplicity is in the patterns—with learning, knowledge, responsibility, values, and wisdom our most effective solution. Seeking to embrace complexity by creating deep knowledge and maximizing our own autobiographies, meta-learning, and learning processes such as collaboration and consilience as part of our growth and development will help us catch up with exponentially exploding complexity while always keeping in mind the need and importance of wisdom.

Appendix E
Examples of Knowledge Capacities

Learning How to Learn (perceiving and representing)

Every individual is unique. Each person has a unique DNA, unique early development history, and adult life experiences and challenges different from all other humans. This uniqueness means that each of us learns differently and, to maximize that learning, we must understand ourselves, how we think and feel about specific subjects and situations, and how we best learn. For example, people who are more visual learners prefer learning through books, movies or databases; those who are more auditory prefer learning through storytelling and dialogue; those who are kinesthetic prefer learning through hands-on approaches such as role-playing.

A first step is to observe ourselves as we learn and assess our efficacy in different learning situations, noting what works well and what doesn't work well. We can also try adding different techniques that aid learning such as journaling, creating songs and stories, or asking others (and ourselves) key questions, then trying to answer those questions, recognizing the importance of emotions and repetitiveness in remembering and understanding. For skills that require body movements, then similar body movements must be included in the learning process. For skills that require mental agility, then mental games or simulations might be involved. In other words, the best way to learn is to understand your preferences and ensure that the learning process is consistent with the skill or knowledge you want to learn.

Undoubtedly, *the most important factor in learning is the desire to learn*, to understand the meaning, ramifications and potential impact of ideas, situations or events. In the present and future CUCA world, learning—that is, the creation and application of knowledge, the capacity to take effective action—is no longer just an advantage. It is a necessity. Because of individual uniqueness, each knowledge worker must learn how they learn best.

There is a relationship between your own learning style preferences and the way you share. Effective facilitation and communication require tailoring learning techniques to the preferred learning styles of your target audience. Applying multiple learning and communication styles enables you to reach target audiences with multiple preferences. Further, exposing multiple learning styles to the larger audience helps expand individual learning capacities, enriching the learning experience.

Shifting Frames of Reference (looking and seeing)

When we find ourselves in confusing situations, ambiguities or paradoxes where we don't know or understand what's happening, it is wise to recognize the limited mental capacity of a single viewpoint or frame of reference. Confusion, paradoxes and riddles are not made by external reality or the situation; they are created by our own limitations in thinking, language and perspective or viewpoint.

The patterns in the mind have strong associations built up through both experience and the developmental structure of the brain. For example, as children we learn to recognize the visual image of a "dog" and with experience associate that visual image with the word "dog". As our experience grows, we identify and learn to recognize attributes of the visual image of "dog" such as large, small, black, brown, head, tail, poodle, Akita, etc. The way we store those in the brain are as associations with the pattern known as "dog" to us, perhaps connected to the particular characteristics of a beloved childhood pet. Thus, when we think of a dog, we immediately associate other attributes to that thought.

Shifting Frames of Reference is the ability to see/perceive situations and their context through different lenses; for example, understanding an organization from the viewpoints of its executives, workforce, customers, etc. The ability to shift frames of reference is enhanced by a diversity of experiences available to networked and interactive knowledge workers. Individuals who are subjected to a wide range of ideas and perspectives through social media are going to be much more attuned to differences, while at the same time becoming engaged through dialogue. This participation with lots of people and interaction with differences helps develop a healthy self-image, and comfortable connections with different situations and people that build a feeling of "capability." Through these interactions, knowledge workers are actively doing things which in and of themselves demonstrate their capability of interacting with the world. Through this broad set of reference experiences individuals can identify those disciplines or dimensions that they are excited about, and capable and competent to develop and grow from. This process can result in better decisions and choices that match their personal needs.

Frames of reference can be both expanding (as introduced above), and focusing and/or limiting, allowing the individual to go deeper in a bounded direction. Learning to consciously shift our frames of reference offers the self the opportunity to take a multidimensional approach in exploring the world around us. As introduced above, one approach is by looking at an issue from the viewpoint of different stakeholders. For example, if you are looking at an organization problem, you might ask the following questions: How would our customers see this problem? How would other employees see this problem? How would senior management see this problem? How would the bank see this problem? As another example, when exploring a system's issue, you might look at it from the inside out as well as the outside in, and then try to understand how you might see it differently from looking at it from the boundaries. Another example is learning to debate both sides of an issue. Still another approach is to look at an issue first as simple, then as complicated, then as complex, and then as chaotic, each yielding a different potential decision set. A unique capability that develops as the self becomes proficient at shifting frames of reference is the ability to extend our visual and auditory sensing perception capabilities by analogy to other time dimensions. For example, having the ability to "see" and "hear" some point in the future that is the result of a decision made today.

An excellent example of shifting frames of reference is the use of Dihedral Group Theory. Thought processes of entrepreneurs like Steve Jobs follow six distinct shifts in perspective which directly correspond to the six permutations of what is known in mathematics as a Dihedral (3) Group. Each of the six models changes the relationship of subject/verb/object, offering the opportunity to discover hidden connections and unique insights, giving rise to faster innovation and potentially more significant breakthroughs. This meaning-making approach also helps individuals understand their personal focus, that is, where their awareness is centered.

Mathematician Tom McCabe's legendary work on algorithm complexity has led to an even more impactful mathematical breakthrough. He has discovered a connection between mathematical group theory and consciousness, directly connecting the mathematical group Dihedral order 6 with different perspectives of our thoughts.[754]

Reversal (looking and seeing)

One of the fun ways to shift our frame of reference is Reversal, that is, the ability to see/perceive situations and their context by turning something inside out, or generally reversing the order of things, whether front and back, or top and bottom, or side to side. There are lots of ways to think about this. For example, during a big Acquisition Reform movement in the U.S. Department of the Navy, part of which was the shift to performance-based standards, there was the need to eliminate thousands of standards that had crept into various contracting vehicles over the years. Given one year and a pot of money to accomplish this task, the DON began down the same path as the other services, holding a mini-trials with each standard, one-by-one, the defendant, such that it had to be "proved" that a standard was not needed. The task was an impossible one; there was always some contractor or contracting officer who felt that each standard was absolutely essential. As the weeks went by, with maybe 5 or 6 standards eliminated out of several thousand needing to be addressed, it was clear this approach was doomed to failure. Embracing the Knowledge Capacity of Reversal, all of the standards were eliminated, and mini-trials were held for those around which contractors and contracting officers had enough energy to bring back to the board and support their reinstatement. This was a game changer; when all was said and done, a couple of hundred standards were important enough to invest the energy necessary to have them reinstated.

Comprehending Diversity (perceiving and representing)

From an internal perspective, quick responses require a diversity of responses from which to draw. Since there is not much time to effectively respond in a CUCA environment, it makes sense to explore and develop a variety of potential responses prior to their need. An example is the use of scenario building, a foresight methodology that has been well-developed and tested in government, business and education. Scenarios are a form of story that can be used to consider possible, plausible, probable and preferable outcomes. Possible outcomes (what might happen) are based on future knowledge; plausible outcomes (what could happen) are based on current knowledge; probable outcomes (what will most likely happen) are based on current trends; and preferable outcomes (what you want to happen) are based on value judgments. For a well-connected knowledge worker, building scenarios can be both fruitful and fun. When facing surprises, scenarios can help in understanding new situations or at least foster a faster response by comparing the surprise with a related scenario.

From an external perspective, Comprehending Diversity means developing a competency in identifying and comprehending a wide variety of situations. For example, if you know nothing about complexity you won't be able to differentiate a complex system from a complicated system, each of which requires different sets of decisions and actions to achieve goals.

A first step is to recognize what you are looking at: the existence of diversity, the situation, and its context. Key questions: Is it diverse? Does it have many aspects that are in play or that may come into play? A second step is to comprehend it. *Vericate*, that is, consult a trusted ally, someone who understands the systems at play. Develop knowledge about a situation to comprehend it within the context of the situation. Move through the value chain of the individual change model to develop knowledge about the diversity, that is, awareness, understanding, meaning, insight, intuition, judgment, creativity, and anticipating the outcome of your decisions and actions.

Orchestrating Drive (acting and being)

There are many wives' tales and beliefs about our personal energy. One is that we just have so much energy in a life, and we just sit down and die when it is spent. Another says the more you give away the more you have. Regardless of whether we refer to this energy as spark, subtle energy (metaphysics), prana (Hindu), chi (Chinese), libido (Freud), orgone energy (Reich), or any other of the numerous other descriptive terms, every individual possesses a life force or, as described by Henri Bergson, a French philosopher, the élan vital, a source of efficient causation and evolution in nature. What we have learned about this energy—both by observation, and confirmed more recently through neuroscience findings—is its relationship to feelings. As Candace Pert, a research professor of physiology and biophysics at Georgetown University Medical Center, asserts, "… this mysterious energy is actually the free-flow of information carried by the biochemicals of emotion, the neuropeptides and their receptors."[772]

While the expression of any strong emotion requires some energy output, the expression of negative emotions generally represents a larger expenditure of energy, and the expression of positive emotions generally represents a generator of energy. For example, consider the crowds following a close-tied football game. While all may be physically tired from the experience, those who supported the loosing team are generally depressed and drag home; those who supported the winning team are generally buoyant, and may well go out and celebrate.

By understanding—and using—the emotions as a personal guidance system and motivator, knowledge workers can orchestrate their energy output. For example, by interacting, working with, and writing about ideas that have personal resonance, a knowledge worker is generating energy while expending energy, thus extending their ability to contribute and influence.

Symbolic Representation (perceiving and representing)

Representations in terms of words and visuals are the tools of trade for facilitating common understanding. The mind/brain does not store exact replicas of past events or memories. Rather, it stores invariant representations that color the meaning or essence of incoming information.[773] There is a hierarchy of information where hierarchy represents "an order of some complexity, in which the elements are distributed along the gradient of importance."[774] This hierarchy of information is analogous to the physical design of the neocortex, "a sheet of cells the size of a dinner napkin as thick as six business cards, where the connections between various regions give the whole thing a hierarchical structure."[775] There are six layers of hierarchical patterns in the architecture of the cortex. While **only documented for the sense of vision**, it appears that the patterns at the lowest level of the cortex are fast changing and spatially specific (highly situation dependent and context sensitive) while the patterns at the highest level are slow changing and spatially invariant.[776] For example, values, theories, beliefs, and assumptions created (repeatedly) through past learning processes represent a higher level of invariant form, one that does not easily change, compared to lower-level patterns.[777]

Thus, once learned, the mind/brain can quickly associate with symbols which can represent large amounts of context yet be immediately understood and interpreted. For example, a cross or menorah carries with it all the myths it represents. "It is an outward sign of an inward belief."[778] As self, symbols are everywhere we look. Mathematics is built on hypotheses and relationships, that is, patterns, assumptions and relationships. Letters represent sounds, notes represent tones, pictures represent thoughts and beliefs, shapes of signs on the highway represent the context of rules, and so on. A whole field of endeavor, *Semiotics*, has emerged around the study of signs.

We use symbols to organize our thoughts. For example, in human face-to-face interactions it has long been recognized that non-verbals and voicing (tone, emphasis) can play a larger role in communication than the words that are exchanged. New patterns are emerging in social media that represent and convey these aspects of communication, helping provide the context and "feeling" for what is being said. In electronic communication, symbols, or emogi, are small icons used to express a concept or emotion. For example, whether on Twitter or email, ":)" immediately conveys a smiley face, so much so that when these keystrokes are entered in MSWord followed by a space, they are immediately translated into ☺. As social media has matured, these symbols have become patterns of patterns, well understood by practicing social networkers and quickly conveying the message they are sending.

Appendix F
Relationship Network Management Assessment Chart

A sample chart is provided on the next page with the following columns:

a. Name and Relationship
b. Length of Relationship
c. Related Expertise and Knowledge
d. Access
e. Willingness to Share
f. Follow-through
g. Your Feelings
h. **Your** Contribution
NOTES AND ACTIONS

A process for using the chart:

(1) On a separate sheet of paper, list the critical knowledge and skill areas which are needed to achieve your goals and create innovative health solutions. Put that aside.

(2) On the same chart, fill out columns a and b (listing the individuals with whom you interact and the groups in which you participate).   Examples of "Relationship" would be: friend, colleague, mentor, manager, etc. Assess columns c through h in terms of a strength scale from 1-10, with 1 being weak and 10 being strong. "Your Feelings" would be rated in terms of respect, trustworthiness and ability to interact. "Contribution" refers to the level or value of knowledge you contribute to the relationship. Positive learning relationships would be those rated above the midpoint (5). Under "NOTES AND ACTIONS" write anything you think may be important to the relationship; for example, "Need to interact more often."

(3) From this simple chart, assess your gaps, that is. circle any number less than 5, and—comparing with your list of knowledge and skill areas— determine the relationships that need to be expanded or relationships that need to be added. For example, if your numbers are low and a specific individual or team is important to accomplishing your goals, then actions must be taken to build/expand that relationship and increase the assessment numbers. (Refer to the RNM key success factors.)   Add the action you plan to take under "NOTES AND ACTIONS".

a. Name and Relationship	b. Length of Relationship	c. Related Expertise and Kin	d. Access	e. Willingness to share	f. Follow-through	g. Your Feelings	h. Your contribution	NOTES AND ACTIONS

Table 4. The Relationship Network Management Assessment Chart.

Appendix G
A Story: An Author's Existential Experience

A few weeks ago, I was working in a support role with a large internationally renowned orchestra. My job was to make sure everyone had the sheet music they needed, run errands for the Maestro, and generally handle the small emergencies that regularly occur when you bring together nearly 60 professional musicians! How I enjoyed my work! During rehearsals I had a folding chair right behind the Maestro. And, over the years, many of these musicians had become personal friends. Just imagine being right in the middle of things as the orchestra recorded Mahler's and Tchaikovsky's symphonies. But it was the operas that tugged at my heart. I've heard it said that the first time someone attends the opera they either love it or hate it. For me, it was definitely the former.

The Maestro was well-respected, a recipient of the National Medal of Arts from the President, and the Gold Baton from the American Symphony Orchestra League. Little wonder that the owner of the theatre was sitting in the front row stage right grinning, thoroughly enjoying the rehearsal. The focus on this particular occasion was Bizet's Carmen; the last hour had been spent reviewing a few hot spots throughout the score. Now, the Maestro wiped his brow, stuck the red handkerchief in his pocket, and announced "No. 1. The Prelude." There was a slight scuffing of chairs, rattling of paper and a soft swish as bows were renewed.

By way of context, all the detail provided above is more of a knowing that all this was happening. And now, the essence of the experience.

The Maestro turned his head to give me a wink, smiled at the theatre owner, then, turning back to me, handed me the baton and said, "Take them through this. I'll be right back." And he was gone, heading out the back of stage right.

From the edge of my eye, I see the owner's mouth hanging open. But I hop right up on the rostrum and raise my arm. I glance quickly at the music. Allegro giocoso fortissimo. The slight smile I know is on my face expands. What fun! 2/4 time. I set the tempo with two beats, and we are off and playing. Perfect staccatos. Four measures in, a perfect trill. Repeat.

Now, piano. Flow, staccato, flow, staccato ... Pianissimo, crescendo molto. My body bounces, softly, with the rhythm, tears move down my face, coming from I know not where since I'm immersed in pure joy. I KNOW this music. I FEEL this music. And near the end we move into ¾ time Andante moderator, expressive movement weaving note-into-note-into-note, building to a final chord that is staccato fortississimo!

My body is shaking while simultaneously rooted in place. Pausa lunga. The Maestro is behind me, smiling. And then I woke up in my bed here at Mountain Quest, eyes wide open, still filled with all the joy and fullness and majesty of the experience, and all the body sensations—tears in my eyes, excitement in my heart, sweat on my brow, down to a rather sore right arm (not used to active conducting). And all the detail there, existing still in my "woken" state.

How could this be? How could an experience so real happen? I began to flick back through my life experiences. Yes, I knew and had the opportunity to work with Maurice Abravanel when he joined the Music Academy of the West for summer programs. Yes, I've sung in the chorus in Carmen productions and studied the role of Micaela, so there is no doubt that at some point I've watched and heard an orchestra play the Prelude to Carmen. Yes, I've conducted choral groups, small bands and small string groups, and even the 7th Fleet Band in Yokosuka once, so the feel of conducting is familiar. STILL, I've never conducted a large orchestra, and certainly never the Prelude to Carmen! Yet, I JUST DID THAT, perfectly, accompanied by all the activity and feelings of the moment. (Bennet, 2018, pp. 54-56)

REFLECT: Repeatedly in the storyteller's words, there is a state of "Joy". In terms of Hawkins' levels of consciousness scale (see Chapter 8), that is, at the level of 540, logarithmically 10^{540} (a number too large to write out—too large for the mind to fathom), this is 10^{340} beyond the balance point between negative and positive influence (200), and 10^{40} beyond the level of love (500). The point is, this was indeed a "lived" experience, complete with the neuronal firings of the mind, the feelings of the heart, and the beauty of belief.[755]

Appendix H
The Expanded Individual Change Model
(In the context of experiential learning)

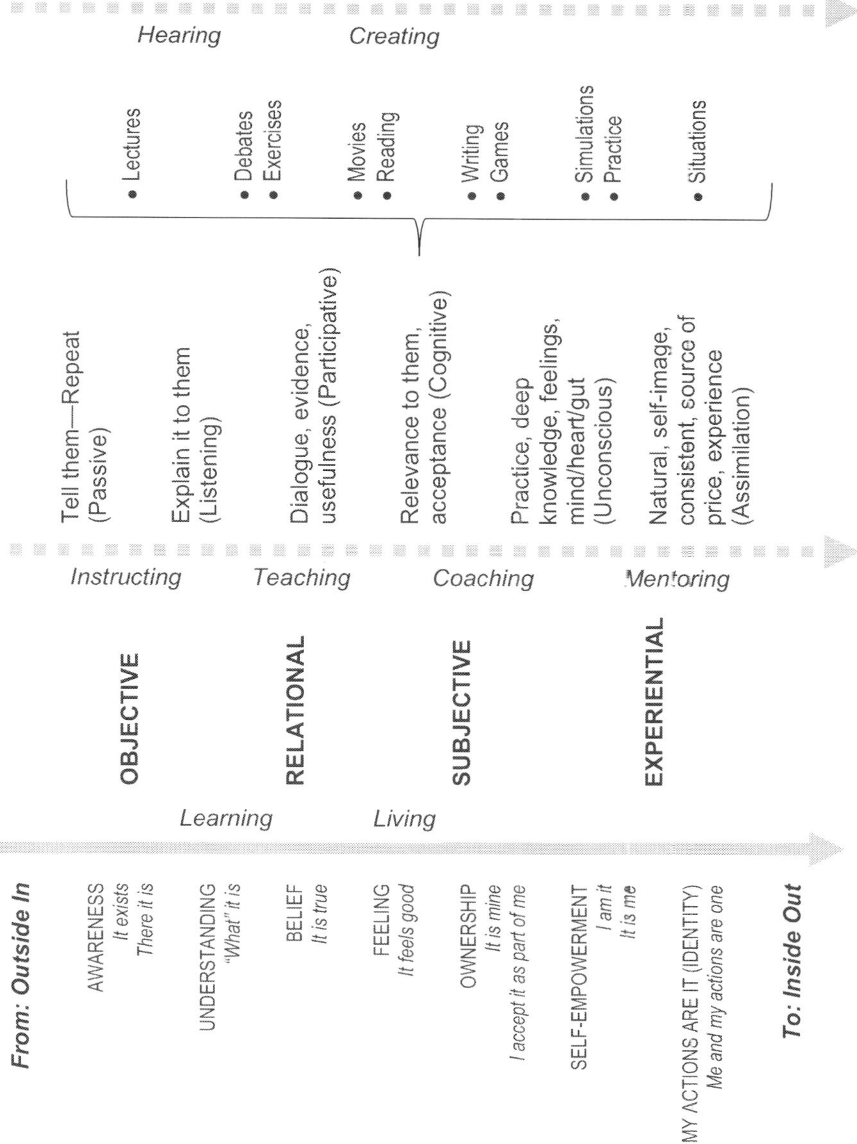

Figure 23. The Expanded Individual Change Model.

Appendix I

A Story: Living a New Frame of Reference

I was snuggled deep into the flannel sheets under several layers of soft blankets and a quilt while outside a light covering of snow reflected a bright night sky. The temperature had plummeted during the night.

In that dreamy state of well-being and warmth there was a soft tapping on my forehead. Tap. Tap. Tap. A gentle movement of my hair. Tap. Tap. Tap. I reached up to brush it off and felt the soft padding of our large tabby's left foot. Cat Walker. My hand reached to his head and rubbed down across his neck. He was not to be dissuaded. Tap. Tap. Tap.

Rolling out of the covers on automatic, I hit the cool rug with my bare feet and ambled around the corner to the cat bowl, opening a small dark blue can and responsibly dishing out a spoonful of his favorite food into his dish. Then, accordingly, a habit of age, went along the top landing for a quick bathroom trip and a drink of water. As I headed back toward the bedroom, I noted that Cat Walker had not touched the food; indeed, a second cat, Zeusi, who had been a starved stray several years ago—but was no longer—was heading toward the bowl. Woops! She was overweight and couldn't stop eating, so this was not a good idea. Where was Cat Walker? Since he was nowhere to be found, I picked up the dish, covered it, and left it in the coolness for later.

Back to bed. The sheets felt wonderful and sleep was immediate. I was on a campus; the school where I had sent my child, not a child I know in this lifetime (mine are now grown), rather a young teen, a daughter? Since I am awake in my dreams, I thought: *This must be my inner child.*

As an educator, I was concerned about the methodologies and curricula of the school. What subjects were required? What activities were available? What standard tests did they use? How did they grade these kids? What did our children know when they graduated? And traveling through my head were so many more of the everyday kind of questions we ask of our school systems. What kind of education was I paying for? Hmmm. Since emersion and experience trump all the words in the world, I decided to live with my kid at school for a while and get some answers. And since it was still night, and still cold, I was cuddled in warm flannel sheets next to my ... daughter?

Tap. Tap. Tap. There it was again. As I woke, my half-grown child had tired of sleeping and gone off to a class, some activity, I didn't know exactly what. But the covers were warm and it was cold outside, so I wrapped them around my face. Tap. Tap. Tap. On the top of my head. There was the cat again. *Oh, no, I get your game now. You aren't really hungry.* Only he stayed and insisted. *Okay, okay.* I swung out of bed and repeated my earlier routine. Again, he was not there, only the second cat, to whom this time I gave a bite of food. A quick bathroom trip, a glass of water, and back into the covers while there was still some darkness left.

Back on campus, my now mid-teen child came into the room. How did those years pass while I was sleeping? Only half awake, I asked where *she* had been. *This really interesting class on brain chemicals and emotions,* she responded. Hmmm. Didn't realize her interests lay in that direction. *Then I went over and polished a skate board I want to try out later today, during afternoon games.* Hmmm. Didn't know she liked skateboarding. They have games every afternoon? *And I stopped by a dance class before playing some numbers games with a really cute guy who likes physics.* Hmmm. Guy?! She's clearly growing up. That's a lot. Had I really been asleep that long? *Are you still tired mom? You're missing all the fun.* My eyes had closed again—so I pulled them open and looked her direction. She was leaving, again. *I'm going to go check out this class on ancient literature ... at least it sounds ancient, a really great writer named Beowolf.* Beowolf? That rang a bell. When did she become interested in literature?

I closed my eyes, but did not go back to sleep. Tap. Tap. Tap. I dragged myself from the bed. That wasn't quite as difficult as I thought it would be. I was still on the campus. Awake now, warmly dressed, I walked outside. It was a beautiful day. The dusting of snow across the campus had melted into the ground and there was laughing, playing children of all ages moving among the buildings and parks and playgrounds. I quickly followed the paths and quietly walked into the back of the common room of the administrative building. An educator facilitator was there with some older children. *Now students,* the facilitator lightly reminded, *there are some very special learning opportunities available to you this semester. Be sure and take advantage of those classes if they offer something you would like to learn. Choose well. You will be graduating in the Spring. And if there is an area that you have passion around that we have not offered, let's have a conversation and we will find additional resources for you. Meanwhile, everyone enjoy the day!*

The children were leaving; only they all seemed older now. College Age? From my point of observation, they seemed to float out the door. How do I mean that? Well, there was a buoyancy about them, a healthiness, an eagerness, and laughter. They interacted continuously, either verbally or silently sending signals to each other, signals that I seemed to be able to pick up and understand. From one, *What do I want to learn today?* From another, *I can't wait to try this creative numbers class out.* From another, *Did you see the new Asian recipe book from Thailand? I'm going to have a cooking experience today.* From another, *There's a new organizational accounting system based on feelings that I want to learn about.* And from another, *They brought in an old 1966 Pontiac GTO to work on.* Excitedly, *We're going to take it apart and put it back together.*

It was then that I saw the Head Master of the school. There was a bright aluminum object in his hand that caught my attention, about ten inches long, cylindrical. He caught my glance and smiled, walking over my direction. *Here,* handing me the object. The cool energy of it caused a quick jolt, and my eyes opened wider. Was I still asleep? No, I don't think so. I was still on campus and the Head Master was right in front of me. I rubbed my hand over the smooth wall of the cylinder and felt energy move into my hand, up my wrist, traveling up my arm. Confused, I looked up at the Head Master. He responded, *A Universal connector that helps us fill the gaps between our mind/body capabilities, the opening of our hearts, and the wonderful learning energies available to us from the Cosmos. When you were attracted to it, I knew you could feel the energies. Are you open to learning, then? It is a choice.*

When had I not been open to learning? For a moment I felt back into the flannel sheets and covered my head. Tap. Tap. Tap. *Wake up and Experience. Wake up and Be.* Then, I was again standing in front of the Head Master.

Open to learning? *Yes,* I silently answered. *Yes, yes, yes.*

Sit with it a moment, said the Master nodding at the Universal connector as he moved toward the Learning Facilitator who was waiting a few feet away, and engaged her in conversation. I sat at a small desk nearby. There was a stack of various sizes of colored paper and writing and drawing utensils atop the desk. Almost reverently, I placed the cylinder against a multi-colored page and drew a line down one side with a wide-tip pen that issued a gold leaf residue. The reflection of the gold bounced against the brightness of the cylinder, highlighting an energy glow that had moved into my awareness and was expanding. Yes, I could feel the connections. I could feel myself opening up. I could feel the energy building. Silently I repeated a favorite mantra, *I am open to receive; I am open to experiencing; I am open to learning*. Only, now I wasn't *saying* it, I was *feeling* it. *It was me*. The energy that had moved up my arm glided around my shoulder and through my chest, rhythmically engaging my heartbeat, then lightly floating upwards into my head. There was a swirling, a dancing, an awakening. Had I ever felt so alive? So hungry to learn?

Through misty eyes I saw my now-grown child come through the front door into the building. She walked up to me smiling. The words from her glowing face answered so many questions. *Hello, mom. Welcome to my world"*.[756]

REFLECT:
Do I tap into the learner I am?
What will the learning environment look like for future generations?

Consilience
Systems
ICALS
Epistemology
Constructivism

Endnotes

[1] Bennet, A., Bennet, D., Shelley, A., Bullard, T., & Lewis, J. (2020a). *The profundity and bifurcation of change part I: Laying the groundwork*. MQIPress.

[2] Ibid.

[3] Bennet, D., Bennet, A., & Turner, R. (2018). *Expanding the self: The intelligent complex adaptive learning system*. A new theory of adult learning. MQIPress.

[4] Bennet, D., & Bennet, A. (2010). Social learning from the inside out: The creation and sharing of knowledge from the mind/brain perspective. In J. Girard & J. Girard (Eds.), *Social knowledge: Using social media to know what you know* (pp. 1-23). IGI Global.

[5] *Oxford English Dictionary* (5th ed.) (2002). Oxford University Press, p. 2963.

[6] Stebbins, L. H. (2010). Development of reality system theory. *Journal of Business & Economics Research, 8*(4).

[7] Bennet, A., & Bennet, D. (2018). *Decision-making in the new reality: Complexity, knowledge and knowing*. MQIPress.

[8] Bennet, A., & Bennet, D. (2004). *Organizational survival in the new world: The intelligent complex adaptive system*. Elsevier.

[9] Johnson, S. (2001). *Emergence*. Scribner.

[10] Axelrod, R., & Cohen, M. D. (1999). *Harnessing complexity: Organizational implications of a scientific frontier*. The Free Press, p. 15.

[11] Marshall, I., & Zohar, D. (1997). *Who's afraid of Schrödinger's cat?* William Morrow and Company, Inc. Also, Prigogine, I. (1996). *The end of certainty: Time, chaos, and the new laws of nature*. The Free Press.

[12] Zohar, D., & Marshall, I. (1994). *The quantum society: Mind, physics, and the new social vision*. William Morrow and Co., Inc.

[13] Wilber, K. (2000). *Integral psychology: Consciousness, spirit, psychology, therapy*. Shambhala Publications, p. 89.

[14] Bennet, A., Bennet, D., & Avedisian, J. (2018). *The course of knowledge: A 21st century theory*. MQIPress.

[15] *Websters Encyclopedic Unabridged Dictionary of the English Language* (1996). Portland House.

[16] Wiig, K. (1993). *Knowledge management methods—Practical approaches to managing knowledge*. Schema Press.

[17] Hawkins, J. C. (2021). *A thousand brains: A new theory of intelligence*. Basic Books, p. 134.

[18] Ibid., p. 130.

[19] Ibid., p. 131-132.

[20] SQ related to Spiritual Quotient, a measure of Spiritual Intelligence, which is beyond the mental (IQ) and emotional (EQ) skills, measuring spiritual aptitude, which includes self-awareness and a level of Otherness (sensitivity to others). At some level all humans are spiritual in nature, and there is a direct positive correlation between spiritual characteristics and human learning. See Bennet, A., & Bennet, D. (2007). The knowledge and knowing of spiritual learning. *VINE: The Journal of Information and Knowledge Management Systems, 37*(1), 27-40.

[21] Zohar & Marshall, Intro.

[22] Bennet, A., Bennet, D., Shelley, A., Bullard, T., & Lewis, J. (2020a).

[23] Csikszentmihalyi, M. (1993). *The evolving self: A psychology for the third millennium*. HarperCollins Publishing, p. 22.

[24] Ibid., p. 216.

[25] Bennet et al. (2020a).

[26] Zimmer, C. (2007). The neurobiology of the self. In F. E. Bloom (Ed.), *Best of the brain from Scientific American: Mind, matter, and tomorrow's brain* (pp. 47-57). Dana Press.

[27] Csikszentmihalyi, p. 23.

[28] LeDoux, J. (1996). *The emotional brain: The mysterious underpinnings of emotional life*. Touchstone.

[29] Bennet, A., Bennet, D., Shelley, A., Bullard, T., & Lewis, J. (2020c). *The profundity and bifurcation of change volume III: Learning in the present*. MQIPress.

[30] Damasio, A. R. (1999). *The feeling of what happens: Body and emotion in the making of consciousness*. Harcourt Brace & Company.

[31] Walsh, N. D. (2009). *When everything changes change everything: In a time of turmoil, a pathway to peace*. EmNin Books.

[32] Buonomano, D. V., & Merzenich, M. M. (1998). Cortical plasticity: From synapses to maps. *Annual Review of Neuroscience, 21*, 149-186.

[33] Cozolino, L., & Sprokay, S. (2006). Neuroscience and adult learning. In S. Johnson & T. Taylor (Eds.), *The neuroscience of adult learning* (pp. 11-19). Jossey-Bass. Also, Cowan, W. M., & Kandel, E. R. (2001). A brief history of synapses and synaptic transmission. In W. C. Cowan, T. C. Sudhof & C. F. Stevens (Eds.), *Synapses*. Johns Hopkins Press. Also, Zhu, X. O., & Waite, P. M. E. (1998). Cholinergic depletion reduces plasticity of barrel field cortex. *Cerebral Cortex, 8*, 63-72.

[34] Medina, J. (2008, 2014). *Brain rules: 12 principles for surviving and thriving at work, home, and school*. Pear Press. Also, Kandel, E. R. (2006). *In search of memory: the emergence of a new science of mind*. W. W. Norton & Company.

[35] Buonomano, D. V., & Merzenich, M. M. (1998). Cortical plasticity: From synapses to maps. *Annual Review of Neuroscience, 21*, 149-186.

[36] *The American Heritage Dictionary of the English Language* (4th ed.) (2000). Houghton-Mifflin, p. 585.

[37] Immordino-Yang, M. H. (2016a). *Emotions, learning, and the brain: Exploring the educational implications of affective neuroscience*, p. 20.

[38] Pert, C. B. (1997). *Molecules of emotion: A science behind mind-body medicine*. Touchstone, p. 179.

[39] Damasio (1999), p. 42.

[40] Bennet, A., & Scott C. L. (2019). *With passion, we live and love*. MQIPress.

[41] Greenfield, S. (2000). *The private life of the brain: Emotions, consciousness, and the secret of the self*. John Wiley & Sons.

[42] Plotkin, H. (1994). *Darwin machines and the nature of knowledge*. Harvard University Press, p. 211.

[43] Mulvihill, M. K. (2003). The Catholic Church in crisis: Will transformative learning lead to social change through the uncovering of emotion? In C. A. Weisner, S. R. Meyers, N. L. Pfhal & P. J. Neaman (Eds.), *Proceedings for the 5th International Conference on Transformative Learning* (pp. 320-325). Teachers College, Columbia University, p. 322.

[44] Christos, G. (2003). *Memory and dreams: The creative human mind*. Rutgers University Press.

[45] Pert.

[46] Lipton, B. (2005). *The biology of belief: Unleashing the power of consciousness.* Hay House, p. 132.

[47] Bennet, A. (2018). *Possibilities that are YOU! volume 14: The emoting guidance system.* MQIPress.

[48] LeDoux (1996).

[49] Damasio, A. R., Grabowski, T. J., Bechara, A., Damasio, H., Ponto, L L., Parvizi, J., & Hichwa, R. D. (2000). Subcortical and cortical brain activity during the feeling of self-generated emotions. *Nature Neuroscience, 3*(10) 1049-1056.

[50] Bennet, A., Bennet, D., Shelley, A., Bullard, T., & Lewis, J. (2020b). *The profundity and bifurcation of change volume II: Learning from the past.* MQIPress.

[51] Irvine, W. B. (2006). *On desire: Why we want what we want.* Oxford University Press.

[52] Heijnen, R. D. (2022). Desire—Life as a drama based game or the thermodynamics of everything. In S. B. Schafer & A. Bennet (Eds.), *The handbook of global media's preternatural influence on global technological singularity, culture and government* (pp. 214-242). IGI Global.

[53] Shand, A. F. (1920). *The foundation of character* (2nd ed.). Macmillan, p. 519.

[54] Holton, R. (n.d.). Belief/desire psychology, the Humean theory of motivation & emotion. *Moral Psychology, 24*(120), 1.

[55] May, R. (1975). *The courage to create.* Bantam.

[56] Warner Brothers (2011). *Green Lantern.*

[57] Excerpts taken from Bennet et al. (2018c).

[58] Bennet, A., Bennet, D., Shelley, A., Bullard, T., & Lewis, J. (2020a-e). *The profundity and bifurcation of change parts 1-5.* MQIPress.

[59] Dave Snowden, Founder and Chief Scientific Officer of Cognitive Edge, in the 2014 Sampler Research Call sponsored by the Mountain Quest Institute and quoted in: Bennet, A., & Bennet, D. (2014). Knowledge, theory and practice in knowledge management: Between associative patterning and context-rich action. In P. Lambe (Ed.), *Knowledge management special issue: Connecting theory and practice, Journal of Entrepreneurship, Management and Innovation, 10*(1), 7-56.

[60] Stonier, T. (1997). *Information and meaning: An evolutionary perspective.* Springer-Verlag. Also, Stonier, T. (1992). *Beyond information: the natural history of intelligence.* Springer-Verlag. Also, Stonier. T. (1990). *Information and the internal structure of the Universe.* Springer-Verlag.

[61] Bennet, D., & Bennet, A. (2008b). Engaging tacit knowledge in support of organizational learning. *The Journal of Information and Knowledge Management Systems, 38*(1), 72-94.

[62] Bennet, A., Bennet, D., & Avedisian, J. (2018). *The course of knowledge : A 21st century theory.* MQIPess.

[63] Weyl, H. (1952). *Symmetry.* Princeton University Press.

[64] Gell-Mann, M. (1994). *The quark and the jaguar: Adventures in the simple and the complex.* W. H. Freeman and Company.

[65] McHale, J. (1977). Futures problems or problems in future studies. In H. A. Linstone & W. H. C. Simmonds (Eds.), *Futures research: New directions.* Addison-Wesley Publishing Company, Inc.

[66] Salk, J. (1973). *The survival of the wisest.* Harper & Row.

[67] Bennet, A. (2018). *Possibilities that are YOU! Volume 8: Me as co-creator.* MQIPress.

[68] Bateson, G. (1972). *Steps to an ecology of the mind.* Ballantine.

[69] Dilts, R. (2003). *From coach to awakener.* Meta Publications.

[70] McWhinney, W. (1997). *Paths of change: Strategic choices for organizations and society.* SAGE Publications, Inc.

[71] Argyris, C., & Schon, D. (1978). *Organizational learning: A theory of action perspective.* Addison-Wesley.

[72] Bateson, pp. 301-305.

[73] Berman, M. (1981). *The reenchantment of the world.* Cornell University Press, p. 346.

[74] Dunning, J. (2014). Discussion of consciousness via the internet. *Incluessence* is that state of Being far beyond the small drop of previous possibility accepted as true, far beyond that which we have known to dream.

[75] Bateson, p. 465.

[76] For a detailed treatment of the ISCJ, see Bennet, A., Bennet, D., Shelley, A., Bullard, T., & Lewis, J. (2018). *The intelligent social change journey: Foundation for the Possibilities that are YOU!* MQIPress. Also, see *The profundity and bifurcation of change* (2020a-e) five-book series by the same authors from which the *Possibilities that are YOU!* series is extracted.

[77] Macdonald, D. (1996). *Toward wisdom: finding our way to inner peace, love, and happiness.* Hampton Roads, p. 1.

[78] Clayton, V., & Birren, J. E. (1980). The development of wisdom across the lifespan: A reexamination of an ancient topic. In P. B. Baltes & O. G. J. Brim (Eds.), *Life span development and behavior* (pp. 104-135). Academic Press.

[79] Holiday, S. G., & Chandler, M. J. (1986). *Wisdom: Explorations in adult competence: Contributions to human development, Vol. 17.* Basel. Also, Erikson, J. M. (1988). *Wisdom and the senses: The way of creativity.* Norton. Also, Sternberg, R. J. (Ed.) (1990). *Wisdom: Its nature, origins and development.* Cambridge University Press. Also, Jarvis, P. (1992). *Paradoxes of learning: On becoming an individual in society.* Jossey-Bass. Also, Kramer, D. A., & Bacelar, W. T. (1994). The educated adult in today's world: Wisdom and the mature learner. In J. D. Sinnott (Ed.), *Interdisciplinary handbook of adult lifespan learning.* Greenwood Press. Also, Bennett-Woods, D. (1997). Reflections on wisdom. (Unpublished paper). University of Northern Colorado.

[80] Merriam, S. B., & Caffarella, R. S. (1999). *Learning in adulthood: A comprehensive guide* (2nd ed.). Jossey-Bass, p. 165.

[81] Erikson, J. M. (1988). *Wisdom and the senses: The way of creativity.* Norton, p. 184.

[82] Costa, J. D. (1995). *Working wisdom: The ultimate value in the new economy.* Stoddart, p. 3.

[83] Macdonald, p. 1.

[84] Bennet, A., & Bennet, D. (2004). *Organizational survival in the new world: The intelligent complex adaptive system.* Elsevier.

[85] Bennet, D., & Bennet, A. (2021). *The QUEST: Where the mountains meet the library.* MQIPress.

[86] Harvey, R. (2013). The ancient thread of authenticity. An interview on Sacred Attentional Therapy. www.sacredattentiontherapy.com/articles.html

[87] Ehrenfeld, D. (1981). *The arrogance of humanism.* Oxford University Press, p. viii.

[88] Bennet et al. (2020c).

[89] Hawkins, D. R. (1995). *Power vs force: The hidden determinants of human behavior.* Veritas.

[90] Ibid., p. 278.

[91] Ibid., p. 54.

[92] Manuel Castells (2022, February 19). *Wikipedia.* www.wikipedia.com

[93] Castells, M. (1996). *The information age: Economy, society and culture, Volume 1: The rise of the network society.* Blackwell Publishers, Ltd.

[94] Ibid. p. 469.

[95] Recounted in Sima Qian (145-ca. 89 BC/1922/1985). Confucius. In Hu Shi (Ed.), *The development of logical methods in ancient China.* Shanghai. Oriental Book Company, p. 125.

[96] Castells, p. xv.

[97] BBC Future (n.d.).

[98] Stuart, K. (2016). Has black mirror episode predicted the future of video games? *The Guardian.*

[99] Hawkins, J. C., p. 120.

[100] Horstman, J. (2010). *The Scientific American brave new brain: How neuroscience, brain-machine interfaces, neuroimaging, psychopharmacology, epigenetics, the internet, and our minds are stimulating and enhancing the future of mental power.* John Wiley & Sons, Inc. and Scientific American.

[101] Bennet, Bennet, & Avedisian.

[102] Bennet, A., & Bennet, D. (2008c). The fallacy of knowledge reuse. *Journal of Knowledge Management, 12*(5), pp. 21-33. Also, Chickering, A. W., Dalton, J. C., & Stamm, L. (2005). *Encouraging authenticity & spirituality in higher education.* Jossey-Bass. Also, Clausing, D. (1994). *Total quality development: A step-by-step guide to world-class concurrent engineering.* Asme Press. Also, National Research Council (2000). Also, Oakes, J., & Lipton, M. (1999). *Teaching to change the world.* McGraw Hill.

[103] Trapnell, P., Sinclair, L. (2012, January). Texting frequency and the moral shallowing hypothesis. Poster presented at the Annual Meeting of the Society for Personality and Social Psychology, San Diego, CA.

[104] See *All about millennials*, https://millennialtype.com/faq

[105] Buzsaki, G. (2006). *Rhythms of the brain.* Oxford University Press, p. vii.

[106] Llinas, R. R. (2001). *I of the vortex: From neurons to self.* The MIT Press.

[107] Damasio (1999), p. 298.

[108] Hawkins, J. & Blakeslee, S. (2004). *On intelligence: How a new understanding of the brain will lead to the creation of truly intelligent machines.* Times Books, p. 88.

[109] Hawkins, J. C.

[110] Ibid.

[111] Ibid., p. 97.

[112] Kolb, D. A. (1984). *Experiential learning: Experience as the source of learning and development.* Prentice-Hall.

[113] Hawkins, & Blakeslee, p. 78.

[114] Kandel, p. 298.

[115] Hawkins, J. C.

[116] Ratey, J. J. (2001). *A user's guide to the brain: Perceptions, attention, and the four theaters of the brain.* Pantheon Books.

[117] Stonier (1997).

[118] Bennet, D., & Bennet, A. (2008a). Associative patterning : The unconscious life of an organization. In. J. P. Girard (Ed.), *Organizational memory* (pp. 201-224). IGI Global.

[119] Hawkins & Blakeslee.

[120] Marton, F., & Booth, S. (1997). *Learning and awareness*. Erlbaum.

[121] Moon, J. A. (2004). *A handbook of reflective and experiential learning: Theory and practice*. Routledge-Falmer.

[122] Bennet, Bennet, & Avedisian.

[123] Avedisian, J., & Bennet, A. (2010). Values as knowledge: A new frame of reference for a new generation of knowledge workers. *On the Horizon, 18*(3), 255-265.

[124] *American Heritage Dictionary*.

[125] Tapscott, D. (2010). *Grown up digital*. McGraw Hill.

[126] Ibid., p. 266.

[127] Leyden, P., Teixeria, R., & Greenberg, E. (2007, June 20). *The progressive politics of the millennial generation*. New Politics Institute, p. 1. www.newpolitics.net

[128] Grimm, R., Dietz, N., Foster-Bey, J., Reingold, D., & Nesbit, R. (2006). *Volunteer growth in America: A review of trends since 1974*. Research Report. Corporation for National and Community Service.

[129] Bennet, A. (2017). The culture of connection. Presentation at Conference on Leading Digital and Cultural Transformation. G-LINK.

[130] *American Heritage Dictionary*.

[131] Heckscher, C. (2007). *The collaborative enterprise*. Yale University Press.

[132] Tapscott, p. 163.

[133] Panetta, S. (2013). *How to lead the millennial generation in the workforce*. Thesis Report. City University of Seattle, p. 51.

[134] Avedisian, & Bennet.

[135] Holm, J. (2013, March). Creating new businesses with open data. Presentation at KM Middle east. DATA.GOV.

[136] For a deeper treatment of this, see Bennet, A. (2022). Seeking global coherence: The waxing and waning of trust in government media. In S. B. Schafer & A. Bennet (Eds.), *Handbook of research on global media's preternatural influence on global technological singularity, culture, and government* (pp. 1-16). IGI Global.

[137] Hutchins, R, (1952). *The Great Conversation: The Substance of a Liberal Education*, pp xiii.

[138] Ibid, p. 3.

[139] Drucker, P. F. (1993). *1990s and later: Post-capitalist society*, p. 3.

[140] See www.weforum.org/agenda/2019/05/how-finland-is-fighting-fake-news-in-the-classroom/

[141] Renn, J. (2020). *The Evolution of Knowledge: Rethinking Science for the Anthropocene*, p. 395

[142] Kurzweil, R. (2005). *The singularity is near: When humans transcend biology*. Viking.

[143] Wilber, K. (1999). *The collected works of Ken Wilber, vols. 1-8*. Shambhala.

[144] Wilber, K. (2000).

[145] Tapscott.

[146] Houston, J. (2000). *Jump time: Shaping your future in a world of radical change*. Penguin Putnam, Inc., p. 1.

[147] Ibid., p. 11.

[148] Ibid, p. 18.

[149] Dunning.

[150] Kissinger, H. A., Schmidt, E., & Huttenlocher, D. (2021). *The age of AI and our human future.* Little, Brown and Company, p. 17.

[151] Autor, D., Mindell, D. A., & Reynolds, E. B. (2021). *The work of the future: Building better jobs in an age of intelligent machines.* The MIT Press, p. 3.

[152] Ibid.

[153] Ibid., p. 7.

[154] Ibid., p. 40.

[155] Ibid., p. 44.

[156] Hess, E. D. (2020). *Hyper-learning: How to adapt to the speed of change.* Berrett-Koehler Publishers, Inc., p. 3.

[157] Bennet, A., Bennet, D., Shelley, A., Bullard, T., Lewis, J., & Panucci, D. (2020e). *The profundity and bifurcation of change part V: Living the future.* MQIPress, p. 156.

[158] Hess, E. D., & Ludwig, K. (2017). *Humility is the new smart: Rethinking human excellence in the smart machine age.* Berrett-Kochler Publishers, Inc., p. 8.

[159] Hook, J. (2015). Blog. https://www.joshuanhook.com/what-is-humility/

[160] Hess & Ludwig, p. 23.

[161] Ibid., p. 5.

[162] Bennet et al. (2020a).

[163] Bennet, D., & Bennet, A. (2010).

[164] Pert.

[165] Silberman, M. (2007). *The handbook of experiential learning.* Pfeiffer, p. 3.

[166] Kolb, p. 21.

[167] Hawkins, J. C., p. vii.

[168] Hyman, S. E. (2007). Diagnosing disorders. In F. E. Bloom (Ed.), *Best of the brain from Scientific American: Mind, matter, and tomorrow's brain* (pp. 132-141). Dana Press.

[169] Nicolelis, M. A. L., & Chapin, J. K. (2007). Controlling robots with the mind. In F. E. Bloom (Ed.), *Best of the brain from Scientific American: Mind, matter, and tomorrow's brain* (pp.197-212). Dana Press.

[170] George, M. S. (2007). Stimulating the brain. In F. E. Bloom (Ed.), *Best of the brain from Scientific American: Mind, matter, and tomorrow's brain* (pp. 20-34). Dana Press.

[171] Goldberg, E. (2001). *The executive brain: Frontal lobes and the civilized mind.* Oxford University Press.

[172] Zull, J. E. (2002). *The art of changing the brain: Enriching the practice of teaching by exploring the biology of learning.* Stylus, pp. 18-29.

[173] Hawkins, J. C., p. 19.

[174] Bennet, Bennet, & Turner (2018).

[175] Battram, A. (1996). *Navigating complexity: The essential guide to complexity theory in business and management.* The Industrial Society, p. 12.

[176] Wiig.

[177] Stonier (1997).

[178] McMaster, M. D. (1996). *The intelligence advantage: Organizing for complexity.* Butterworth-Heinemann.

[179] Hawkins, & Blakeslee, p. 12.

[180] Bennet, Bennet, & Avedisian.

[181] For an in-depth treatment of intelligent behavior, see Bennet, A., & Bennet D., (2004). *Organizational survival in the new world: The intelligent complex adaptive system*. Elsevier.
[182] Ibid., p. 291.
[183] Battram, p. 145.
[184] Pert.
[185] Buzsaki.
[186] Edelman, G. M., & Mountcastle, V. B. (1982). *The mindful brain: Cortical organization and the group-selective theory of higher brain function*. MIT Press.
[187] Edelman & Mountcastle.
[188] Hawkins, J. C., p. 19.
[189] Ibid., p. 15.
[190] Novak, J. D. (1998). *Learning. Creating and using knowledge: Concept maps as facilitative tools in schools and corporations*. Lawrence Erlbaum Associates. Also, Novak, J. D., & Gowin, D. B. (1996). *Learning how to learn*. Cambridge University Press.
[191] Ausubel, D. (1968). *Educational psychology: A cognitive view*. Holt, Rinehart Winston.
[192] Kolb.
[193] Zull.
[194] Cozolino, L. J. (2006). *The neuroscience of human relationships: Attachment and developing social brain*. W. W. Norton, p. 3.
[195] Kolb.
[196] Bennet, Bennet, & Avedisian.
[197] Kolb.
[198] Bennet, Bennet, & Avedisian.
[199] Bennet, D., & Bennet, A. (2008a).
[200] Bennet, D. (2001). Loosening the world knot. Unpublished paper. Mountain Quest Institute. www.mountainquestinstitute.com
[201] Dunning.
[202] James, W. (1890/1980). *The principles of psychology, Vol. I*. Holt, Rinehar & Winston.
[203] Kolb, p. 27.
[204] Kolb.
[205] Ibid., p. 31.
[206] Edelman & Mountcastle.
[207] Hawkins, J., p. 26.
[208] Ibid., p. 34.
[209] Maturana, H. R., & Varela, F. J. (1987). *The tree of knowledge: The biological roots of human understanding*. Shambhala.
[210] Andreason, N. C. (2005). *The creating brain: The neuroscience of genius*. The Dana Foundation.
[211] While "knowing" can be defined in many ways (and has been), in the Bennet body of work it is considered as a sense—driven by the unconscious as an integrated unit—gained from experience that resides in the subconscious part of the mind, AND the energetic connection our mind enjoys with the superconscious. Poetically, knowing can be described as *seeing beyond images, hearing beyond words, sensing beyond*

appearances, and feeling beyond emotions. See Bennet, A., & Bennet, D. (2018). *Decision-making in the new reality: Complexity, knowledge and knowing.* MQIPress.

[212] Maturana & Varela. Also, Battram. Also, von Krogh, G., & Roos, J. (1995). *Organizational epistemology.* St. Martin's Press.

[213] Zull.

[214] Buzsaki, p. 21.

[215] Csikszentmihalyi, p. 42.

[216] Ibid.

[217] The introduction is excerpted from Bennet, A., Bennet, D., Shelley, A., Bullard, T., & Lewis, J. (2020d). *The profundity and bifurcation of change part IV: Co-creating the future.* MQIPress.

[218] Tallis, F. (2002). *Hidden minds: A history of the unconscious.* Arcade, p. 182.

[219] Lakoff, G., & Johnson, M. (1999). *Philosophy in the flesh: The embodied mind and its challenge to Western thought.* Basic Books, p. 13.

[220] Epstein, M. (1995). *Thoughts without a thinker.* Basic Books, p. ix.

[221] Besant, A., & Leadbeater, C. W. (1999). *Thought-forms.* The Theosophical Publishing House.

[222] Templeton, Sir John (2002). *Wisdom from world religions: Pathways toward heaven on earth.* Templeton Foundation Press.

[223] Butler, J. (1906). *The analogy of religion, natural & revealed.* J. M. Dent & Company.

[224] Armstrong, D. M. (1989). *Universals: An opinionated introduction.* Westview Press.

[225] Locke, J. (2016). Essay concerning human understanding, book III, p. 377.

[226] Armstrong.

[227] Gomide, W. (2022). Transreal numbers and sentient logic. In S. B. Schafer & A. Bennet (Eds.), *Handbook of research on global media's preternatural influence on global singularity, culture and government* (pp. 269-278). IGI Global.

[228] For an in-depth treatment of decision-making in the new reality, see Bennet, A., & Bennet, D. (2018).

[229] Source unknown.

[230] Halpern, D. F. (2996). *Thought and knowledge: An introduction to critical thinking.* Erbaum Associates, p. 117.

[231] Minsky, M. (2006). *The emotion machine: Commonsense thinking, artificial intelligence, and the future of the human mind.* Simon and Schuster, p. 6.

[232] Middlebrook, P. N. (1974). *Social psychology and modern life.* Alfred A. Knoph.

[233] This is addressed extensively in Bennet, Bennet, & Avedisian.

[234] Brown, M. Y. (1979). *The art of guiding: The psychosynthesis approach to individual counselling and psychology.* Johnson College, University of Redlands.

[235] Rowan, J. (1990). *Subpersonalities: The people within us.* Routledge, p. 7.

[236] Minsky, p. 24.

[237] Gomide.

[238] Minsky, pp. 226-228.

[239] Ibid., p. 227.

[240] Ibid., p. 228.

[241] An in-depth discussion of "Engaging Tacit Knowledge" is chapter 5 in Bennet, Bennet, & Avedisian.

[242] Goldberg, E. (2005). *The wisdom paradox: How your mind can grow stronger as your brain grows older.* Gotham Books.

243 Csikszentmihalyi, M. (1990). *Flow: The psychology of optimal experience*. Harper Perennial.

244 Goswami, A. (2014). *Quantum creativity: Think quantum. Be creative*. Hay House, Inc.

245 Sheldrake, R. (1989). *The presence of the past: Morphic resonance and the habits of nature*. Vintage Books.

246 Warmington, E. H. (Ed.) (1984). *Great dialogues of Plato* (P. G. Rouse, Trans.). Mentor, p. 18.

247 Andreasen.

248 Boden, M. (1991). *The creative mind, myths & mysticism*. Basic Books.

249 Hobson, J. A. (1999). *Consciousness*. Scientific American Library.

250 Stonier (1997).

251 Begley, S. (2007). *Train your mind change your brain: How a new science reveals our extraordinary potential to transform ourselves*. Ballantine Books, p. 214.

252 Goldberg (2005).

253 Kipling, R. (1985/1937).1 Working-tools in Ghiselin, B. (Ed.), *The creative process: A symposium*. University of California Press, pp. 161-163.

254 de Bono, E. (1992). *Serious creativity: Using the power of lateral thinking to create new ideas*. HarperColllins.

255 Freud, S. (1908/1959). The relation of the poet to day-dreaming. In *Collected papers 4* (pp. 173-183). Hogarth Press.

256 Guilford, J. P. (190). Creativity. *American Psychologist, 5*, 444-454.

257 Torrance, E. P. (1974). *Torrance test of creative thinking*. Personnel Press.

258 Finke, R. A., Ward, T. B., & Smith, S. M. (1992). *Creative cognition: Theory, research, and applications*. MIT Press.

259 Amabile, T. M. (1983). The social psychology of *creativity*. Springer.

260 Eysenck, H. J. (1993). Creativity and personality: A theoretical perspective. *Psychological Inquiry, 4*, 147-178.

261 Campbell, D. T. (1960). Blind variation and selective retention in creative thought and other knowledge processes. *Psychological Review, 67*, 380-400.

262 Perkins, D. N. (1995). *Outsmarting IQ: The emerging science of learnable intelligence*. Free Press.

263 Simonton, D. K. (1999). Talent and its development: An emergenic and epigenetic mode. *Psychological Review, 106*, 435-457.

264 Csikszentmihalyi, M., & Csikszentmihalyi, I. S. (Eds.) (1988). *Optimal experience: Psychological studies of flow in consciousness*. Cambridge University Press.

265 Amabile, T. M. (1996). *Creativity in context*. Westview.

266 Gruber, H. E. (1989). The evolving systems approach to creative work. In D. B. Wallace & H. E. Gruber (Eds.), *Creative people at work: Twelve cognitive case studies* (pp. 3-24). Oxford University Press.

267 Andreasen.

268 Byrnes. J. P. (2001). *Minds, brains, and learning: Understanding the psychological and educational relevance of neuroscientific research*. The Guilford Press.

269 Begley. Also, Jensen, E. (1998). *Teaching with the brain in mind*. Association for Supervision and Curriculum Development.

270 Bennet, A., Bennet, D., & Lewis, J. (2015). *Leading with the future in mind: Knowledge and emergent leadership*. MQIPress.

271 Gell-Mann, p. 264.

[272] Andreasen.

[273] Christos.

[274] Poincaré, H. (2001). *The value of science: Essential writings of Henri Pincaré.* Random House (Modern Library Science).

[275] Poincaré, H. ([1902,1904, 1908] 1982) Science and hypothesis, The value of science, science and method. In *The foundations of science.* University Press of America, p. 389.

[276] Ibid.

[277] Tchaikovsky, P. E. (2016). *The life and letters of Pete Ilich Tchaikovsky.* University of Toronto Press.

[278] Crandall, B., Klein, G., & Hoffman, R. R. (2006). *Working minds: A practitioner's guide to cognitive task analysis.* The MIT Press.

[279] Ibid.

[280] Christos.

[281] Ibid., p. 90.

[282] Sternberg, R. J. (2003). *Wisdom, intelligence, and creativity synthesized.* Cambridge University Press, p. 125.

[283] Ibid., p. 188.

[284] Andreasen.

[285] Grudin, R. (1984). The ethics of inspiration. In *The phenomenon of CHANGE.* Rizzoli for Cooper-Hewitt Museum, The Smithsonian Institution's National Museum of Design.

[286] Ibid., p. 15.

[287] Green, R. D. (2016). Conversation from *Huffington Post.* http://www.huffingtonpost.com/r-kay-green/giving-back_b_3298691.html

[288] Machlup, F. (1962). *The production and distribution of knowledge in the United States.* Princeton University Press, p. 179.

[289] Bennet et al. (2020a).

[290] Csikszentmihalyi (1993), p. 22.

[291] Ibid., p. 216.

[292] Ibid., p. 28.

[293] Zimmer, pp. 47-57.

[294] Csikszentmihalyi (1993), p. 23.

[295] Lipton, p. 15.

[296] Zull.

[297] Byrnes, J. P. (2007). Some ways in which neuroscientific research can be relevant to education. In D. Coch, K. W. Fischer & G. Dawson (Eds.), *Human behavior, learning and the developing brain: Typical development.* The Guilford Press.

[298] Lipton, p. 17.

[299] Ross, C. A. (2006). Brain self-repair in psychotherapy: Implications for education. In. S. Johnson & K. Taylor (Eds.), *The neuroscience of adult learning: New directions for adult and continuing education.* Jossey-Bass, p. 32.

[300] Reik, W., & Walter, J. (2001). Genomic imprinting: Parental influence on the genome. *Nature Reviews Genetics, 2,* 21+. Also, Surani, M. A. (2001). Reprogramming of genome function through epigenetic inheritance. *Nature, 414,* 122-128.

[301] Bownds, M. D. (1999). *The biology of mind: Origins and structures of mind, brain, and consciousness.* Fitzgerald Science Press. Also, Lipton. Also, Rose, S. (2005) *The*

future of the brain: The promise and perils of tomorrow's neuroscience. Oxford University Press. Also, Begley.

[302] Lipton, p. 143.

[303] Reik & Walter.

[304] Jensen, E. (2006). *Enriching the brain: How to maximize every learner's potential.* Jossey-Bass.

[305] Begley.

[306] Lipton, p. 143.

[307] Lipton.

[308] Ibid., p. 143.

[309] Bennet, Bennet, & Turner. Also, Bownds. Also, Begley.

[310] Gibson, J. J. (1979). *The ecological approach to visual perception.* Houghton Mifflin.

[311] Damasio, A. R. (2010). *Self comes to mind: constructing the conscious brain.* Vintage Books.

[312] Immordino-Yang, M. H., & Singh, v. (2016). Perspectives from social and affective neuroscience on the design of digital learning technologies. In. M. H. Immordino-Yang (Ed.), *Emotions, learning and the brain: Exploring the educational implications of affective neuroscience* (pp. 181-190). Norton.

[313] Hawkins, J. C.

[314] Bennet, Bennet, & Turner.

[315] Hawkins, J. C., p. 51.

[316] Eilan, N., Marcel, A., & Bermudez, J. L. (1995). Self-consciousness and the body: An interdisciplinary introduction. In J. L. Bermudez, A., Marcel & N. Eilan (Eds.), *The body and the self.* MIT Press.

[317] Bennet, Bennet, & Turner.

[318] Damasio (2010), p. 24.

[319] Ibid.

[320] Mulvihill, p. 322.

[321] Buks, E., Schuster, R., Heiblum, M., Mahalu, D., & Umansky, V. (1998). Dephasing in electron interference by a "which-path" detector. *Nature, 391,* 871-874.

[322] Damasio (2010).

[323] LeDoux.

[324] Schacter, D. L. (1996). *Searching for memory: The brain, the mind, and the past.* Basic Books.

[325] Laughlin, S. B. (2004). The implications of metabolic energy requirements for the representation of information in neurons. In M. S. Gazzaniga (Ed.), *The cognitive neuroscience III.* The MIT Press, 187-196.

[326] Ibid., p. 188.

[327] Pert.

[328] Bargh, J. A. (2005). Bypassing the will: Towards demystifying the nonconscious control of social behavior. In R. R. Hassan, J. S. Uleman, & J. A. Bargh (Eds.), *The new unconscious.* Oxford University Press, p. 39.

[329] Rock, A. (2004). *The mind at night: The new science of how and why we dream.* Perseus Books Group. Also, Schacter.

[330] Jensen (2006). Also, Taylor, K. (2006). Brain function and adult learning: Implications for practice. In S. Johnson & K. Taylor (Eds.), *The neuroscience of adult learning: New directions for adult and continuing education* (pp. 71-86). Jossey-Bass.

[331] Talllis.

[332] Bohm, D. (1992). *Thought as a system*. Routledge.

[333] Andreasen, p. 165.

[334] Pinker, S. (2007). *The stuff of thought: Language as window into human nature*. Viking.

[335] Bennet, A., & Bennet, D. (2008a). The human knowledge system: Music and brain coherence. *VINE: The Journal of Information and Knowledge Management Systems, 38*(3), 277-295.

[336] Oster, G. (1973). Auditory beats in the brain. *Scientific American, 229*, 94-102.

[337] Swann, R., Bosanko, S., Cohen, R., Midgley, R., & Seed, K. M. (1982). *The brain: A user's manual*. G. P. Putnam & Sons.

[338] Hink, R. F., Kodera, K., Yamada, O., Kaga, K., & Suzuki, J. (1980). Binaural interaction of a beating frequency following response. *Audiology, 19*, 36-43. Also, Marsh, J. T., Brown, W. S., & Smith, J. C. (1975). Far-field recorded frequency-following response: Correlates of low pitch auditory perception in humans. *Electroencephalography and Clinical Neurophysiology, 38*, 113-119. Also, Smith, J. C., Marsh, J. T., Greenberg, S., & Brown, W. S. (1978). Human auditory frequency-following responses to a missing fundamental. *Science, 201*, 639-641.

[339] Ritchey, D. (2003). *The H.I.S.S. of the A.S.P.: Understanding the anomalously sensitive person*. Headline Books.

[340] Atwater, F. H. (2004). *The Hemi-Sync® Process*. The Monroe Institute. Also, Delmonte, M. M. (1984). Electrocortical activity and related phenomena associated with mediation practice: A literature review. *International Journal of Neuroscience, 24*, 217-231. Also, Fischer, R. (1971). A cartography of ecstatic and meditative states. *Science, 174*(4012), 897-904.

[341] Kandel.

[342] Ibid., p. 281.

[343] Engle, R. W., Cantor, J., & Carullo, J. J. (1993). Individual difference in working memory and comprehension: A test of four hypotheses. *Journal of Experimental Psychology: Learning, memory, and Cognition, 19*, 972-992. Also, Laming, D. (1992). Analysis of short-term retention: Models for Brown-Peterson Experiments. *Journal of Experimental Psychology: Learning, Memory, and Cognition, 18*, 1342-1365.

[344] Miller, G. A. (1956). The magical number seven, plus or minus two: Some limits on our capacity for processing information. *Psychological Review, 63*, 81-96.

[345] Baddeley, A. D. (1997). *Human memory: Theory and Practice* (Rev.). Psychology Press.

[346] Christos.

[347] Miller.

[348] Ross, P. E. (2006, August). The expert mind. *Scientific American*, 64-71.

[349] Lesch, M. F., & Pollatsek, A. (1993). Automatic access of semantic information by phonological codes to visual word recognition. *Journal of Experimental Psychology: Learning, Memory and Cognition, 19*, 285-294.

[350] Goldringer, S. D. (1996). Words and voices: Episodic traces in spoken word identification and recognition memory. *Journal of Experimental Psychology: Learning, Memory, and Cognition, 22*, 1166-1183.

[351] Haberlandt, K. (1998). *Human memory: Exploration and application*. Allyn & Bacon.

[352] Ibid.

[353] Kluwe, R. H., Luer, G., & Rosler, F. (Eds.) (2003). *Principles of learning and memory*. Birkhauser Verlag. Also, Markowitsch, H. (1999), Functional neuroimaging correlates of functional amnesia. *Memory, 8*, 567-583. Also, Schacter.

[354] Bennet, A., & Bennet, D. (2004). Also, Hawkins & Blakeslee.

[355] Christos.

[356] Amen, D. G. (2005). *Making a good brain great*. Harmony Books. Also, Begley. Also, Lipton. Also, Medina.

[357] Christos.

[358] Hawkins, & Blakeslee. Also, Ratey, J. J. (2001). *A user's guide to the brain: Perceptions, attention, and the four theaters of the brain*. Pantheon Books.

[359] Hawkins, & Blakeslee.

[360] Edelman, G., & Tononi, G. (2000). *A Universe of consciousness: How matter becomes imagination*. Basic Books.

[361] Hawkins, & Blakeslee, p. 82.

[362] Ibid., p. 71.

[363] Bennet et al. (2018a).

[364] Hawkins & Blakeslee, p. 70.

[365] Schank, R. E. (1995). *Tell me a story: Narrative and intelligence*. Northwestern University Press, p. 11.

[366] Bennet, A., Bennet, D., & Lee, S. L. (2009). Exploring the military contribution to KBD through leadership and values. *Journal of Knowledge Management, 14*(2), p. 326.

[367] Hawkins, pp. 71-72.

[368] Bloom, F. E. (2007). *Best of the brain from Scientific American: Mind, matter and tomorrow's brain*. Dana Press, pp. 63-64.

[369] Begley. Also, Hawkins.

[370] Hawkins.

[371] Scripps Research Institute (2012, May 9). Scientists identify neurotransmitters that lead to forgetting. *ScienceDaily*.

[372] LeDoux.

[373] *American Heritage Dictionary of the English Language*.

[374] Connerton, P. (2008). Seven types of forgetting. http://mss.sagepub.com/content/1/1/49.short?rss=1&ssource=mfr

[375] Ebbinghaus, H. (188). Memory: A contribution to experimental psychology. http://nwkpsych.rutgers.edu/

[376] Schacter.

[377] Brown, J. (1958). Some tests of the decay theory of immediate memory. *Quarterly Journal of Experimental Psychology*, 10, 12-21. Also, Peterson, L. R., & Peterson, M. J. (1959). Short-term retention of individual verbal items. *Journal of Experimental Psychology, 58*, 193-198.

[378] Andrews, B., Brewin, C. R., Ochera, J., Morton, J., Bekerian, D.A., Davies, G. M., et al. (1999, August). Characteristics, context and consequences of memory recovery among adults in therapy. *British Journal of Psychiatry, 175*, 141-146.

[379] Lief, H., & Fetkewicz, J. (1995). Retractors of false memories: The evolution of pseudomemories. *Journal of Psychiatry Law, 23*, 411-35.

[380] Ceci, S. J. (1995). False beliefs: Some developmental and clinical considerations. In D. L. Schacter (Ed.), *Memory distortion: How minds, brains and societies reconstruct the past* (pp. 92-128). Harvard University Press.

[381] See Bennet et al. (2018c) for a more in-depth treatment of forgetting.

[382] Wickelgren, I. (2012, January 1). Forgetting is key to a healthy mind. *Scientific American Mind*.

[383] Details of the Sedona Method are included at www.sedona.com

[384] Argyris, C. (1990). *Overcoming organizational defenses: Facilitating organizational learning*. Prentice Hall.

[385] Bennet et al. (2018a).

[386] Belasco, J. A. (1997). Leading the new organizations. In K. Shelton (Ed.), *A new paradigm of leadership: Visions of excellence for 21ˢᵗ century organizations*. Executive Excellent Publishing, pp. 37-38.

[387] Davenport, T. H. (2001). Knowledge work and the future of management. In W. Bennis, G. M. Spreitzer & T. G. Cummings (Eds.), *The future of leadership: Today's top leadership thinkers speak to tomorrow's leaders*. Jossey-Bass, p. 58.

[388] Bennis, W. (1995). *Beyond leadership: Balancing economics, ethics and ecology*. Blackwell Publishers, pp. 37-38.

[389] Davenport, p. 47.

[390] Bullard, B., & Bennet, A. (2013/2020). *Remembrance: Pathways to expanded learning with music and meta-music®*. MQIPress, p. 27.

[391] Graziano, M. S. A.; Kastner, S (2011). Human consciousness and its relationship to social neuroscience: A novel hypothesis. *Cognitive Neuroscience 2* (2), 98-113.

[392] Portante, T., & Tarro, R. (1997, September). Paying attention. *Wired Magazine*.

[393] Lanham, R. (1993). The electronic word: Democracy, technology, and the arts. University of Chicago Press.

[394] Davenport, T. H., & Beck, J. C. (2001). *The attention economy*. Harvard Business Review Press.

[395] Ibid., p. 7.

[396] Csikszentmihalyi (1993), p. 217.

[397] Amen, p. 115.

[398] Medina.

[399] Sapolsky, R. M. (1999). Glucocorticoids, stress, and their adverse neurological effects: Relevance to gaining. *Experimental Gerontology, 34*, 721-732.

[400] Carrasco, M., Ling, S., & Read, S. (2004). Attention alters appearance. *Nature Neuroscience, 7*, 308-313.

[401] Jensen (2006), p. 10.

[402] Maunsell, J. H. R., & Treue, S. (2006). Feature-based attention in visual cortex. *Trends in Neuroscience, 29*, 317-322. Also, Asplund, C. L., Fougnie, D., Zughni, S., Martin, J. W., & Marois, R. (2014, January 16). The attentional blink reveals the probabilistic nature of discrete conscious perception. *Psychological Science*. http://pss.sagepub.com/

[403] Zull, p. 141.

[404] Byrnes.

[405] Begley, p. 158.

[406] Akil, H., Campeau, S., Cullinan, W., Lechang, R., Toni, R., Watson, S., & Moore, R. (1999). Neuroendocrine system 1: Overview—thyroid and adrenal axis. In M. Zigmond, F. Bloom, S. Landis, J. Roberts, & L. Squire (Eds.), *Fundamentals of neuroscience*. Academic Press, pp 1146.

[407] Zull (2002).

[408] Cozolino & Sprokay, p. 14.

[409] Ibid, p. 62.

410 Merry, U. (1995). *Coping with uncertainty: Insights from the new sciences of chaos, self-organization and complexity.* Praeger.
411 Begley.
412 Ibid.
413 Selye, H. (1978). *The stress of life.* McGraw-Hill, 2nd ed., pp. 64, 74.
414 Thompson, R. F. (2000). *The brain: A neuroscience primer.* Worth, p. 210.
415 Developed by Dr. Donna Panucci, Blue Ray of Hope.
416 Ramon, S. (1997). *Earthly cycles: How past lives and soul patterns shape your life,* Pepperwood Press, p. 48.
417 Searle, J. R. (1983). *Intentionality: An essay in the philosophy of mind.* Cambridge University Press, p. ix.
418 Searle.
419 Prinz, W. (2005). An ideomotor approach to imitation. In S. Hurley & N. Chater, *Perspectives on imitation: From neuroscience to social science Vol. 1: Mechanisms of imitation and imitation in animals* (pp. 141-156). MIT Press.
420 Searle.
421 Jahn, R. G., & Dunne, B. J. (2011) *Consciousness and the source of reality: The PEAR odyssey.* ICRL Press.
422 Pribram, K. H. (1991). *Brain and perception: Holonomy and structure in figural processing.* Lawrence Erlbaum. Also, Pribram, K. H. (1993). Rethinking neural networks: Quantum fields and biological data. *Proceedings of the First Appalachian Conference on Behavioral Neurodynamics.* Lawrence Erlbaum. Also, Pribram, IK. H. (1998). Autobiography in anecdote: The founding of experimental neuropsychology. In R. Bilder (Ed.), *The history of neuroscience in autobiography* (pp. 306-349). California Academic Press.
423 Popp, F. A. (2002). Biophotonics: A powerful tool for investigating and understanding life. In H. P. Durr, F. A. Popp, and W. Schommers (Eds.), *What is life? Scientific approaches and philosophical positions.* (Series on the Foundations of Natural Science and Technology). World Scientific.
424 McTaggart, L. (2002). *The field: The quest for the secret force of the Universe.* Harper Perennial.
425 Brown & Ryan.
426 Creswell, J. D. (2017). Mindfulness interventions. *Annual Review of Psychology, 68*(1), 491-516, p. 492.
427 Kabat-Zinn, J. (2003). Mindfulness-based interventions in context: Past, present, and future. *Clinical Psychology: Science and Practice, 10*(2), 144-156.
428 Siegel, R. D., Germer, C. K., & Olendzki, A. (2009). Mindfulness: what is it? Where did it come from? In F. Didonna (Ed.), *Clinical handbook of mindfulness* (pp. 17-36). Springer.
429 Worawichayavongsa, W. (2020). The influence of mindfulness practice at work on self-leadership on Thai non-managerial employees: The lived experience in an international K12 educational environment, p. 50. [Unpublished doctoral dissertation]. Southeast Asia Institute of Knowledge and Innovation, Bangkok University, Thailand.
430 Senge, P. M., Scharmer, C. O., Jaworski, J., & Flowers, B. S. (2004). *Presence: Exploring profound change in people, organizations and society.* Random House, Inc., p. 133.
431 Senge, P. M. (1990). *The fifth discipline.* Doubleday.

432 Bower, J. L. (2000). The purpose of change: A commentary on Jensen and Senge. In M. K. Beer & N. Nohria (Eds.), *Breaking the code of change* (pp. 93-95). Harvard Business School Press.
433 Hawkins, J. C., p. 137.
434 Senge et al.
435 Ibid.
436 Scharmer, C. O. (2009). *Theory U: Leading from the future as it emerges*. Berrett-Koehler.
437 Forwarded in the January 08, 2017, daily quote from Abraham-Hicks Publications,
438 Medina.
439 Amen (2005).
440 Sousa, D. A. (2006). *How the brain learns*. Corwin Press, p. 19.
441 Stonier (1997).
442 Medina, p. 14.
443 *American Heritage Dictionary*, p. 1928.
444 Begley.
445 Ibid., p. 69.
446 Amen (2005). Also, Goldberg (2005). Also, Sapolsky.
447 Amen (2005).
448 Diamond, M. (2001, March 10). Successful aging of the healthy brain. Presentation at the American Society on Aging Conference, New Orleans.
449 Ibid.
450 Amen (2005).
451 Ibid., p. 114.
452 Begley, p. 58.
453 Kempermann, G., Kuhn, H. G., & Gage, F. H. (1997). More hippocampal neurons in adult mice living in an enriched environment. *Nature, 386*, 493-95.
454 Begley, p. 58.
455 Ibid.
456 Ibid.
457 Byrnes (2001). Also, Jensen (1998).
458 Skoyles, J. R., & Sagan, D. (2002). Up from dragons: The evolution of human intelligence. McGraw-Hill.
459 Ibid.
460 *The American Heritage Dictionary of the English Language,* p. 585.
461 Immordino-Yang (2016a), p. 20.
462 Adolphs, R. (2004). Processing of emotional and social information by the human *amygdala*. In M. S. Gazzaniga (Ed.), *The cognitive neurosciences III* (pp. 1017-1030)The Bradford Press.
463 LeDoux.
464 Haberlandt.
465 Mulvihill, p. 322.
466 Ralston, F. (1995). *Hidden dynamics: How emotions affect business performance & how you can harness their power for positive results*. American Management Association.
467 Bennet & Scott.
468 McCraty, R. (2015). *Science of the heart: Exploring the role of the heart in human performance volume 2*. HeartMath Institute.

[469] Strogatz, S. H., l& Stew-art, I. (2020/1993). Coupled oscillators & biological synchronization. *Scientific American, 269*(6), 102-109.

[470] McCraty, R., & Childre, D. (2010, July/August). Coherence: bridging personal, social, and global health. *Alternative Therapies, 16*(4), p. 10.

[471] Ibid.

[472] McCraty.

[473] LeDoux. Also, Smith, C. A., & Ellsworth, P. C. (1985). Patterns of cognitive appraisal in emotion. *Journal of Personality and Social Psychology, 56*, 339-53.

[474] Zull.

[475] National Research Council (2000). *How people learn: Brain, mind, experience, and school*. National Academy Press.

[476] Lipton.

[477] Hawkins, J., p. 174.

[478] Pert.

[479] Christos. Also, Kluwe, Luer, & Rosler.

[480] Immordino-Yang (2016a), p. 18.

[481] Christos.

[482] Plotkin.

[483] Carroll, S. (2016). *The big picture: On the origins of life, meaning, and the Universe itself*. Dutton.

[484] Walsch.

[485] Bennet & Scott.

[486] LeDoux.

[487] Hawkins (1995), p. 70.

[488] Ibid., p. 235.

[489] Cooper, R. (2005). Austinian truth, attitudes and type theory. *Research on Language and Computation 5*, 333-362, p. 42.

[490] Walsh.

[491] Hawkins (1995), p. 235.

[492] Gomide (2022). Also, Gomide, W. (2021a). Non-causality and transreal numbers: The world seen at speed of light. [Unpublished]. UFMT/Brazil. Also, Gomide, W. (2021b). Metaphysical aggregates and their insertion into Cantor's thought. [Unpublished]. UFMT/Brazil.

[493] Anderson, J., & Dos Reis, T. (2014). Transdifferential and transintegral calculus. In H. K. Kim, M. A. Amouzegar, & S. Ao (Eds.), *Transactions on engineering technologies: World congress on engineering and computer science 2014*, 92-96.

[494] Gomide (2022), Conclusion.

[495] Schafer, S. B. (2022). Gamers as homeopathic media therapy: Electromagnetic antibodies in the toxic media-field. In S. B. Schafer & A. Bennet (Eds.), *Handbook of research on global media's preternatural influence on global singularity, culture and government* (pp. 397-425). IGI Global. Also, Cools, J. (2022). Counter-hacking the subconscious mind antidote for faulty human perceptions from the media sphere. In S. B. Schafer & A. Bennet (Eds.), *Handbook of research on global media's preternatural influence on global singularity, culture and government* (pp. 372-396). IGI Global. Also, Luca, M. (2022). Does the algorithm heal a business organization? In S. B. Schafer & A. Bennet (Eds.), *Handbook of research on global media's preternatural influence on global singularity, culture and government* (pp. 345-371). IGI Global.

[496] Gomide (2022).

[497] Bennet, A., & Bennet, D. (2008b). Engaging tacit knowledge in support of organizational learning. *VINE: The Journal of Information and Knowledge Management Systems, 38*(1), 72-94.

[498] Strogatz, S. (2003). *Sync: How order emerges from chaos in the Universe, nature, and daily life.* Hyperion Books.

[499] Turmel, R. (1997, Summer). Resonant frequencies and the human brain. *The Vaults of EROWID 1.*

[500] Damasio.

[501] Bechara, A. (2004). The role of emotion in decision-making: Evidence from neurological patients with orbitofrontal damage. *Brain Cognition 55*(1), 30-40.

[502] Christos. Also, Kluwe et al.

[503] LeDoux.

[504] Begley.

[505] Damasio.

[506] Cooper, p. 28.

[507] Pert.

[508] Lambrou, P., & Pratt, G. (2000). *Acupressure for the emotions: Instant emotional healing.* Broadway Books, p. 52.

[509] LeDoux (1996).

[510] Damasio (1994), p. 139.

[511] LeDoux, (1996), p. 19

[512] LeDoux.

[513] Goleman, D. (2015). *A force for good: The Dalai Lama's vision for our world.* Bantam Books.

[514] Goleman, D. (1995). *Emotional intelligence.* Bantam Books.

[515] Pert, p. 145.

[516] Hyman.

[517] Nicolelis & Chapin.

[518] George.

[519] Cozolino, p. 3.

[520] Johnson, S. (2006). The neuroscience of the mentor-learner relationship. In S. Johnson & K. Taylor (eds.), *The neuroscience of adult learning: New directions for adult and continuing education* (pp. 63-70). Jossey-Bass.

[521] Ibid., p. 65.

[522] Stern, D. N. (2004). *The present moment in psychotherapy and everyday life.* Norton.

[523] Buzsaki.

[524] Cozolino, p. 213.

[525] Bownds.

[526] LeDoux, p. 33.

[527] Tallis, p. 129.

[528] Dewey, J. (1938/1997). Experience and education. Simon & Schuster, p. 39.

[529] Johnson (2006), p 65.

[530] Brookfield, S. D. (1987). *Developing critical thinkers.* Jossey-Bass. Also, Daloz, L. (1986). Effective teaching and mentoring. Jossey-Bass. Also, Daloz, L. (1999). *Mentor: Guiding the journey of adult learners.* Jossey-Bass. Also, Mezirow, J. (1991). *Transformative dimensions of adult learning.* Jossey-Bass. Also, Perry, W. G.

(1970/1988). *Forms of ethical and intellectual development in the college years.* Jossey-Bass.

[531] Cozolino, p. 291.

[532] Taylor, p. 82.

[533] Ibid., pp. 213, 293.

[534] Cozolino & Sprokay, p. 13.

[535] Caine, G., & Caine, R. N. (2006). Meaningful learning and the executive functions of the brain. In S. Johnson & K. Taylor (Eds.), *The neuroscience of adult learning: New directions for adult and continuing education.* Jossey-Bass, p. 54.

[536] Amen.

[537] Chung, S., Tack, G., Lee, B., Eom, G., Lee, S., & Sohn, J. (2004). The effect of 30 oxygen on visuospatial performance and brain activation: An fMRI study. *Brain and Cognition, 56*, 279-285. Also, Scholey, A., Moss, M., Naeve, N., & Wesnes, K. (1999). Cognitive performance, hyperoxia, and heart rate following oxygen administration in healthy young adults. *Physiological Behavior, 67*, 783-789.

[538] Sousa, D. A. (2006). *How the brain learns.* Corwin Press.

[539] Cozolino.

[540] Iacoboni, M. (2008). *The new science of how we connect with others: Mirroring people.* Farrar, Straus & Giroux.

[541] Rizzolatti, G. (2006, November). Mirrors in the mind. *Scientific American*, p. 56.

[542] Bennet, Bennet, & Turner.

[543] Rizzolatti.

[544] Immordino-Yang, M. H. (2016b). The smoke around mirror neurons: Goals as sociocultural and emotional organizers of perception and action in learning. In M. H. Immordino-Yang (Ed.), *Emotions, learning, and the brain: Exploring the educational implications of affective neuroscience.* Norton.

[545] Iacoboni.

[546] Rizzolatti, p. 61.

[547] Zull.

[548] Ibid., p. 156.

[549] Ibid.

[550] Nelson, C. A., deHaan, M., & Thomas, K. M. (2006). *Neuroscience of cognitive development: The role of experience and the developing brain.* John Wiley & Sons.

[551] Stern, p. 76.

[552] Cozolino, p. 203.

[553] Riggio, R. E. (2015). Are you empathic? 3 types of empathy and what they mean. *Psychology Today.*

[554] Masserman, J., Wechkin, S., & Terris, W. (1964, December). Altruistic behavior in Rhesus monkeys. *American Journal of Psychiatry, 121*, 584-585.

[555] Cozolino.

[556] Singer, T., Seyour, B., O'Doherty, J., Kaube, H., Dolan, R. J., & Frith, C. D. (2004). Empathy for pain involves the affective but not sensory components of pain. *Science, 303*, 1157-1162. Also, Jackson, P. L., Meltzoff, A. N., & Decety, J. (2005). How do we perceive the pain of others? A window into the neural processes involved in empathy. *NeuroImage, 24*, 771-779.

[557] deWaal, R. (2009). *The age of empathy: Nature's lessons for a kinder society.* Crown Publishing Group.

[558] Cozolino, p. 206.

[559] Carr, L., Iacoboni, M., Dubeau, M. c., Mazziotta, J. C., & Lenzi, G. L. (2003). Neural mechanisms of empathy in humans: A relay from neural systems for imitation to limbic areas. *Proceedings of the National Academy of Science, 100*, 5497-5502. Also, Kawashina, R., Sugiura, M., Kato, T., Nakamura, A., Hatano, K., Ito, K., et al. (1999). The human amygdala plays an important role in gaze monitoring: A PET study. *Brain, 122*, 779-783.

[560] Andersson, J. L., Lilja, A., Hartvig, P., Langstrom, B., Gordh, T., Handwerker, H., et al. (1997). Somatotopic organization along the central sulcus, for pain localization in humans, as revealed by positron emission tomography. *Experimental Brain Research, 117*, 192-199.

[561] Immordino-Yang, M. H., Christodoulou, J. A., & Singh, V. (2016). 'Rest is not idleness': Implications of the brain's default mode for human development and education. In M. H. Immordino-Yang (Ed.), *Emotions, learning, and the brain: Exploring the educational implications and affective neuroscience* (pp. 43-68). Norton.

[562] Bennet, et al. (2020d), p. 167.

[563] Shelley, A. (2007). *Organizational zoo: A survival guide to work place behavior.* Aslan Publishing.

[564] Bennet & Bennet (2004).

[565] Vickers, G. (1977). The future of culture. In H. A. Linstone & W. H. C. Simmonds (Eds.), *Futures research: New directions.* Addison-Wesley Publishing Company, Inc.

[566] Schore (1994) as cited in Cozolino, p. 191.

[567] Bennet et al. (2020a).

[568] *Oxford English Dictionary* 5th Ed., p. 2928.

[569] Csikszentmihalyi, M. (2003). *Good business: Leadership flow and the making of meaning.* Viking, p. 19.

[570] Inglehart, R., Basanez, M., & Moreno, A. (1998). *Human values and beliefs: A cross-cultural sourcebook.* The University of Michigan Press, p. 168.

[571] Haider, P. (n.d.). 60+ percent of all people believe in the soul. Sivana East. http://www.blog.sivanaspirit.com

[572] Beauregard, M., & O'Leary, D. (2007). The spiritual brain: A neuroscientist's case for the existence of the soul. HarperOne, p. x.

[573] Ibid.

[574] Baier. A. (2004). Demoralization, trust, and the virtues. In C. Calhoun (Ed.), *Setting the moral compass: Essays by women philosophers* (pp. 176-189). Oxford University Press, p. 180.

[575] *Oxford English Dictionary* 5th Ed., p. 2963,

[576] *American Heritage Dictionary* 4th Ed., p. 910.

[577] *Oxford English Dictionary* 5th Ed., p. 2963.

[578] Zohar, D., & Marshall, I. (2003). Spiritual capital: Wealth we can live by. Berrett-Kochler Publishers, Inc., p. 27.

[579] Bennet, A., & Bennet, D. (2006). Learning as associative patterning. *VINE: The Journal of Information and Knowledge Management Systems, 36*(4), 371-376.

[580] Teasdale, W., & the Dali Lama (1999). *The mystic heart: Discovering al universal spirituality in the world's religions.* New World Library, pp. 17-18.

[581] Cousins, E. (1999). Spirituality on the evening of a new millennium: An overview. *Chicago Studies, 38*(1), 5-14. Also, Shafer, I. H. (1999). World commission on global consciousness and spirituality.

https://www.sourcewatch.org/index.php/World_Commission_on_Global_Consciousnes s_and_Spirituality
[582] English, L. M., & Gillen, M. A. (Eds.) (2000). *Addressing the spiritual dimensions of adult learning: What educators can do.* Jossey-Bass Publishers.
[583] *Merriam Webster* (2016). http://www.merriam-webster.com/dictionary/virtue
[584] Virtue (2016). Google.
[585] See Chapter 34 in Bennet, A., Bennet, D., Shelley, A., Bullard, T., Lewis, J., & Panucci, D. (2020e).
[586] Gardner, H. (2011). *Truth, beauty and goodness reframed: Educating for the virtues in the age of truthiness and twitter.* Basic Books.
[587] "From the point of view of formative causation, even our thoughts and attitudes influence other people, and we're in turn influenced by countless other people's thoughts and attitudes. So, our minds are not insulated and separate in the way many modern people like to imagine, they're far more permeable to each other …. Formative causation makes us see societies and social groups as having al wholeness, a unity, and enables us to think more about the reality of the social forms in which we live." Sheldrake, p. 168-169.
[588] Gomide (2022).
[589] Kant, I. (1993). *Groundwork of the metaphysics of morals* 3rd ed. (J. W. Ellington, Trans.). Hackett.
[590] Pelegrinis, T. N. (1980). *Kant's conceptions of the categorical imperative and the will.* Zeno Publishers.
[591] Haidt, J. (2003). Elevation and the positive psychology of morality. In C. L. M. Keyes & J. Haidt (Eds.), *Flourishing positive psychology and the life well-lived* (pp. 275-289). American Psychological Association.
[592] Immordino-Yang, M. H., McColl, A., Damasio, H., & Dalmasio, A. (2009). Neural correlates of admiration and compassion. *Proceedings of the National Academy of Sciences, 106*(19), 8021-8026.
[593] Immordino-Yang, M. H., & Sylvan, L. (2016). Admiration for virtue: Neuroscientific perspectives on a motivating emotion. In M. H. Immordino-Yang (Ed.), *Emotions, learning, and the brain: Exploring the educational implications of affective neuroscience* (pp. 165-179). W. W. Norton.
[594] Schacter.
[595] Laughlin, S. B. (2004). The implications of metabolic energy requirements for the representation of information in neurons. In M. S. Gazzaniga (Ed.), *The cognitive neuroscience III* (187-196). The MIT Press.
[596] Begley, p. 214.
[597] Schafer, R. M. (1993). *The soundscape: Our sonic environment and the tuning of the world.* Destiny Books.
[598] The Monroe Institute furthers the experience and exploration of consciousness, expanded awareness and discovery of self through technology and education. See www.monroeinstitute.org/
[599] Seeley, W. W., Menon, V., Schatzberg, A. F., Keller, J., Glover, G. H., Kenna, H., … & Greicius, M. D. (2007). Dissociable intrinsic connectivity networks for salience processing and executive control. *Journal of Neuroscience, 27*(9), 2349-2356.
[600] Buckner, R. L., & Vincent, J. L. (2007). Unrest at rest: Default activity and spontaneous network correlations. *NeuroImage 37*(4), 1091-1096. Also, Raichle, M. E., MacLeod, A. M., Snyder, A. Z., Powers, W. J., Gusnard, D. A., & Shulman, G. L.

(2001). A default mode of brain function. *Proceedings of the National Academy of Sciences, 98*(2), 676-682.

[601] Immordino-Yang, Christodoulou, & Singh, p. 46.

[602] Greicius, M. D., Krasnow, B., Reiss, A. L., & Menon, V. (2003). Functional connectivity in the resting brain: A network analysis of the default mode hypothesis. *Proceedings of the National Academy of Sciences 100*(1), 253-258.

[603] Esposito, F., Bertolino, A., Scarabino, T., Latoffe, V., Blasi, G., Popolizio, T., ... Di Salle, F. (2006). Independent component model of the default-ode brain function: Assessing the impact of active thinking. *Brain Research Bulletin, 70*(4-6), 263-269. Also, Fox, M. D., Snyder, A. Z., Vincent, J. L., Corbetta, M., Van Essen, D. C., & Raichle, M. E. (2005). The human brain is intrinsically organized into dynamic, anti-correlated functional networks. Proceedings of the *National Academy of Sciences, 102*(27), 9673-9678.

[604] Immordino-Yang, Christodoulou, & Singh, p. 48.

[605] Andreasen.

[606] Ibid., p. 164.

[607] Smallwood, J., Obsonsawin, M., & Heim, D. (2003). Task unrelated thought: The role of distributed processing. *Consciousness and Cognition, 12*(2), 169-189.

[608] Buckner, R. L., Andrews-Hanna, J. R., & Schacter, D. L. (2008). The brain's default network: Anatomy, function, and relevance to disease., *Annals of the New York Academy of Sciences 1124*, 1-38. Also, Gilbert, D. T., & Wilson, T. D. (2007). Prospection: Experiencing the future. *Science, 317*(5843), 1351-1354. Also, Spreng, R. N., & Grady, C. L. (2010) Patterns of brain activity supporting auto-biographical memory, prospection, and theory of mind, and their relationship to the default mode network. *Journal of Cognitive Neuroscience, 22*(6), 1112-1123.

[609] Immordino-Yang, Christodoulou, & Senge, pp. 44-45.

[610] Bennet, Bennet, & Turner (2018).

[611] Pert.

[612] McCraty.

[613] Besant & Leadbeater, p. 47.

[614] Epstein, p. ix.

[615] Besant & Leadbeater, p. xx.

[616] Long, T. A. (1986). Narrative unity and clinical judgment. *Theoretical Medicine 7*, 75-92.

[617] Gell-Mann.

[618] Gardner, H. (2006). *Multiple intelligences: New horizons in theory and practice.* Basic Books, p. 46.

[619] Ross, P. E., pp. 64-71.

[620] Lipton, B., & Bhaerman, S. (2009). *Spontaneous evolution: Our positive future (and a way to get there from here).* Hay House, p. 217.

[621] Hawkins, J. C.

[622] Mulford, P. (2007). *Thoughts are things.* Barnes & Noble. p. 72.

[623] Alberts, D. S., & Hayes, R. E. (2005). *Power to the edge: Command, control in the information age.* Command & Control Research Program.

[624] Ibid., p. 48.

[625] Michael, D. N. (1977). Planning's challenge to the systems approach. In H. A. Linstone & W. H. C. Simmonds (Eds.), *Futures research: New directions.* Addison-Wesley Publishing Company, Inc.

[626] Meyer, C., & Davis, S. (2003). *It's alive: The coming convergence of information, biology, and business*. Crown Business.

[627] Alberts & Hayes, p. 227.

[628] Ibid., p. 228.

[629] Bennet & Bennet (2004).

[630] Mulford, p. 72.

[631] Ibid.

[632] Demasio, p. 570.

[633] Hawkins, J. C.

[634] Bennet, Bennet, & Lewis.

[635] The tenets of Appreciative Inquiry developed by Hammond and Hall in 1996 are: (1) In every society, organization, or group (and within every individual) something works. (2) What we focus on becomes our reality. (3) Reality is created in the moment and there are multiple realities. (4) The act of asking questions of an organization or group (or individual) influences the group (or individual) in some way. (5) People have more confidence and comfort to journey into the future (the unknown) when they carry forward parts of the past (the known). (6) If we carry parts of the past forward, they should bel what is best about the past. (7) It is important to value differences. (8) The language we use creates our reality. See Bennet, A., Bennet, D., Shelley, A., Bullard, T., & Lewis, J. (2020b). *The profundity and bifurcation of change part II: Learning from the past*. MQIPress, p. 64.

[636] Wade, W. (2012). *Scenario planning: A field guide to the future*. John Wiley & Sons, Inc.

[637] Churchman, C. W. (1977). A philosophy for complexity. In H. A. Linstone & W. H. C. Simmonds (Eds.), *Futures research: New directions*. Addison-Wesley Publishing Company, Inc. Also, von Forester, H. (1977). Objects: Tokens for (eigen-) behaviors. In B. Inheler, R. Gracia, & J. Voneche (Eds.), *Hommage a Jean Piaget: Epistemologie Genetique et Eqilibration*. Delachaux et Niestel.

[638] Worawichayavongsa, & Bennet.

[639] Ibid.

[640] Bennet, A. & Bennet, D. (2008).

[641] Oster.

[642] Ritchey. Headline Books.

[643] Schafer, S. B. (2022).

[644] Schafer, R. M. (1993)..

[645] Hodges, D. (2000). Implications of music and brain research. *Music Educators Journal, 87*(2), 17-22.

[646] Weinberger, N. M. (1995). Non musical outcomes of music education. *Musical Journal, II*(2), 6.

[647] Wilson as quoted in Hodges, p. 18.

[648] Sousa.

[649] Swain, J. (1997). *Musical languages*. W. W. Norton.

[650] Fagan, J., Prigot, J., Carroll, M., Pioli, L., Stein, A., & Franco, A. (1997). Auditory context and memory retrieval in young infants. *Child Development, 68*, 1057-1066.

[651] Weinberger, N. M. (2004). Music and the brain. *Scientific American, 291*, 89-95. Also, Hannon, E. E., & Johnson, S. P. (2005). Infants use meter to categorize rhythms and melodies: Implications for musical structure learning. *Cognitive Psychology, 50*, 354-377.

652 Sousa.

653 Nakamura, S., Sadato, N., Oohashi, T., Nishina, E., Fuwamotc, Y., & Yonekura, Y. (1999). Analysis of music-brain interaction with simultaneous measurement of regional cerebral blood flow and electroencephalogram beta rhythm in human subjects. *Neuroscience Letters, 275*, 222-226.

654 Berendt, J. E. (1992, April 1). *The third ear: On listening to the world* (T. Nevill, Trans.). Henry Holt & Co.

655 Begley, p. 108.

656 Ibid.

657 Edelman & Tononi.

658 Zull.

659 Chugani, H. T. (1998). Biological basis of emotion: brain systems and brain development. *Pediatrics 102*(5), 1225-1229.

660 Zull.

661 Ibid. Also, Edelman.

662 Jensen, E., (2000). *Brain based learning: The new science of teaching and training.* The Brain Store, p. 246.

663 Bullard, B. (2003). Metamusic: Music for inner space. *Hemi-Sync Journal, XXI*(3, 4), i-v.

664 Sousa, p. 224.

665 Rauscher, F. H., Shaw, G. L., & Ky, K. N. (1993). Music and spatial task performance. *Nature, 365*, 611.

666 Weinberger, N. M. (1999, Spring). Can music really improve the mind? The question of transfer effects. *MuSICA Research Notes: V VI*, 12.

667 Coch, D., Fischer, K. W., & Dawson, G. (2007). *Human behavior, learning, and the developing brain: Typical development.* The Guilford Press.

668 Hetland, L. (2000). Listening to music enhances spatial-temporal reasoning: Evidence for the Mozart Effect. *Journal of Aesthetic Education, 34*, 105-148.

669 Jensen.

670 Webb, D., & Webb, T. (1990). *Accelerated learning with music.* Accelerated Learning Systems.

671 Hodges, p. 18.

672 Bennet, A. (2018c). *Possibilities that are YOU! Volume 2: Grounding.* MQIPress.

673 Scott, S. (2000). *From pride to humility: A biblical perspective.* Focus Publishing, Inc., p. 6.

674 Bennet, A. (2018d). *Possibilities that are YOU! Volume 20: The humanness of humility.* MQIPress.

675 Sriastva, S., & Cooperrider (D. L. (Eds.) (1990). *Appreciative management and leadership.* Jossey-Bass.

676 Hammond, S. A., & Hall, J. (1996). *The thin book of appreciative inquiry.* Thin Book Publishing.

677 Murray, A. (2016). *Humility: The beauty of holiness.* Aneko Press, p. 45.

678 Bridges, J. (2016). *The blessing of humility.* NavPress.

679 Murray.

680 von Hildebrand, D. (1997). *Humility: Wellspring of virtue.* Scphia Institute Press, p. 5.

681 Collins, O. (Ed.) (1998). *Speeches that changed the world.* Westminister John Knox Press, p. 32.

[682] Williams, P., & Denney, J. (2016). *Humility: The secret ingredient of success.* Shiloh Run Press, pp. 191-192.

[683] Bennet, A. (2018e). *Possibilities that are YOU! Volume 4: Conscious compassion.* MQIPress.

[684] Bridges, p. 3.

[685] Kornfield, J. (1993). *A path with heart.* Bantam Books.

[686] Davis, D., Rice, K., McElroy, S., DeBlaere, C., Choe, E., van Tongeren, D. R., & Hook, J. N. (2016). Distinguishing intellectual humility and general humility. *Journal of Positive Psychology, 11*(3), Abstract.

[687] Shelley, xiii.

[688] Forbes, J. (n.d.). The confidence gap in men and women and why it matters and how to overcome it. www.forbes.com/sites/jackzenger,2018/04/08the-confidence-gap-in-men-and-women-why-it-matters-and-how-to-overcome-it/#8b03e9663bfa

[689] Vaynerchuk, G. (2018). Blog. www.garyvaynerchuk.com/confidence-humility/

[690] Quoted from Garry Vaynerchuk's blog. www.garyvaynerchuk.com/confidence-huility/

[691] Hardwick, C. J. (2018). Blog. https://cynthiahardwick.com/humble-curiosity/

[692] Ibid.

[693] Hook, J. (2018). Blog. https://www.joshuanhook.com/what-is-huility/

[694] Cozolino.

[695] Hess & Ludwig, p. 92.

[696] Athlos Academies (20170. Reinforcing performance character growth at home: Humility. https://athlosacademies.org/humility-performance-character-growth/

[697] Kruss, E., Chancellor, J., Ruberton, P. M., & Lyubomirsky, S. (2014). An upward spiral between gratitude and humility. *Social Psychological and Personality Science, 1-10*(1).

[698] See Bloom, Paul. Psychology Course: Humility and Optimism. https://www.coursera.org/lecture/introduction-psychology/humility-and-optimism-8t96r

[699] The daily quote from Abraham-Hicks Publications (January 06, 2017).

[700] Srivastva & Cooperrider.

[701] Kruss et al., p. 8.

[702] Gardner, H. (2006b). *Five minds for the future.* Harvard Business Press.

[703] PBC Part 1, p. 68.

[704] Rescher, N. (2001). *Paradoxes: Their roots, range, and resolution.* Open Court.

[705] Salinsbury, R. M. (1988). *Paradoxes.* Cambridge University Press.

[706] van Heijenoort, J. (Ed.) (1967). *From Frege to Gödel: A source book in mathematical logic, 1879-1931.* Harvard University Press.

[707] Bennet, A. (2022, June 17). Expanding consciousness through paradox to stimulate system change. 5th Paradox & Plurality Annual Meeting, Cascais, Nova SBE, Portugal.

[708] Clark, M. (2002). *Paradoxes from A to Z* (Third Ed.). Routledge,

[709] Quine, W. V. (1966). *Ways of paradox and other essays.* Random House.

[710] Schafer (2022).

[711] Hollmen, J. (1996). *Self-organizing map (SOM).* https://users.ics.aalto.fi/jhollmen/dippa/node9.html

[712] Luca.

[713] Gomide (2022), (2021a), (2021b).

[714] Schafer, S.B. (2022, June). Personal email.

[715] Ibid.

[716] Luca.

[717] Greenfield.

[718] Sriastva, S., & Cooperrider, D. L. (Eds.) (1990). *Appreciative management and leadership*. Jossey-Bass.

[719] Hammond, S. A., & Hall, J. (1996). *The thin book of appreciative inquiry*. Thin Book Publishing.

[720] LeDoux.

[721] Bruchac, J. (1997, September). Storytelling and the sacred: On the uses of Native American stories. *Storytelling Magazine*. Special Edition.

[722] Denning, S. (2000). *The springboard: How storytelling ignites action in knowledge-era organizations*. Butterworth Heinemann, p. xx.

[723] Ibid.

[724] Ray Ban (2016). MQI research study.

[725] Wenger, E. (2000). *Communities of practice: Learning, meaning, and identity*. Cambridge University Press.

[726] Bennet et al., (2020a-e).

[727] Shelley, A. (2007). *Organizational zoo: A survival guide to work place behavior*. Aslan Publishing. Also, Shelley, A. (2017). *KNOWledge SUCCESSion: Sustained performance and capability growth through strategic knowledge projects*. Business Expert Press.

[728] Bennet et al., (2020b). This model and examples are extracted from that text.

[729] Bennet et al. (2020a).

[730] Ibid.

[731] Bennet et al. (2020c).

[732] Senge et al. (2004).

[733] Bennet & Bennet (2004).

[734] Bennet et al. (2020b).

[735] Bennet et al. (2020c).

[736] Bennet et al. (2020d).

[737] Hess & Ludwig.

[738] Bennet, Bennet, & Avedisian.

[739] Hawkins, J. C., p. 122.

[740] Ibid.

[741] Ibid, p. 96

[742] Bennet, A., & Bennet, D. (2014). Knowledge, theory and practice in knowledge management: Between associative patterning and context-rich action. In P. Lambe (Ed.), Knowledge management special issue: Connecting theory and practice, *Journal of Entrepreneurship, Management and Innovation 10*(1), 7-56.

[743] Bohm, D. (1980). *Wholeness and the implicate order*. Routledge & Kegal Paul.

[744] Surinder Kumar Batra in the Sampler Call. Bennet & Bennet (2014).

[745] Kissinger et al., p. 194.

[746] Ibid., p. 202.

[747] See https://www.humanbrainproject.eu/en/about/overview/ for an overview of the European Human Brain Project (HBP).

[748] See https://braininitiative.nih.gov for the NIH BRAIN Initiative.

[749] See https://medlineplus.gov/neurologicdiseases.html for U.S. National Library of Medicine Information site for Neurologic Diseases.

[750] Haines, D. E. (2019). *Neuroanatomy atlas in clinical context: Structures, sections, systems, and syndrome*. Wolters Kluwer.

[751] James Grier Miller (1916-2002) is the originator of the living systems theory. He is also credited with developing the term "behavioral science" and founded and directed the Mental Health Research Institute at the University of Michigan. See Miller, J. G. (1978). Living systems. McGraw-Hill.

[752] Faulkner, W. (1950, December 10). The Nobel Prize in Literature 1949 banquet speech, given at City Hall in Stockholm. www.nobelprize.org

[753] Wilson, E. O. (1998). *Consilience: The unity of knowledge*. Alfred A. Knopf.

[754] Ibid. Also, Bennet, & Bennet (2004).

[755] Wilson (1998).

[756] Snow in Wilson (1998), p. 11.

[757] Stonier (1997), p. 14.

[758] Ibid. Also, Stonier (1990).

[759] Bennet, Bennet & Avedisian.

[760] Edelman & Tononi, p. 13.

[761] Merriam, S. B., Caffarella, A. S., & Baumgartner, L. M. (2007). *Learning in adulthood: A comprehensive guide*. Jossey-Bass. Also, Mezirow, J., & Associates (2000). *Learning at transformation: Critical perspectives on a theory in progress*. Jossey-Bass.

[762] Mezirow (2000).

[763] Jarvis, P. (2004). *Adult education and lifelong learning: Theory and practice*. RoutledgeFalmer. Also, Merriam et al. (2007). Also, Moon.

[764] Candy, P. C. (1991). *Self-direction for lifelong learning*. Jossey-Bass. Also, Dewey (1938/197, 1958). Also, Lave, J. (1988). *Cognition in practice: Mind, mathematics and culture in everyday life*. Cambridge University Press. Also, Piaget, J. (1951). *Play, dreams and imitation in childhood*. W. W. Norton. Also, Piaget, J. (1968). *Structuralism*. Harper Torchbooks. Also, Piaget, J. (1971). *Psychology and epistemology*. Penguin Books. Also, Vygotsky, L. S. (1978). *Mind in society: The development of higher psychological processes*. Harvard University Press.

[765] Jarvis. Also, Merriam (2007). Also, Mezirow (1991). Also, Moon.

[766] Miller (1978).

[767] Senge (1990).

[768] Battram.

[769] Ibid. Also, Buchanan, M. (1004, Spring). Power laws and the new science of complexity management. *Strategy + Business, 34*, 70-79. Also, Gell-Mann. Also, Gladwell, M. (2000). *The tipping point: How little things can make a big difference*. Little, Brown.

[770] Bennet, A. (2018f). *Possibilities that are YOU ! Volume 15: Seeking wisdom*. MQIPress.

[771] Bennet & Bennet (2004).

[772] Wong, C. Y. L., Choi, C. J., & Millar, C. C. J. M. (2006). The case of Singapore as a knowledge-based city. In F. J. Carrillo (Ed.), *Knowledge cities: Approaches, experiences, and perspectives*. Elsevier, p. 91.

[773] Lipton.

[774] Ibid., p. 144.

[775] Bennet, & Bennet (2008b); Bennet, Bennet, & Avedisian.

[776] Church, D. (2006). *The genie in your genes: Epigenetic medicine and the new biology of intention*. Elite Books, p. 144.
[777] Meacham, J. A. (1995). The loss of wisdom. In R. J. Sternberg (Ed.), *Wisdom, its nature, origins, and development*. Cambridge University Press.
[778] *American Heritage Dictionary*.
[779] Rock.
[780] Kandel.
[781] Marshall, S. P. (2005, September-November). A decidedly different mind. *Shift: At the frontiers of Consciousness*.
[782] Bennet, Bennet, & Avedisian.
[783] Sternberg, R. J. (Ed.) (1998). *Wisdom: Its nature, origins, and development*. Cambridge University Press.
[784] Goldberg (2005). Also, Goldberg, E. (2009). *The new executive brain: Frontal lobes in a complex world*. Oxford University Press.
[785] *American Heritage Dictionary*.
[786] Bennet, Bennet, & Avedisian.
[787] Bennet, & Bennet (2013).
[788] See Expanded Consciousness Institute: www.expanded-consciousness.com.
[789] Pert.
[790] Hawkins (2004).
[791] Kuntz (1968), p. 162.
[792] Hawkins (2004), p. 109.
[793] Ibid.
[794] Bennet, & Bennet (2013).
[795] Bennet, A. (2018g). *Possibilities that are YOU! Volume 22: Beyond action*. MQIPress, pp. 54-56.
[796] Bennet, Bennet, & Turner (2018).

Bibliography

Adams, J. K. (1957). Laboratory studies of behavior without awareness. *Psychological Bulletin, 54*, 383-405.

Adolphs, R. (2004). Processing of emotional and social information by the human amygdala. In M. S. Gazzaniga (Ed.), *The cognitive neurosciences III* (pp. 1017-1030). The Bradford Press.

Akil, H., Campeau, S., Cullinan, W., Lechang, R., Toni, R., Watson, S., & Moore, R. (1999). Neuroendocrine system 1: Overview—thyroid and adrenal axis. In M. Zigmond, F. Bloom, S. Landis, J. Roberts & L. Squire (Eds.), *Fundamentals of neuroscience* (p. 1146). Academic Press.

Alberts, D. S., & Hayes, R. E. (2005). *Power to the edge: Command, control in the information age.* Command & Control Research Program.

Amabile, T. M. (1996). *Creativity in context.* Westview.

Amabile, T. M. (1983). *The social psychology of creativity.* Springer.

Amen, D. G. (2005). *Making a good brain great.* Harmony Books.

American Heritage Dictionary (4th ed.). (2006). Houghton Mifflin Co.

Anderson, J., & Dos Reis, T. (2014). Transdifferential and transintegral calculus. In H. K. Kim, M. A. Amouzegar & S. Ao (Eds.), *Transactions on engineering technologies: World congress on engineering and computer science 2014* (pp. 92-96).

Anderson, J. A. (1988). Cognitive styles and multicultural populations. *Journal of Teacher Education, 39*(1), 2-9.

Anderson, J. R. (2004). *Cognitive psychology and its implications.* Worth.

Andersson, J. L., Lilja, A., Hartvig, P., Langstrom, B., Gordh, T., Handwerker, H., et al. (1997). Somatotopic organization along the central sulcus, for pain localization in humans, as revealed by positron emission tomography. *Experimental Brain Research, 117*, 192-199.

Andreason, N. C. (2005). *The creating brain: The neuroscience of genius.* The Dana Foundation.

Andrews, B., Brewin, C. R., Ochera, J., Morton, J., Bekerian, D.A., Davies, G. M., & Mollon, P. (1999, August). Characteristics, context and consequences of memory recovery among adults in therapy. *British Journal of Psychiatry, 175*, 141-146.

Argyris, C. (1990). *Overcoming organizational defenses: Facilitating organizational learning.* Prentice Hall.

Argyris, C., & Schon, D. (1978). *Organizational learning: A theory of action perspective.* Addison-Wesley.

Armstrong, D. M. (1989). *Universals: An opinionated introduction.* Westview Press.

Ascoli, G. A. (Ed.). (2002). *Computational neuroanatomy: Principles and methods.* Humana Press.

Asplund, C. L., Fougnie, D., Zughni, S., Martin, J. W., & Marois, R. (2014, January 16). The attentional blink reveals the probabilistic nature of discrete conscious perception. *Psychological Science.* http://pss.sagepub.com/

Atherton, J. S. (2005). *Learning and teaching: Reflection and reflective practice.* http://www.learningandteaching.info/learning/reflecti.htm

Athlos Academies (2017). *Reinforcing performance character growth at home: Humility.* https://athlosacademies.org/humility-performance-character-growth/

Atwater, F. H. (2004). *The Hemi-Sync® Process.* The Monroe Institute.

Ausubel, D. (1968). *Educational psychology: A cognitive view*. Holt, Rinehart Winston.

Autor, D., Mindell, D. A., & Reynolds, E. B. (2021). *The work of the future: Building better jobs in an age of intelligent machines*. The MIT Press.

Avedisian, J., & Bennet, A. (2010). Values as knowledge: A new frame of reference for a new generation of knowledge workers. *On the Horizon, 18*(3), 255-265.

Axelrod, R., & Cohen, M. D. (1999). *Harnessing complexity: Organizational implications of a scientific frontier*. The Free Press.

Baier. A. (2004). Demoralization, trust, and the virtues. In C. Calhoun (Ed.), *Setting the moral compass: Essays by women philosophers* (176-189). Oxford University Press.

Bargh, J. A. (2005). Bypassing the will: Towards demystifying the nonconscious control of social behavior. In R. R. Hassan, J. S. Uleman, & J. A. Bargh (Eds.), *The new unconscious*. Oxford University Press.

Bateson, G. (1972). *Steps to an ecology of the mind*. Ballantine.

Battram, A. (1996). *Navigating complexity: The essential guide to complexity theory in business and management*. The Industrial Society.

Bear, M. F., Connors, B. W., & Paradiso, M. A. (2001). *Neuroscience: Exploring the brain* (2nd ed.). Lippincott Williams & Wilkins.

Beauregard, M., & O'Leary, D. (2007). *The spiritual brain : A neuroscientist's case for the existence of the soul*. HarperOne.

Bechara, A. (2004). The role of emotion in decision-making: Evidence from neurological patients with orbitofrontal damage. *Brain Cognition 55*(1), 30-40.

Beggs, J. M., Brown, T. J., Byrne, J. H., Crow, T., LeDoux, J. E., LeBar, K., & Thompson, R. J. (1999). Learning and memory; Basic mechanisms. In M. Zigmond, F. Bloom, S. Landis, J. Roberts & L. Squire (Eds.), *Fundamental neuroscience* (1411-1454). Academic Press.

Begley, S. (2007). *Train your mind change your brain: How a new science reveals our extraordinary potential to transform ourselves*. Ballantine Books.

Belasco, J. A. (1997). Leading the new organizations. In K. Shelton (Ed.), *A new paradigm of leadership: Visions of excellence for 21st century organizations* (37-38). Executive Excellent Publishing.

Bennet, A. (2022, June 17). *Expanding consciousness through paradox to stimulate system change*. [Presentation]. 5th Paradox & Plurality Annual Meeting, Cascais, Nova SBE, Portugal.

Bennet, A. (2022). Seeking global coherence: The waxing and waning of trust in government media. In S. B. Schafer & A. Bennet (Eds.), *Handbook of research on global media's preternatural influence on global technological singularity, culture, and government* (pp. 1-16). IGI Global.

Bennet, A. (2018a). *Possibilities that are YOU! Volume 14: The emoting guidance system*. MQIPress.

Bennet, A. (2018b). *Possibilities that are YOU! Volume 8: Me as co-creator*. MQIPress.

Bennet, A. (2018c). *Possibilities that are YOU! Volume 2: Grounding*. MQIPress.

Bennet, A. (2018d). *Possibilities that are YOU! Volume 20: The humanness of humility*. MQIPress.

Bennet, A. (2018e). *Possibilities that are YOU! Volume 4: Conscious compassion*. MQIPress.

Bennet, A. (2018f). *Possibilities that are YOU! Volume 15: Seeking wisdom.* MQIPress.

Bennet, A. (2018g). *Possibilities that are YOU! Volume 22: Beyond action.* MQIPress.

Bennet, A. (2017). *The culture of connection.* [Presentation]. Leading Digital and Cultural Transformation, G-LINK, Bangkok, Thailand.

Bennet, A. (2005). *Exploring aspects of knowledge management that contribute to the passion expressed by its thought leaders.* Self-published. www.mountainquestinstitute.com

Bennet, A., & Bennet, D. (2018). *Decision-making in the new reality: Complexity, knowledge and knowing.* MQIPress.

Bennet, A., & Bennet, D. (2014). Knowledge, theory and practice in knowledge management: Between associative patterning and context-rich action. In P. Lambe (Ed.), *Knowledge management special issue: Connecting theory and practice, Journal of Entrepreneurship, Management and Innovation, 10*(1), 7-56.

Bennet, A., & Bennet, D. (2008a). The human knowledge system: Music and brain coherence. *VINE: The Journal of Information and Knowledge Management Systems, 38*(3), 277-295.

Bennet, A., & Bennet, D. (2008b). Engaging tacit knowledge in support of organizational learning. *VINE: The Journal of Information and Knowledge Management Systems, 38*(1), 72-94.

Bennet, A., & Bennet, D. (2008c). The fallacy of knowledge reuse. *Journal of Knowledge Management, 12*(5), 21-33.

Bennet, A., & Bennet, D. (2008). The decision-making process for complex situations in a complex environment. In C.W. Holsapple & F. Burstein (Eds.), *Handbook on decision support systems* (3-20). Springer-Verlag.

Bennet, A., & Bennet, D. (2007). The knowledge and knowing of spiritual learning. *VINE: The Journal of Information and Knowledge Management Systems, 37*(1), 27-40.

Bennet, A., & Bennet, D. (2007). *Knowledge mobilization in the social sciences and humanities: Moving from research to action.* MQIPress.

Bennet, A., & Bennet, D. (2006). Learning as associative patterning. *VINE: The Journal of Information and Knowledge Management Systems, 36*(4), 371-376.

Bennet, A., & Bennet, D. (2004). *Organizational survival in the new world: The intelligent complex adaptive system.* Elsevier.

Bennet, A., Bennet, D., & Avedisian, J. (2018). *The course of knowledge: A 21st century theory.* MQIPress.

Bennet, A., Bennet, D., & Lee, S. L. (2009). Exploring the military contribution to KBD through leadership and values. *Journal of Knowledge Management, 14*(2), 326.

Bennet, A., Bennet, D., & Lewis, J. (2015). *Leading with the future in mind: Knowledge and emergent leadership.* MQIPress.

Bennet, A., Bennet, D., Shelley, A., Bullard, T., & Lewis, J. (2020a). *The profundity and bifurcation of change part I: Laying the groundwork.* MQIPress.

Bennet, A., Bennet, D., Shelley, A., Bullard, T., & Lewis, J. (2020b). *The profundity and bifurcation of change part II: Learning from the past.* MQIPress.

Bennet, A., Bennet, D., Shelley, A., Bullard, T., & Lewis, J. (2020c). *The profundity and bifurcation of change volume III: Learning in the present.* MQIPress.

Bennet, A., Bennet, D., Shelley, A., Bullard, T., & Lewis, J. (2020d). *The profundity and bifurcation of change part IV: Co-creating the future.* MQIPress.

Bennet, A., Bennet, D., Shelley, A., Bullard, T., Lewis, J., & Panucci, D. (2020e). *The Profundity and Bifurcation part V: Living the future.* MQIPress.

Bennet, A., Bennet, D., Shelley, A., Bullard, T., & Lewis, J. (2018). *The intelligent social change journey: Foundation for the Possibilities that are YOU!* MQIPress.

Bennet, A., & Scott C. L. (2019). *With passion, we live and love.* MQIPress.

Bennet, D. (2006). Expanding the knowledge paradigm. *VINE: The Journal of Information and Knowledge Management Systems, 36*(2), 175-181.

Bennet, D. (2001). Loosening the world knot. Unpublished paper. Mountain Quest Institute. www.mountainquestinstitute.com

Bennet, D., & Bennet, A. (2021). *The QUEST: Where the mountains meet the library.* MQIPress.

Bennet, D., & Bennet, A. (2010). Social learning from the inside out: The creation and sharing of knowledge from the mind/brain perspective. In J. Girard & J. Girard (Eds.), *Social knowledge: Using social media to know what you know* (pp. 1-23). IBI Global.

Bennet, D., & Bennet, A. (2008a). Associative patterning: The unconscious life of an organization. In. J. P. Girard (Ed.), *Organizational memory* (pp. 201-224). IGI Global.

Bennet, D., & Bennet, A. (2008b). Engaging tacit knowledge in support of organizational learning. *The Journal of Information and Knowledge Management Systems, 38*(1), 72-94.

Bennet, D., Bennet, A., & Turner, R. (2018). *Expanding the self: The intelligent complex adaptive learning system. A new theory of adult learning.* MQIPress.

Bennett-Woods, D. (1997). Reflections on wisdom. (Unpublished paper). University of Northern Colorado.

Bennis, W. (1995). *Beyond leadership: Balancing economics, ethics and ecology.* Blackwell Publishers.

Berendt, J. E. (1992, April 1). *The third ear: On listening to the world* (T. Nevill, Trans.). Henry Holt & Co.

Berman, M. (1981). *The reenchantment of the world.* Cornell University Press.

Berry, D. C., & Broadbent, D. E. (1984). On the relationship between task performance and associated verbalizable knowledge. *Quarterly Journal of Experimental Psychology, 36A,* 209-231.

Besant, A., & Leadbeater, C. W. (1999). *Thought-forms.* The Theosophical Publishing House.

Blackmore, S. (2004). *Consciousness: An introduction.* Oxford University Press.

Blackmore, S. (1999). *The meme machine.* Oxford University Press.

Blakemore, S., & Frith, Y. (2005). *The learning brain: Lessons for education.* Blackwell.

Bloom, F. E. (2007). *Best of the brain from Scientific American: Mind, matter and tomorrow's brain.* Dana Press.

Bloom, F. E., Fischer, B. A., Landis, S. C., Roberts, J. L., Squire, L. R., & Zigmond, M. J. (1999). Fundamentals of neuroscience. In M. J. Zigmond, F. E. Bloom, B. A. Landis, S. C. Roberts, & L. R. Squire (Eds.), *Fundamental neuroscience* (3-8). Academic Press.

Boden, M. (1991). *The creative mind, myths & mysticism.* Basic Books.

Bohm, D. (1992). *Thought as a system.* Routledge.

Boud, D., Keogh, R., & Walker, D. (Eds.) (1994). *Reflection: Turning experience into learning.* Routledge Falmer.

Bower, J. L. (2000). The purpose of change: A commentary on Jensen and Senge. In M. K. Beer & N. Nohria (Eds.), *Breaking the code of change* (93-95). Harvard Business School Press.

Bownds, M. D. (1999). *The biology of mind: Origins and structures of mind, brain, and consciousness.* Wiley.

Brainnet (2008). *Decade of the brain 1990-1999. Proclamation by the President of the United States of America.* http://www.brainnet.org/proclaim.htm

Bridges, J. (2016). *The blessing of humility.* NavPress.

Brookfield, S. D. (1987). *Developing critical thinkers.* Jossey-Bass.

Brown, J. (1958). Some tests of the decay theory of immediate memory. *Quarterly Journal of Experimental Psychology, 10,* 12-21.

Brown, K. W., & Ryan, R. M. (2003). The benefits of being present: Mindfulness and its role in psychological well-being. *Journal of Personality and Social Psychology, 84*(4), 822-848.

Brown, M. Y. (1979). *The art of guiding: The psychosynthesis approach to individual counselling and psychology.* Johnson College, University of Redlands.

Bruchac, J. (1997, September). Storytelling and the sacred: On the uses of Native American stories. *Storytelling Magazine.* Special Edition.

Buchanan, M. (2004, Spring). Power laws and the new science of complexity management. *Strategy + Business, 34,* 70-79.

Buckner, R. L., Andrews-Hanna, J. R., & Schacter, D. L. (2008). The brain's default network: Anatomy, function, and relevance to disease., *Annals of the New York Academy of Sciences 1124,* 1-38.

Buckner, R. L., & Vincent, J. L. (2007). Unrest at rest : Default activity and spontaneous network correlations. *NeuroImage 37*(4), 1091-1096.

Buks, E., Schuster, R., Heiblum, M., Mahalu, D., & Umansky, V. (2998). Dephasing in electron interference by a "which-path" detector. *Nature, 391,* 871-874.

Bullard, B. (2003). Metamusic: Music for inner space. *Hemi-Sync Journal, XXI*(3, 4), i-v.

Bullard, B., & Bennet, A. (2013/2020). *Remembrance: Pathways to expanded learning with music and meta-music®.* MQIPress.

Buonomano, D. V., & Merzenich, M. M. (1998). Cortical plasticity: From synapses to maps. *Annual Review of Neuroscience, 21,* 149-186.

Butler, J. (1906). *The analogy of religion, natural & revealed.* J. M. Dent & Company.

Buzsaki, G. (2006). *Rhythms of the brain.* Oxford University Press.

Byrnes, J. P. (2007). Some ways in which neuroscientific research can be relevant to education. In D. Coch, K. W. Fischer & G. Dawson (Eds.), *Human behavior, learning and the developing brain: typical development.* The Guilford Press.

Byrnes, J. P. (2001). *Minds, brains, and learning: Understanding the psychological and educational relevance of neuroscientific research.* The Guilford Press.

Caine, G., & Caine, R. N. (2006). Meaningful learning and the executive functions of the brain. In S. Johnson & K. Taylor (Eds.), *The neuroscience of adult learning: New directions for adult and continuing education.* Jossey-Bass.

Calaprice, A. (2000). *The expanded quotable Einstein.* Princeton University Press.

Campbell, D. T. (1960). Blind variation and selective retention in creative thought and other knowledge processes. *Psychological Review, 67*, 380-400.

Candy, P. C. (1991). *Self-direction for lifelong learning.* Jossey-Bass.

Cannon, W. B. (1929). *Bodily changes in pain, hunger, fear and rage.* (Vol. 2). Appleton.

Carlsson, B., Keane, P., & Martin, J. B. (1976, Spring). R&D organizations as learning systems. *Sloan Management Review,* 21-31.

Carr, L., Iacoboni, M., Dubeau, M. c., Mazziotta, J. C., & Lenzi, G. L. (2003). Neural mechanisms of empathy in humans: A relay from neural systems for imitation to limbic areas. *Proceedings of the National Academy of Science, 100,* 5497-5502.

Carrasco, M., Ling, S., & Read, S. (2004). Attention alters appearance. *Nature Neuroscience, 7,* 308-313.

Carroll, S. (2016). *The big picture: On the origins of life, meaning and the universe itself.* Dutton.

Carter, R. (2002). *Exploring consciousness.* University of California Press.

Castells, M. (1996). *The information age: Economy, society and culture, Volume 1: The rise of the network society.* Blackwell Publishers, Ltd.

Ceci, S. J. (1995). False beliefs: Some developmental and clinical considerations. In D. L. Schacter (Ed.), *Memory distortion: How minds, brains and societies reconstruct the past* (92-128). Harvard University Press.

Checkland, P., & Holwell, S. (1998). *Information, systems and information systems:Making sense of the field.* John Wiley & Sons.

Chickering, A. W. (1977). *Experience and learning: An introduction to experiential learning.* Change Magazine Press.

Chickering, A. W., Dalton, J. C., & Stamm, L. (2005). *Encouraging authenticity & spirituality in higher education.* Jossey-Bass.

Christos, G. (2003). *Memory and dreams: The creative human mind.* Rutgers University Press.

Chugani, H. T. (1998). Biological basis of emotion: Brain systems and brain development. *Pediatrics 102*(5), 1225-1229.

Chung, S., Tack, G., Lee, B., Eom, G., Lee, S., & Sohn, J. (2004). The effect of 30 oxygen on visuospatial performance and brain activation: An fMRI study. *Brain and Cognition, 56,* 279-285.

Church, D. (2006). *The genie in your genes: Epigenetic medicine and the new biology of intention.* Elite Books.

Churchland, P. S. (2002). *Brain-wise studies in neurophilosophy.* The MIT Press.

Churchman, C. W. (1977). A philosophy for complexity. In H. A. Linstone & W. H. C. Simmonds (Eds.), *Futures research: New directions.* Addison-Wesley Publishing Company, Inc.

Clark, M. (2002). *Paradoxes from A to Z* (Third Ed.). Routledge.

Clausing, D. (1994). *Total quality development: A step-by-step guide to world-class concurrent engineering.* Asme Press.

Clayton, V., & Birren, J. E. (1980). The development of wisdom across the lifespan: A reexamination of an ancient topic. In P. B. Baltes & O. G. J. Brim (Eds.), *Life span development and behavior* (104-135). Academic Press.

Coch, D., Fischer, K. W., & Dawson, G. (2007). *Human behavior, learning, and the developing brain: Typical development.* The Guilford Press.

Collins, O. (Ed.) (1998). *Speeches that changed the world*. Westminister John Knox Press.

Connerton, P. (2008). *Seven types of forgetting*. http://mss.sagepub.com/content/1/1/49.short?rss=1&ssource=mfr

Cools, J. (2022). Counter-hacking the subconscious mind antidote for faulty human perceptions from the media sphere. In S. B. Schafer & A. Bennet (Eds.), *Handbook of research on global media's preternatural influence on global singularity, culture and government* (372-396). IGI Global.

Cooper, R. (2005). Austinian truth, attitudes and type theory. *Research on Language and Computation 5*, 333-362.

Costa, J. D. (1995). *Working wisdom: The ultimate value in the new economy*. Stoddart.

Cousins, E. (1999). Spirituality on the evening of a new millennium: An overview. *Chicago Studies, 38*(1), 5-14.

Cowan, W. M., & Kandel, E. R. (2001). A brief history of synapses and synaptic transmission. In W. C. Cowan, T. C., Sudhof & C. F. Stevens (Eds.), *Synapses*. Johns Hopkins Press.

Cozolino, L. J. (2006). *The neuroscience of human relationships: Attachment and developing social brain*. W. W. Norton.

Cozolino, L. J. (2002). *The neuroscience of psychotherapy: Building and rebuilding the human brain*. Norton.

Cozolino, L., & Sprokay, S. (2006). Neuroscience and adult learning. In S. Johnson & T. Taylor (Eds.), *The neuroscience of adult learning* (11-19). Jossey-Bass.

Crandall, B., Klein, G., & Hoffman, R. R. (2006). *Working minds: A practitioner's guide to cognitive task analysis*. The MIT Press.

Creswell, J. D. (2017). Mindfulness interventions. *Annual Review of Psychology, 68*(1), 491-516.

Csikszentmihalyi, M. (2003). *Good business: Leadership, flow and the making of meaning*. Viking.

Csikszentmihalyi, M. (1993). *The evolving self: A psychology for the third millennium*. HarperCollins Publishing.

Csikszentmihalyi, M. (1990). *Flow: The psychology of optimal experience*. Harper Perennial.

Csikszentmihalyi, M., & Csikszentmihalyi, I. S. (Eds.) (1988). *Optimal experience: Psychological studies of flow in consciousness*. Cambridge University Press.

Daloz, L. (1999). *Mentor: Guiding the journey of adult learners*. Jossey-Bass.

Daloz, L. (1986). *Effective teaching and mentoring*. Jossey-Bass.

Damasio, A. R. (2010). *Self comes to mind: Constructing the conscious brain*. Vintage Books.

Damasio, A. R. (2007). How the brain creates the mind. In F. E. Bloom (Ed.), *Best of the brain from Scientific American: Mind, matter, and tomorrow's brain* (pp. 58-67). Dana Press.

Damasio, A. R. (1999). *The feeling of what happens: Body and emotion in the making of consciousness*. Harcourt Brace & Company.

Damasio, A. R. (1994). *Descartes' error: Emotion, reason, and the human brain*. G.P. Putnam's Sons.

Damasio, A. R., Grabowski, T. J., Bechara, A., Damasio, H., Ponto, L. L., Parvizi, J., & Hichwa, R. D. (2000). Subcortical and cortical brain activity during the feeling of self-generated emotions. *Nature Neuroscience, 3*(10), 1049-1056.

Davenport, T. H. (2001). Knowledge work and the future of management. In W. Bennis, G. M. Spreitzer & T. G. Cummings (Eds.), *The future of leadership: Today's top leadership thinkers speak to tomorrow's leaders*. Jossey-Bass.

Davenport, T. H., & Beck, J. C. (2001). *The attention economy.* Harvard Business Review Press.

Davis, D., Rice, K., McElroy, S., DeBlaere, C., Choe, E., van Tongeren, D. R., & Hook, J. N. (2016). Distinguishing intellectual humility and general humility. *Journal of Positive Psychology, 11*(3), Abstract.

de Bono, E. (2015). *Lateral thinking: Creativity step by step.* Harper Colophon.

Delmonte, M. M. (1984). Electrocortical activity and related phenomena associated with mediation practice: A literature review. *International Journal of Neuroscience, 24*, 217-231.

Denning, S. (2000). *The springboard: How storytelling ignites action in knowledge-era organizations.* Butterworth Heinemann.

deWaal, R. (2009). *The age of empathy: Nature's lessons for a kinder society.* Crown Publishing Group.

Dewey, J. (1958). *Experience and nature.* Dover Publications.

Dewey, J. (1938/1997). *Experience and education.* Simon & Schuster.

Diamond, M. (2001, March 10). Successful aging of the healthy brain. [Presentation]. American Society on Aging Conference, New Orleans.

Dilts, R. (2003). *From coach to awakener.* Meta Publications.

Dobbs, D. (2007). Turning off depression. In F. E. Bloom (Ed.), *Best of the brain from Scientific American: Mind, matter, and tomorrow's brain* (pp. 169-178). Dana Press.

Dorfman, J., Shames, V. A., & Kihlstrom, J. F. (1996). Intuition, incubation, and insight: Implicit cognition in problem-solving. In G. Underwood (Ed.), *Implicit cognition*. Oxford University Press.

Dunning, J. (2014). Discussion of consciousness via the internet on December 13.

Ebbinghaus, H. (188). Memory: A contribution to experimental psychology. http://nwkpsych.rutgers.edu/

Edelman, G. M. (1992). *Bright air, brilliant fire: On the matter of the mind.* Basic Books.

Edelman, G., & Tononi, G. (2000). *A universe of consciousness: How matter becomes imagination.* Basic Books.

Ehrenfeld, D. (1981). *The arrogance of humanism.* Oxford University Press.

Eich, E., Kihlstrom, J. F., Bower, G. H., Forgas, J. P., & Niedenthal, P. M. (2000). *Cognition and emotion.* Oxford University Press.

Eilan, N., Marcel, A., & Bermudez, J. L. (1995). Self-consciousness and the body: An interdisciplinary introduction. In J. L. Bermudez, A., Marcel & N. Eilan (Eds.), *The body and the self.* MIT Press.

Einstein, A., & Infeld, L. (1950). *The evolution of physics.* Simon & Schuster.

Engle, R. W., Cantor, J., & Carullo, J. J. (1993). Individual difference in working memory and comprehension: A test of four hypotheses. *Journal of Experimental Psychology: Learning, memory, and Cognition, 19*, 972-992.

English, L. M., & Gillen, M. A. (Eds.) (2000). *Addressing the spiritual dimensions of adult learning: What educators can do.* Jossey-Bass Publishers.

Epstein, M. (1995). *Thoughts without a thinker.* Basic Books.

Ericsson, K. A., Charness, N., Feltovich, P. J., & Hoffman, R. R. (Eds.) (2006). *The Cambridge handbook of expertise and expert performance.* Cambridge University Press.

Erikson, J. M. (1988). *Wisdom and the senses: The way of creativity.* Norton.

Esposito, F., Bertolino, A., Scarabino, T., Latoffe, V., Blasi, G., Popolizio, T., … Di Salle, F. (2006). Independent component model of the default-ode brain function: Assessing the impact of active thinking. *Brain Research Bulletin, 70*(4-6), 263-269.

Eysenck, H. J. (1993). Creativity and personality: A theoretical perspective. *Psychological Inquiry, 4,* 147-178.

Fagan, J., Prigot, J., Carroll, M., Pioli, L., Stein, A., & Franco, A. (1997). Auditory context and memory retrieval in young infants. *Child Development, 68,* 1057-1066.

Feigl, H. (1958). The mental and the physical. In *Concepts, theories and the mind-body problem* (pp. 370-497). University of Minnesota Press.

Feldman, A. (1996). Enhancing the practice of physics teachers. *Journal of Research in Science Teaching, 33*(5), 513-540.

Finke, R. A., Ward, T. B., & Smith, S. M. (1992). *Creative cognition: Theory, research, and applications.* MIT Press.

Fischer, R. (1971). A cartography of ecstatic and meditative states. *Science, 174*(4012), 897-904.

Forbes, J (n.d.). *The confidence gap in men and women and why it matters and how to overcome it.* www.forbes.com/sites/jackzenger,2018/04/08the-confidence-gap-in-men-and-women-why-it-matters-and-how-to-overcome-it/#8b03e9663bfa

Fox, M. D., Snyder, A. Z., Vincent, J. L., Corbetta, M., Van Essen, D. C., & Raichle, M. E. (2005). The human brain is intrinsically organized into dynamic, anti-correlated functional networks. Proceedings of the *National Academy of Sciences, 102*(27), 9673-9678.

Freud, S. (1908/1959). The relation of the poet to day-dreaming. In *Collected papers 4* (173-183). Hogarth Press.

Friedman, T. L. (2005). *The world is flat: A brief history of the twenty-first century.* Farrar, Straus & Giroux.

Frijda, N. H. (2000). The psychologists' point of view. In M. Lewis & J. M. Haviland-Jones (Eds.), *Handbook of emotions.* The Guilford Press.

Frith, C., & Wolpert, D. (2003). *The neuroscience of social interaction: Decoding, imitating, and influencing the actions of others.* Oxford University Press.

Gardner, H. (2011). *Truth, beauty and goodness reframed: Educating for the virtues in the age of truthiness and twitter.* Basic Books.

Gardner, H. (2006a). *Multiple intelligences: New horizons in theory and practice.* Basic Books.

Gardner, H. (2006b) *Five minds for the future.* Harvard Business Press.

Gell-Mann, M. (1994). *The quark and the jaguar: Adventures in the simple and the complex.* W. H. Freeman and Company.

George, M. S. (2007). Stimulating the brain. In F. E. Bloom (Ed.), *Best of the brain from Scientific American: Mind, matter, and tomorrow's brain* (pp. 20-34). Dana Press.

Gibson, J. J. (1979). *The ecological approach to visual perception.* Houghton Mifflin.

Gigerenzer, G. (2007). *Gut feelings: The intelligence of the unconscious.* Penguin Books.

Gilbert, D. T., & Wilson, T. D. (2007). Prospection: Experiencing the future. *Science, 317*(5843), 1351-1354.

Gladwell, M. (2005). *Blink: The power of thinking without thinking.* Little, Brown.

Gladwell, M. (2000). *The tipping point: How little things can make a big difference.* Little, Brown.

Goldberg, E. (2009). *The new executive brain: Frontal lobes in a complex world.* Oxford University Press.

Goldberg, E. (2005). *The wisdom paradox: How your mind can grow stronger as your brain grows older.* Gotham Books.

Goldberg, E. (2001). *The executive brain: Frontal lobes and the civilized mind.* Oxford University Press.

Goldringer, S. D. (1996). Words and voices: Episodic traces in spoken word identification and recognition memory. *Journal of Experimental Psychology: Learning, Memory, and Cognition, 22*, 1166-1183.

Goleman, D. (2015). *A force for good: The Dalai Lama's vision for our world.* Bantam Books.

Gomide, W. (2022). Transreal numbers and sentient logic. In S. B. Schafer & A. Bennet (Eds.), *Handbook of research on global media's preternatural influence on global singularity, culture and government* (pp. 269-278) IGI Global.

Gomide, W. (2021a). Non-causality and transreal numbers: The world seen at speed of light. [Unpublished]. UFMT/Brazil.

Gomide, W. (2021b). Metaphysical aggregates and their insertion into Cantor's thought. [Unpublished]. UFMT/Brazil.

Goswami, A. (2014). *Quantum creativity: Think quantum. Be creative.* Hay House, Inc.

Green, R. D. (2016). Conversation from *Huffington Post.* http://www.huffingtonpost.com/r-kay-green/giving-back_b_3298691.html

Greenfield, S. (2000). *The private life of the brain: Emotions, consciousness, and the secret of the self.* John Wiley & Sons.

Greicius, M. D., Krasnow, B., Reiss, A. L., & Menon, V. (2003). Functional connectivity in the resting brain: A network analysis of the default mode hypothesis. *Proceedings of the National Academy of Sciences 100*(1), 253-258.

Grimm, R., Dietz, N., Foster-Bey, J., Reingold, D., & Nesbit, R. (2006). *Volunteer growth in America: A review of trends since 1974.* [Research Report]. Corporation for National and Community Service.

Gruber, H. E. (1989). The evolving systems approach to creative work. In D. B. Wallace & H. E. Gruber (Eds.), *Creative people at work: Twelve cognitive case studies* (pp. 3-24). Oxford University Press.

Grudin, R. (1984). The ethics of inspiration. In *The phenomenon of CHANGE.* Rizzoli for Cooper-Hewitt Museum, The Smithsonian Institution's National Museum of Design.

Guilford, J. P. (190). Creativity. *American Psychologist, 5*, 444-454.

Gusnard, D. A., & Raichle, M. E. (2004). Functional imaging, neurophysiology, and the resting state of the human brain. *The cognitive neurosciences III* (pp. 1267-1280). Cambridge, MA: MIT Press.

Haberlandt, K. (1998). *Human memory: Exploration and application.* Allyn & Bacon.

Haider, P. (n.d.). *60+ percent of all people believe in the soul.* Sivana East. http://www.blog.sivanaspirit.com

Haidt, J. (2003). Elevation and the positive psychology of morality. In C. L. M. Keyes & J. Haidt (Eds.), *Flourishing positive psychology and the life well-lived* (pp. 275-289). American Psychological Association.

Haines, D. E. (2019). *Neuroanatomy atlas in clinical context: Structures, sections, systems, and syndrome.* Wolters Kluwer.

Halpern, D. F. (2996). *Thought and knowledge: An introduction to critical thinking.* Erbaum Associates.

Hammond, S. A., & Hall, J. (1996). *The thin book of appreciative inquiry.* Thin Book Publishing.

Hannon, E. E., & Johnson, S. P. (2005). Infants use meter to categorize rhythms and melodies: Implications for musical structure learning. *Cognitive Psychology, 50,* 354-377.

Hardwick, C. J. (2018). Blog. https://cynthiahardwick.com/humble-curiosity/

Harvey, R. (2013). The ancient thread of authenticity. An interview on Sacred Attentional Therapy. www.sacredattentiontherapy.com/articles.html

Hawkins, D. R. (1995). *Power vs force: The hidden determinants of human behavior.* Veritas.

Hawkins, J. with Blakeslee, S. (2004). *On intelligence: How a new understanding of the brain will lead to the creation of truly intelligent machines.* Times Books.

Heckscher, C. (2007). *The collaborative enterprise.* Yale University Press.

van Heijenoort, J. (Ed.) (1967). *From Frege to Gödel: A source book in mathematical logic, 1879-1931.* Harvard University Press.

Hess, E. D. (2020). *Hyper-learning: How to adapt to the speed of change.* Berrett-Koehler Publishers, Inc.

Hess, E. D., & Ludwig, K. (2017). *Humility is the new smart: Rethinking human excellence in the smart machine age.* Berrett-Kochler Publishers, Inc.

Heijnen, R. D. (2022). Desire—Life as a drama based game or the thermodynamics of everything. In S. B. Schafer & A. Bennet (Eds.), *The handbook of global media's preternatural influence on global technological singularity, culture and government* (214-242). IGI Global.

Hetland, L. (2000). Listening to music enhances spatial-temporal reasoning: Evidence for the Mozart Effect. *Journal of Aesthetic Education, 34,* 105-148.

Hink, R. F., Kodera, K., Yamada, O., Kaga, K., & Suzuki, J. (1980). Binaural interaction of a beating frequency following response. *Audiology, 19,* 36-43.

Hobson, J. A. (1999). *Consciousness.* Scientific American Library.

Hodges, D. (2000). Implications of music and brain research. *Music Educators Journal, 87*(2), 17-22.

Holiday, S. G., & Chandler, M. J. (1986). *Wisdom: Explorations in adult competence: Contributions to human development, Vol. 17.*

Hollmen, J. (1996). *Self-organizing map (SOM).* https://users.ics.aalto.fi/jhollmen/dippa/node9.html

Holm, J. (2013, March). Creating new businesses with open data. Presentation at KM Middle east. DATA.GOV.

Holman, D., Pavlica, K., & Thorpe, R. (1997). Rethinking Kolb's theory of experiential learning: The contribution of social constructivism and activity theory. *Management Learning, 28*, 135-148.

Holton, R. (n.d.). Belief/desire psychology, the Humean theory of motivation & emotion. *Moral Psychology, 24*(120).

Hook, J. (2015). Blog. https://www.joshuanhook.com/what-is-humility/

Hopkins, R. (1993). David Kolb's learning machine. *Journal of Phenomenological Psychology, 24*, 46-62.

Horstman, J. (2010). *The Scientific American brave new brain: How neuroscience, brain-machine interfaces, neuroimaging, psychopharmacology, epigenetics, the internet, and our minds are stimulating and enhancing the future of mental power.* John Wiley & Sons, Inc. and Scientific American.

Houston, J. (2000). *Jump time: Shaping your future in a world of radical change.* Penguin Putnam, Inc.

Huberman, A. M. & Miles, M. B. (2002). *The qualitative researcher's companion.* Sage Publications.

Hyman, S. E. (2007). Diagnosing disorders. In F. E. Bloom (Ed.), *Best of the brain from Scientific American: Mind, matter, and tomorrow's brain* (pp. 132-141). Dana Press.

Iacoboni, M. (2008). *The new science of how we connect with others: Mirroring people.* Farrar, Straus & Giroux.

Illeris, K. (2004). *Three dimensions of learning.* Roskilde University Press.

Immordino-Yang, M. H. (2016a). *Emotions, learning, and the brain: Exploring the educational implications of affective neuroscience.* Norton.

Immordino-Yang, M. H. (2016b). The smoke around mirror neurons: Goals as sociocultural and emotional organizers of perception and action in learning. In M. H. Immordino-Yang (Ed.), *Emotions, learning, and the brain: Exploring the educational implications of affective neuroscience* (pp. 151-164) Norton.

Immordino-Yang, M. H., Christodoulou, J. A., & Singh, V. (2016). 'Rest is not idleness': Implications of the brain's default mode for human development and education. In M. H. Immordino-Yang (Ed.), *Emotions, learning, and the brain: Exploring the educational implications and affective neuroscience* (pp. 43-68). Norton.

Immordino-Yang, M. H., McColl, A., Damasio, H., & Dalmasio, A. (2009). Neural correlates of admiration and compassion. *Proceedings of the National Academy of Sciences, 106*(19), 8021-8026.

Immordino-Yang, M. H., & Singh, v. (2016). Perspectives from social and affective neuroscience on the design of digital learning technologies. In. M. H. Immordino-Yang (Ed.), *Emotions, learning and the brain: Exploring the educational implications of affective neuroscience* (pp. 181-190). Norton.

Inglehart, R., Basanez, M., & Moreno, A. (1998). *Human values and beliefs: A cross-cultural sourcebook.* The University of Michigan Press.

Irvine, W. B. (2006). *On desire: Why we want what we want.* Oxford University Press.

Jablonka, E. & Lamb, M. (1995). *Epigenetic inheritance and evolution: The Lamarckian dimension.* Oxford University Press.

Jackson, P. L., Meltzoff, A. N., & Decety, J. (2005). How do we perceive the pain of others? A window into the neural processes involved in empathy. *NeuroImage, 24,* 771-779.

Jahn, R. G., & Dunne, B. J. (2011) *Consciousness and the source of reality: The PEAR odyssey.* ICRL Press.

James, J. (1996). *Thinking in the future tense: A workout for the mind.* Touchstone.

James, W. (1890/1980). *The principles of psychology, Vol. I.* Holt, Rinehar & Winston.

Jarvis, P. (2004). *Adult education and lifelong learning: Theory and practice.* RoutledgeFalmer.

Jarvis, P. (1992). *Paradoxes of learning: On becoming an individual in society.* Jossey-Bass.

Jarvis, P. (1987). *Adult learning in the social context.* Croom Helm.

Jensen, E. (2006). *Enriching the brain: How to maximize every learner's potential.* Jossey-Bass.

Jensen, E. (2005). *Teaching with the brain in mind.* Association for Supervision and Curriculum Development.

Jensen, E., (2000). *Brain based learning: The new science of teaching and training.* The Brain Store.

Jensen, E. (1998). *Teaching with the brain in mind.* Association for Supervision and Curriculum Development.

Johnson, S. (2006). The neuroscience of the mentor-learner relationship. In S. Johnson & K. Taylor (Eds.), *The neuroscience of adult learning: New directions for adult and continuing education* (pp. 63-70). Jossey-Bass.

Johnson, S. (2001). *Emergence.* Scribner.

Johnson, S., & Taylor, K. (2006). *The neuroscience of adult learning: New directions for adult and continuing education.* Jossey-Bass.

Jung, C. G. (1990). *The undiscovered self with symbols and the interpretation of dreams* (R. F. Hull, Trans.). Princeton University Press.

Kabat-Zinn, J. (2003). Mindfulness-based interventions in context: Past, present, and future. *Clinical Psychology: Science and Practice, 10*(2), 144-156.

Kandel, E. R. (2007). The new science of mind. In F. E. Bloom (Ed.), *Best of the brain from Scientific American: Mind, matter, and tomorrow's brain* (pp. 68-78). Dana Press.

Kandel, E. R. (2006). *In search of memory: The emergence of a new science of mind.* W. W. Norton & Company.

Kant, I. (1993). *Groundwork of the metaphysics of morals* (3rd ed.) (J. W. Ellington, Trans.). Hackett.

Katzenbach, J. R. & Smith, D. K. (1993). *The wisdom of teams: Creating the high-performance organization.* Harvard Business School Press.

Kayes, D. C. (2002). Experiential learning and its critics: Preserving the role of experience in management learning and education. *Academy of Management Learning and Education, 1*(2), 137-149.

Kawashina, R., Sugiura, M., Kato, T., Nakamura, A., Hatano, K., Ito, K., et al. (1999). The human amygdala plays an important role in gaze monitoring: A PET study. *Brain, 122,* 779-783.

Keenan, J. P., McCutcheon, B., Freund, S., Gallup, G. G. Jr., Sanders, G., & Pascual-Leone, A. (1999). Left hand advantage in self-face recognition task. *Neuropsychologia, 37,* 1421-1425.

Kempermann, G., Kuhn, H. G., & Gage, F. H. (1997). More hippocampal neurons in adult mice living in an enriched environment. *Nature, 386*, 493-95.

Kihlstrom, J. F. (1987). The cognitive unconscious. *Science, 237*, 1445-1452.

Kihlstrom, J. F., Shames, V. A., & Dorfman, J. (1996). Imitations of memory and thought. In L. Reber (Ed.), *Implicit memory and metacognition*. Erlbaum.

Kipling, R. (1985/1937). Working-tools. In B. Ghiselin (Ed.), *The creative process: A symposium* (pp. 161-163). University of California Press.

Kissinger, H. A., Schmidt, E., & Huttenlocher, D. (2021). *The age of AI and our human future*. Little, Brown and Company.

Klein, G. (2003). *Intuition at work: Why developing your gut instincts will make you better at what you do*. Doubleday.

Kluwe, R. H., Luer, G., & Rosler, F. (Eds.) (2003). *Principles of learning and memory*. Birkhauser Verlag.

Knowles, M. S., Holton, E. F., & Swanson, R. A. (1998). *The adult learner: The definitive classic in adult education and human resource development*. Gulf.

Kolb, D. A. (1984). *Experiential learning: Experience as the source of learning and development*. Prentice-Hall.

Kornfield, J. (1993). *A path with heart*. Bantam Books.

Kramer, D. A., & Bacelar, W. T. (1994). The educated adult in today's world: Wisdom and the mature learner. In J. D. Sinnott (Ed.), *Interdisciplinary handbook of adult lifespan learning*. Greenwood Press.

Kruss, E., Chancellor, J., Ruberton, P. M., & Lyubomirsky, S. (2014). An upward spiral between gratitude and humility. *Social Psychological and Personality Science, 1-10*(1).

Kuntz, P. G. (1968). *The concept of order*. University of Washington Press.

Kurzweil, R. (2005). *The singularity is near: When humans transcend biology*. Viking.

Lakoff, G., & Johnson, M. (1999). *Philosophy in the flesh: The embodied mind and its challenge to Western thought*. Basic Books.

Laming, D. (1992). Analysis of short-term retention: Models for Brown-Peterson Experiments. *Journal of Experimental Psychology: Learning, Memory, and Cognition, 18*, 1342-1365.

Laszlo, E. (1972). *The systems view of the world: The natural philosophy of the new developments in the sciences*. George Braziller.

Laughlin, S. B. (2004). The implications of metabolic energy requirements for the representation of information in neurons. In M. S. Gazzaniga (Ed.), *The cognitive neuroscience III* (pp. 187-196). The MIT Press.

Lave, J. (1988). *Cognition in practice: Mind, mathematics and culture in everyday life*. Cambridge University Press.

Layzer, D. (1990). *Cosmo genesis*. Oxford University Press.

Lazarus, R. S. (1991). Cognition and motivation in emotion. *American Psychologist, 46*(4), 352-67.

LeDoux, J. (2002). *Synaptic self: How our brains become who we are*. Viking.

LeDoux, J. (1996). *The emotional brain: The mysterious underpinnings of emotional life*. Touchstone.

Leonard, D., & Swap, W. (2004). *Deep smarts: How to cultivate and transfer enduring business wisdom*. Harvard Business School Press.

Lesch, M. F., & Pollatsek, A. (1993). Automatic access of semantic information by phonological codes to visual word recognition. *Journal of Experimental Psychology: Learning, Memory and Cognition, 19*, 285-294.

Lewicki, P., Hill, T., & Bizot, E. (1988). Acquisition of procedural knowledge about a pattern of stimuli that cannot be articulated. *Cognitive Psychology, 20*, 24-37.

Lewicki, P., Hill, T., & Czyzewska, M. (1992). Nonconscious acquisition of information. *American Psychologist, 47*, 796-801.

Lewis, J. (2019). *Story Thinking: Transforming organizations for the fourth industrial revolution.* Amazon KDP.

Leyden, P., Teixeria, R., & Greenberg, E. (2007, June 20). *The progressive politics of the millennial generation.* New Politics Institute. www.newpolitics.net

Lief, H., & Fetkewicz, J. (1995). Retractors of false memories: The evolution of pseudomemories. *Journal of Psychiatry Law, 23*, 411-35.

Liggan, D. Y., & Kay, J. (1999). Some neurobiological aspects of psychotherapy: A review. *Journal of Psychotherapy Practice and Research, 8*, 103-114.

Lindeman, E. C. (1961). *The meaning of adult education in the United States.* Harvest House.

Lindsay, G. (2021). *Models of the mind: How physics, engineering and mathematics have shaped our understanding of the brain.* Bloomsbury.

Lipton, B. (2005). *The biology of belief: Unleashing the power of consciousness.* Hay House.

Lipton, B., & Bhaerman, S. (2009). *Spontaneous evolution: Our positive future (and a way to get there from here).* Hay House.

Llinas, R. R. (2001). *I of the vortex: From neurons to self.* The MIT Press.

Locke, J. (2016). Essay concerning human understanding, book III.

Long, T. A. l(1986). Narrative unity and clinical judgment. *Theoretical Medicine 7*, 75-92.

Luca, M. (2022). Does the algorithm heal a business organization? In S. B. Schafer & A. Bennet (Eds.), *Handbook of research on global media's preternatural influence on global singularity, culture and government* (pp. 345-371). IGI Global.

Macdonald, D. (1996). *Toward wisdom : finding our way to inner peace, love, and happiness.* Hampton Roads.

MacGregor, R. J. (2006). *On the contexts of things human: An integrative view of brain, consciousness, and freedom of will.* World Scientific.

Machlup, F. (1962). *The production and distribution of knowledge in the United States.* Princeton University Press.

Mandel, M. (2007, August 20) Which way to the future? *BusinessWeek, 46.*

Manuel Castells (2022, February 19). *Wikipedia.* www.wikipedia.com

Maples, M. F., & Webster, J. M. (1980). Thorndike's connectionism. In G. B. Gazda & R. J. Corsini (Eds.), *Theories of learning* (pp. 1-280). Peacock.

Marchese, T. J. (1998). The new conversations about learning: Insights from neuroscience and anthropology, cognitive science and workplace studies. *New Horizons for Learning.* http://www.newhorizons.org/lifelong/higher_ed/marchese.htm

Marion, R. (1999). *The edge of organizations: Chaos and complexity theories of formal social systems.* CA: SAGE Publications.

Markowitsch, H. (2000). The anatomical basis of memory. In M. S. Gazzaniga (Ed.), *The new cognitive neurosciences*. MIT Press.

Markowitsch, H. (1999), Functional neuroimaging correlates of functional amnesia. *Memory, 8*, 567-583.

Marquardt, M. J. (1999). *Action learning in action: Transforming problems and people for world-class organizational learning*. Davies-Black.

Marrow, A. J. (1969). *The practical theorist*. Basic Books.

Marsh, J. T., Brown, W. S., & Smith, J. C. (1975). Far-field recorded frequency-following response: Correlates of low pitch auditory perception in humans. *Electroencephalography and Clinical Neurophysiology, 38*, 113-119.

Marshall, S. P. (2005, September-November). A decidedly different mind. *Shift: At the frontiers of Consciousness*.

Marshall, I., & Zohar, D. (1997). *Who's afraid of Schrödinger's cat?* William Morrow and Company, Inc.

Marsick, V. J., & Watkins, K. E. (2001). Informal and incidental learning. In S. B. Merriam (Ed.), *The new update on adult learning theory*. Jossey-Bass.

Martin, J. (2006). *The meaning of the 21st century: A vital blueprint for ensuring our future*. Riverhead Books.

Marton, F., & Booth, S. (1997). *Learning and awareness*. Erlbaum.

Masserman, J., Wechkin, S., & Terris, W. (1964, December). Altruistic behavior in Rhesus monkeys. *American Journal of Psychiatry, 121*, 584-585.

Maturana, H. R., & Varela, F. J. (1987). *The tree of knowledge: The biological roots of human understanding*. Shambhala.

Maunsell, J. H. R., & Treue, S. (2006). Feature-based attention in visual cortex. *Trends in Neuroscience, 29*, 317-322.

May, R. (1975). *The courage to create*. Bantam.

McCraty, R. (2015). *Science of the heart: Exploring the role of the heart in human performance volume 2*. HeartMath Institute.

McCraty, R., & Childre, D. (2010, July/August). Coherence: bridging personal, social, and global health. *Alternative Therapies, 16*(4), 10.

McHale, J. (1977). Futures problems or problems in future studies. In H. A. Linstone & W. H. C. Simmonds (Eds.), *Futures research: New directions*. Addison-Wesley Publishing Company, Inc.

McMaster, M. D. (1996). *The intelligence advantage: Organizing for complexity*. Butterworth-Heinemann.

McTaggart, L. (2002). *The field: The quest for the secret force of the universe*. Harper Perennial.

McWhinney, W. (1997). *Paths of change: Strategic choices for organizations and society*. SAGE Publications, Inc.

Meacham, J. A. (1995). The loss of wisdom. In R. J. Sternberg (Ed.), *Wisdom, its nature, origins, and development*. Cambridge University Press.

Medina, J. (2008, 2014). *Brain rules: 12 principles for surviving and thriving at work, home, and school*. Pear Press.

Merriam, S. B., & Caffarella, R. S. (1999). *Learning in adulthood: A comprehensive guide* (2nd ed.). Jossey-Bass.

Merriam Webster (2016). http://www.merriam-webster.com/dictionary/virtue

Merry, U. (1995). *Coping with uncertainty: Insights from the new sciences of chaos, self-organization and complexity*. Praeger.

Medina, J. (2008). *Brain rules: 12 principles for surviving and thriving at work, home, and school*. Pear Press.

Meltzoff, A. N., & Decety, J. (2002). What imitation tells us about social cognition: A rapproachement between developmental psychology and cognitive neuroscience. *Philosophical Transactions of the Royal Society of London—Series B: Biological Sciences, 358*, 491-500.

Mental illness in America: More than a quarter of adults are afflicted. (2008, February/March 15). *Scientific American Mind*.

Merriam, S. B., Caffarella, A. S., & Baumgartner, L. M. (2007). *Learning in adulthood: A comprehensive guide.* Jossey-Bass.

Meyer, C., & Davis, S. (2003). *It's alive: The coming convergence of information, biology, and business*. Crown Business.

Mezirow, J. (1991). *Transformative dimensions of adult learning*. Jossey-Bass.

Mezirow, J., & Associates (2000). *Learning at transformation: Critical perspectives on a theory in progress*. Jossey-Bass.

Michael, D. N. (1977). Planning's challenge to the systems approach. In H. A. Linstone & W. H. C. Simmonds (Eds.), *Futures research: New directions*. Addison-Wesley Publishing Company, Inc.

Middlebrook, P. N. (1974). *Social psychology and modern life*. Alfred A. Knoph.

Miettinen, R. (1998). About the legacy of experiential learning. *Lifelong Learning in Europe, 3*, 165-171.

Miller, G. A. (1956). The magical number seven, plus or minus two: Some limits on our capacity for processing information. *Psychological Review, 63*, pp. 81-96.

Miller, J. G. (1978). Living systems. New York: McGraw-Hill.

Miller, N. (2000). Learning from experience in adult education. In A. L. Wilson & E. R. Hayes (Eds.), *Handbook of adult and continuing education*. Jossey-Bass.

Minsky, M. (2006). *The emotion machine: Commonsense thinking, artificial intelligence, and the future of the human mind*. Simon and Schuster.

Moon, J. A. (2004). *A handbook of reflective and experiential learning: Theory and practice*. Routledge-Falmer.

Morowitz, H. J., & Singer, J. L. (Eds.). (1995) *The mind, the brain, and complex adaptive systems*. Addison-Wesley.

Mulford, P. (2007). *Thoughts are things*. Barnes & Noble.

Mulvihill, M. K. (2003). The Catholic Church in crisis: Will transformative learning lead to social change through the uncovering of emotion? In C. A. Weisner, S. R. Meyers, N. L. Pfhal & P. J. Neaman (Eds.), *Proceedings for the 5th International Conference on Transformative Learning* (pp. 320-325). Teachers College, Columbia University.

Murray, A. (2016). *Humility: The beauty of holiness*. Aneko Press.

Nakamura, S., Sadato, N., Oohashi, T., Nishina, E., Fuwamoto, Y., & Yonekura, Y. (1999). Analysis of music-brain interaction with simultaneous measurement of regional cerebral blood flow and electroencephalogram beta rhythm in human subjects. *Neuroscience Letters, 275*, 222-226.

Naisbitt, J. (2006). *Mind set! Reset your thinking and see the future*. Collins.

National Research Council (2000). *How people learn: Brain, mind, experience, and school*. National Academy Press.

National Research Council. (1994). *Learning, remembering, believing: Enhancing human performance*. National Academy Press.

Neal, A., & Hesketh, B. (1997). Episodic knowledge and implicit learning. *Psychonomic Bulletin & Review, 4*, 24-37.

Nelson, C. A., deHaan, M., & Thomas, K. M. (2006). *Neuroscience of cognitive development: The role of experience and the developing brain*. John Wiley & Sons.

Nicolelis, M. A. L., & Chapin, J. K. (2007). Controlling robots with the mind. In F. E. Bloom (Ed.), *Best of the brain from Scientific American: Mind, matter, and tomorrow's brain* (pp.197-212). Dana Press.

Novak, J. D. (1998). *Learning. Creating and using knowledge: Concept maps as facilitative tools in schools and corporations*. Lawrence Erlbaum Associates. Also, Novak, J. D., & Gowin, D. B. (1996). *Learning how to learn*. Cambridge University Press.

Oakes, J., & Lipton, M. (1999). *Teaching to change the world*. McGraw Hill.

Oakshott, M. (1933). *Experience and its modes*. Cambridge University Press.

Ormrod, J. E. (1999). *Human learning* (3rd ed.). Merrill.

Orrell, D. (2007). *Apollo's arrow: The science of prediction and the future of everything*. Harper Collins.

Oster, G. (1973). Auditory beats in the brain. *Scientific American, 229*, 94-102.

Oxford English Dictionary (5th ed.) (2002). Oxford University Press.

Paradiso, S., Johnson, D. L., Andreasen, N. C., O'Leary, D. S., Watkins, G. L., Ponto, L. L. B. et al. (1999). Cerebral blood flow changes associated with attribution of emotional valence to pleasant, unpleasant, and neutral visual stimuli in a PET study of normal subjects. *American Journal of Psychiatry, 156*, 1618-1629.

Pelegrinis, T. N. (1980). *Kant's conceptions of the categorical imperative and the will*. Zeno Publishers.

Panetta, S. (2013). *How to lead the millennial generation in the workforce*. Thesis Report. City University of Seattle.

Perkins, D. N. (1995). *Outsmarting IQ: The emerging science of learnable intelligence*. Free Press.

Perkins, M. (1971). Matter, sensation and understanding. *American Philosophical Quarterly, 8*, 1-12.

Perry, W. G. (1970/1988). *Forms of ethical and intellectual development in the college years*. Jossey-Bass.

Pert, C. B. (1997). *Molecules of emotion: A science behind mind-body medicine*. Touchstone.

Peterson, L. R., & Peterson, M. J. (1959). Short-term retention of individual verbal items. *Journal of Experimental Psychology, 58*, 193-198.

Phelps, E. A. (2004). The human amygdala and awareness: Interactions of the amygdale and hippocampal complex. *Current Opinion in Neurobiology, 14*, 198-202.

Phelps, E. A., O'Connor, K. J., Gatenby, J. C., Grillon, C., Gore, J.C., & Davis, M. (2001). Activation of the left amygdala to a cognitive representation of fear. *Natural Neuroscience, 4*, 437-441.

Piaget, J. (1971). *Psychology and epistemology*. Penguin Books.

Piaget, J. (1968). *Structuralism*. Harper Torchbooks.

Piaget, J. (1951). *Play, dreams and imitation in childhood*. W. W. Norton.

Pickles, T. (2007). Experiential learning ... on the web. http://www.reviewing.co.uk/research/experiential.learning.htm

Pinker, S. (2007). *The stuff of thought: Language as window into human nature.* Viking.

Plotkin, H. (1994). *Darwin machines and the nature of knowledge.* Harvard University Press.

Poincaré, H. ([1902,1904, 1908] 1982) Science and hypothesis, The value of science, science and method. In *The foundations of science.* University Press of America.

Polanyi, M. (1958). *Personal knowledge: Towards a post-critical philosophy.* The University of Chicago Press.

Popp, F. A. (2002). Biophotonics: A powerful tool for investigating and understanding life. In H. P. Durr, F. A. Popp, and W. Schommers (Eds.), *What is life? Scientific approaches and philosophical positions.* (Series on the Foundations of Natural Science and Technology). World Scientific.

Pribram, K. H. (1998). Autobiography in anecdote: The founding of experimental neuropsychology. In R. Bilder (Ed.), *The history of neuroscience in autobiography* (pp. 306-349). California Academic Press.

Pribram, K. H. (1993). Rethinking neural networks: Quantum fields and biological data. *Proceedings of the First Appalachian Conference on Behavioral Neurodynamics.* Lawrence Erlbaum.

Pribram, K. H. (1991). *Brain and perception: Holonomy and structure in figural processing.* Lawrence Erlbaum

Portante, T., & Tarro, R. (1997, September). Paying attention. *Wired Magazine.*

Prigogine, I. (1996). *The end of certainty: Time, chaos, and the new laws of nature.* The Free Press.

Prinz, W. (2005). An ideomotor approach to imitation. In S. Hurley & N. Chater (Eds.), *Perspectives on imitation: From neuroscience to social science Vol. 1: Mechanisms of imitation and imitation in animals* (pp. 141-156). MIT Press.

Quine, W. V. (1966). *Ways of paradox and other essays.* Random House.

Race, P. (2005). *Making learning happen: A guide for post-compulsory education.* SAGE Publications.

Raichle, M. E., MacLeod, A. M., Snyder, A. Z., Powers, W. J., Gusnard, D. A., & Shulman, G. L. (2001). A default mode of brain function. *Proceedings of the National Academy of Sciences, 98*(2), 676-682.

Ralston, F. (1995). *Hidden dynamics: How emotions affect business performance & how you can harness their power for positive results.* American Management Association.

Ramon, S. (1997). *Earthly cycles: How past lives and soul patterns shape your life.* Pepperwood Press.

Ratey, J. J. (2001). *A user's guide to the brain: Perceptions, attention, and the four theaters of the brain.* Pantheon Books.

Rauscher, F. H., Shaw, G. L., & Ky, K. N. (1993). Music and spatial task performance. *Nature, 365,* 611.

Reber, A. S. (1993). *Implicit learning and tacit knowledge: An essay on the cognitive unconscious.* Oxford University Press.

Reik, W., & Walter, J. (2001). Genomic imprinting: Parental influence on the genome. *Nature Reviews Genetics, 2,* 21+.

Rescher, N. (2001). *Paradoxes: Their roots, range, and resolution.* Open Court.

Revans, R. W. (1980). *Action learning.* Frederick Muller.

Reynolds, M. (1999). Critical reflection and management education: Rehabilitating less hierarchical approaches. *Journal of Management Education, 23,* 537-553.

Riggio, R. E. (2015). Are you empathic? 3 types of empathy and what they mean. *Psychology Today.*

Rischard, J. F. (2002). *High noon 2020: 20 global problems, 20 years to solve them.* Basic Books.

Ritchey, D. (2003). *The H.I.S.S. of the A.S.P.: Understanding the anomalously sensitive person.* Headline Books.

Rizzolatti, G. (2006, November). Mirrors in the mind. *Scientific American.*

Robbins, J. (2000). *A symphony in the brain.* Grove Press.

Rock, A. (2004). *The mind at night: The new science of how and why we dream.* Perseus Books Group.

Rose, S. (2005) *The future of the brain: The promise and perils of tomorrow's neuroscience.* Oxford University Press

Rosenfield, I. (1988). *The invention of memory.* Basic Books.

Ross, C. A. (2006). Brain self-repair in psychotherapy: Implications for education. In S. Johnson & K. Taylor (Eds.), *The neuroscience of adult learning: New directions for adult and continuing education.* Jossey-Bass.

Ross, P. E. (2006, August). The expert mind. *Scientific American,* 64-71.

Rowan, J. (1990). *Subpersonalities: The people within us.* Routledge.

Salk, J. (1973). *The survival of the wisest.* Harper & Row.

Salinsbury, R. M. (1988). *Paradoxes.* Cambridge University Press.

Sapolsky, R. M. (1999). Glucocorticoids, stress, and their adverse neurological effects: Relevance to gaining. *Experimental Gerontology, 34,* 721-732.

Schafer, S. B. (2022). Gamers as homeopathic media therapy: Electromagnetic antibodies in the toxic media-field. In S. B. Schafer & A. Bennet (Eds.), *Handbook of research on global media's preternatural influence on global singularity, culture and government* (pp. 397-425). IGI Global.

Schafer, R. M. (1993). *The soundscape: Our sonic environment and the tuning of the world.* Destiny Books.

Schacter, D. L. (1996). *Searching for memory: The brain, the mind, and the past.* Basic Books.

Scharmer, C. O. (2009). *Theory U: Leading from the future as it emerges.* Berrett-Koehler.

Schank, R. E. (1995). *Tell me a story: Narrative and intelligence.* Northwestern University Press.

Scholey, A., Moss, M., Naeve, N., & Wesnes, K. (1999). Cognitive performance, hyperoxia, and heart rate following oxygen administration in healthy young adults. *Physiological Behavior, 67,* 783-789.

Schon, D. (1983). *The reflective practitioner: How professionals think in action.* Basic.

Schore, A. N. (1994). *Affect regulation and the origin of the self: The neurobiology of emotional development.* Erlbaum.

Schore, A. N. (2002). Dysregulation of the right brain: A fundamental mechanism of traumatic attachment and the psychopathogenesis of posttraumatic stress disorder. *Australian and New Zealand Journal of Psychiatry, 36,* 9-30.

Schwartz, P. (2003). *Inevitable surprises: Thinking ahead in a time of turbulence.* Penguin Group.

Schwarzlose, R. (2021). *Brainscapes: An atlas of your life on earth.* Profile Books.

Scott, S. (2000). *From pride to humility: A biblical perspective.* Focus Publishing, Inc.

Scripps Research Institute (2012, May 9). Scientists identify neurotransmitters that lead to forgetting. *ScienceDaily.*

Searle, J. R. (1983). *Intentionality: An essay in the philosophy of mind.* Cambridge University Press.

Seeley, W. W., Menon, V., Schatzberg, A. F., Keller, J., glover, G. H., Kenna, H., ... & Greicius, M. D. (2007). Dissociable intrinsic connectivity networks for salience processing and executive control. *Journal of Neuroscience, 27*(9), 2349-2356.

Selye, H. (1978). *The stress of life* (2nd ed.). McGraw-Hill.

Senge, P. M. (1990). *The fifth discipline.* Doubleday.

Senge, P. M., Scharmer, C. O., Jaworski, J., & Flowers, B. S. (2004). *Presence: Exploring profound change in people, organizations and society.* Random House, Inc.

Shafer, I. H. (1999). World commission on global consciousness and spirituality. https://www.sourcewatch.org/index.php/World_Commission_on_Global_Conscio usness_and_Spirituality

Shand, A. F. (1920). *The foundation of character* (2nd ed.). Macmillan.

Sheldrake, R. (1989). *The presence of the past: Morphic resonance and the habits of nature.* Vintage Books.

Shelley, A. (2021). *Becoming adaptable: Creative facilitation to develop yourself and transform cultures.* Intelligent Answers Pty Ltd.

Shelley, A. (2017). *KNOWledge SUCCESSion: Sustained performance and capability growth through strategic knowledge projects.* Business Expert Press.

Shelley, A. (2007). *Organizational zoo: A survival guide to work place behavior.* Aslan Publishing.

Siegel, D. J. (2016). *Mind: A journey to the heart of being human.* W. W. Norton & Co.

Siegel, R. D., Germer, C. K., & Olendzki, A. (2009). Mindfulness: what is it? Where did it come from? In F. Didonna (Ed.), *Clinical handbook of mindfulness* (pp. 17-36). Springer.

Silberman, M. (2007). *The handbook of experiential learning.* Pfeiffer.

Sima Qian (145-ca. 89 BC/1922/1985). Confucius. In Hu Shi (Ed.), *The development of logical methods in ancient China.* Shanghai. Oriental Book Company.

Simonton, D. K. (1999). Talent and its development: An emergenic and epigenetic mode. *Psychological Review, 106*, 435-457.

Singer, T., Seyour, B., O'Doherty, J., Kaube, H., Dolan, R. J., & Frith, C. D. (2004). Empathy for pain involves the affective but not sensory components of pain. *Science, 303*, 1157-1162.

Skoyles, J. R., & Sagan, D. (2002). Up from dragons : The evolution of human intelligence. McGraw-Hill.

Smallwood, J., Obsonsawin, M., & Heim, D. (2003). Task unrelated thought: The role of distributed processing. *Consciousness and Cognition, 12*(2), 169-189.

Smith, C. A., & Ellsworth, P. C. (1985). Patterns of cognitive appraisal in emotion. *Journal of Personality and Social Psychology, 56*, 339-53.

Smith, J. C., Marsh, J. T., Greenberg, S., & Brown, W. S. (1978). Human auditory frequency-following responses to a missing fundamental. *Science, 201*, 639-641

Snow, C. P. (1959/1993). *The two cultures.* Cambridge University Press.

Sousa, D. A. (2006). *How the brain learns.* Corwin Press.

Spreng, R. N., & Grady, C. L. (2010) Patterns of brain activity supporting auto-biographical memory, prospection, and theory of mind, and their relationship to the default mode network. *Journal of Cognitive Neuroscience, 22*(6), 1112-1123.

Squire, L. R., & Knowlton, B. (1995). Memory, hippocampus, and brain systems. In M. S. Gazzaniga (Ed.), *The cognitive neurosciences*. The MIT Press.

Sriastva, S., & Cooperrider (D. L. (Eds.) (1990). *Appreciative management and leadership*. Jossey-Bass.

Stacy, R. D. (1996). *Complexity and creativity in organizations*. Berrett-Koehler Publishers.

Stacy, R. D., Griffin, D., & Shaw, P. (2000). *Complexity and management: Fad or radical challenge to systems thinking?* Routledge.

Stern, D. N. (2004). *The present moment in psychotherapy and everyday life*. Norton.

Sternberg, R. J. (2003). *Wisdom, intelligence, and creativity synthesized*. Cambridge University Press.

Sternberg, R. J. (Ed.) (1998). *Wisdom: Its nature, origins, and development*. Cambridge University Press.

Sternberg, R. J., Ed. (1995) *Wisdom: Its nature, origins, and development*. Cambridge University Press.

Sternberg, R. J. (Ed.) (1990). *Wisdom: Its nature, origins and development*. Cambridge University Press

Strogatz, S. (2003). *Sync: How order emerges from chaos in the universe, nature, and daily life*. Hyperion Books.

Stonier, T. (1997). *Information and meaning: An evolutionary perspective*. Springer-Verlag.

Stonier, T. (1992). *Beyond information: the natural history of intelligence*. Springer-Verlag.

Stonier. T. (1990). *Information and the internal structure of the universe*. Springer-Verlag.

Strogatz, S. H., & Stew-art, I. (2020/1993). Coupled oscillators & biological synchronization. *Scientific American, 269*(6), 102-109.

Surani, M. A. (2001). Reprogramming of genome function through epigenetic inheritance. *Nature, 414*, 122-128.

Sriastva, S., & Cooperrider, D. L. (Eds.) (1990). *Appreciative management and leadership*. Jossey-Bass.

Surani, M. A. (2001). Reprogramming of genome function through epigenetic inheritance. *Nature, 414,* 122+.

Swain, J. (1997). *Musical languages*. W. W. Norton.

Swann, R., Bosanko, S., Cohen, R., Midgley, R., & Seed, K. M. (1982). *The brain: A user's manual*. G. P. Putnam & Sons.

Tallis, F. (2002). *Hidden minds: A history of the unconscious*. Arcade.

Tan, O. (Ed.). (2004). *Enhancing thinking through problem-based learning approaches: International perspectives*. Thompson Learning.

Tapscott, D. (2010). *Grown up digital*. McGraw Hill

Taylor, K. (2006). Brain function and adult learning: Implications for practice. In S. Johnson & K. Taylor (Eds.), *The neuroscience of adult learning: New directions for adult and continuing education* (pp. 71-86). Jossey-Bass.

Tchaikovsky, P. e. (2016). *The life and letters of Pete Ilich Tchaikovsky*. University of Toronto Press.

Teasdale, W., & the Dali Lama (1999). *The mystic heart: Discovering al universal spirituality in the world's religions*. New World Library.

Templeton, Sir John (2002). *Wisdom from world religions: Pathways toward heaven on earth*. Templeton Foundation Press.

Tennant, M. (1997). *Psychology and adult learning*. Routledge.

Thompson, R. F. (2000). *The brain: A neuroscience primer*. Worth.

Tooby, J., & Cosmides, L. (1990). The past explains the present: Emotional adaptations and the structure of ancestral environments. *Ethological Sociobiology 11*, 375-424.

Torrance, E. P. (1974). *Torrance test of creative thinking*. Personnel Press.

Turmel, R. (1997, Summer). Resonant frequencies and the human brain. *The Vaults of EROWID 1*.

Uleman, J. (2005). Becoming aware of the new unconscious. In R. R. Hassin, J. S. Uleman & J. A. Bargh (Eds.), *The new unconscious*. Oxford University Press.

Vaynerchuk, G. (2018). Blog. www.garyvaynerchuk.com/confidence-humility/

Vickers, G. (1977). The future of culture. In H. A. Linstone & W. H. C. Simmonds (Eds.), *Futures research: New directions*. Addison-Wesley Publishing Company.

Vince, R. (1998). Behind and beyond Kolb's learning cycle. *Journal of Management Education, 22*, 304-319.

Virtue (2016). Google.

Volk, T. (1998). *Gaia's body: Toward a physiology of earth*. Springer-Verlag.

von Forester, H. (1977). Objects: Tokens for (eigen-) behaviors. In B. Inheler, R. Gracia, & J. Voneche (Eds.), *Hommage a Jean Piaget: Epistemologie Genetique et Eqilibration*. Delachaux et Niestel.

von Hildebrand, D. (1997). *Humility: Wellspring of virtue*. Sophia Institute Press.

von Krogh, G., & Roos, J. (1995). *Organizational epistemology*. St. Martin's Press.

Vygotsky, L. S. (1978). *Mind in society: The development of higher psychological processes*. Harvard University Press.

Wade, W. (2012). *Scenario planning: A field guide to the future*. John Wiley & Sons, Inc.

Walsh, N. D. (2009). *When everything changes change everything: In a time of turmoil, a pathway to peace*. EmNin Books.

Warmington, E. H. (Ed.) (1984). *Great dialogues of Plato* (P. G. Rouse, Trans.). Mentor.

Wang, X., Merzenich, M. M., Sameshima, K., & Jenkins, W. M. (1995). Remodeling of hand representation in adult cortex determined by timing of tactile stimulation. *Nature, 378*, 71-75.

Ward, J. (2006). *The student's guide to cognitive neuroscience*. Psychology Press.

Warner Brothers (2011). *Green Lantern*.

Webb, D., & Webb, T. (1990). *Accelerated learning with music*. Accelerated Learning Systems.

Websters Encyclopedic Unabridged Dictionary of the English Language (1996). Portland House.

Weinberger, N. M. (2004). Music and the brain. *Scientific American, 291*, 89-95.

Weinberger, N. M. (1999, Spring). Can music really improve the mind? The question of transfer effects. *MuSICA Research Notes: V VI*, 12.

Weinberger, N. M. (1995). Non musical outcomes of music education. *Musical Journal, II*(2), 6.

Wenger, E. (2000). *Communities of practice: Learning, meaning, and identity.* Cambridge University Press.

Weyl, H. (1952). *Symmetry.* Princeton University Press.

Wickelgren, I. (2012, January 1). Forgetting is key to a healthy mind. *Scientific American Mind.*

Wilber, K. (2000). *Integral psychology: Consciousness, spirit, psychology, therapy.* Shambhala Publications.

Wilber, K. (1999). *The collected works of Ken Wilber, vols. 1-8.* Shambhala.

Wild, B., Rodden, F. A., Grodd, W., & Ruch, W. (2003). Neural correlates of laughter and humor. *Brain, 126,* 2121-2138.

Williams, P., & Denney, J. (2016). *Humility: The secret ingredient of success.* Shiloh Run Press.

Wilson, A. L., & Hayes, E. R. (Eds.) (2000). *Handbook of adult and continuing education.* Jossey-Bass.

Wilson, E. O. (1998). *Consilience: The unity of knowledge.* Alfred A. Knopf.

Wiig, K. (1993). *Knowledge management methods—Practical approaches to managing knowledge.* Schema Press.

Wolfe, P. (2006). The role of meaning and emotion in learning. In S. Johnson & K. Taylor (Eds.), *The neuroscience of adult learning: New directions for adult and continuing education* (pp. 35-420. Jossey-Bass.

Wong, C. Y. L., Choi, C. J., & Millar, C. C. J. M. (2006). The case of Singapore as a knowledge-based city. In F. J. Carrillo (Ed.), *Knowledge cities: Approaches, experiences, and perspectives.* Elsevier.

Worawichayavongsa, W. (2020). The influence of mindfulness practice at work on self-leadership on Thai non-managerial employees: The lived experience in an international K12 educational environment [Unpublished doctoral dissertation]. Southeast Asia Institute of Knowledge and Innovation, Bangkok University, Thailand.

Zhu, X. O., & Waite, P. M. E. (1998). Cholinergic depletion reduces plasticity of barrel field cortex. *Cerebral Cortex, 8,* 63-72.

Zigmond, M. J., Bloom, F. E., Landis, S. C., Roberts, J. L., & Squire, L. R. (Eds.) (1999). *Fundamental neuroscience.* Academic Press.

Zimmer, C. (2007). The neurobiology of the self. In F. E. Bloom (Ed.), *Best of the brain from Scientific American: Mind, matter, and tomorrow's brain* (pp. 47-57). Dana Press.

Zohar, D., & Marshall, I. (2003). Spiritual capital: Wealth we can live by. Berrett-Kochler Publishers, Inc.

Zohar, D., & Marshall, I. (1994). *The quantum society: Mind, physics, and the new social vision.* William Morrow and Co., Inc.

Zull, J. E. (2002). *The art of changing the brain: Enriching the practice of teaching by exploring the biology of learning.* Stylus.

Subject Index

Index by ICALS Findings

The findings are organized in 13 areas: Aging, Anticipating the Future, Creativity, Emotions, Epigenetics, Exercise and Health, Memory, Mirror Neurons, Plasticity, Social Support, Social Interaction, Stress, and the Unconscious. For purposes of this index, when ICALS findings fit into two categories, they are listed in both of those categories.

Genes are not destiny. The environment can change the actions of genes via non-expression. 114

Genes are operating options modulated by inputs from the environment, resulting in behavior. 114

Exercise and Health

Exercise increases brainpower. 146

Volition is necessary for benefit. Forced exercise does not promote neurogenesis. 158

Meditation and other mental exercises can change feelings, attitudes and mindsets. 103

Meditation quiets the mind. 103, 217

Positive and negative beliefs affect every aspect of life. 117

Physical and mental exercise and social bonding are significant sources of stimulation of the brain. 193

Physical activity increases the number (and health) of neurons. 159

Less than seven hours of sleep may impair memory. 135

Memory

Memory is scattered throughout the entire cortex; it is not stored locally. 129

Working memory is limited. 127

Emotional tags impact memory. 173

The brain stores only a part of the meaningful incoming sensory information. The gaps are filled in (re-created) when the memory is recalled. 130

Memory stores invariant forms (used to predict the future). 130

Memories are re-created each time they are recalled and therefore never the same. 129

Memory recall is improved through temporal sequences of associated patterns, that is, stories and songs. 132, 273

Repetition increases memory recall. 146

Emotional tags influence memory recall. 9, 173

Memory patterns cannot be erased at will. 133

Memory patterns decay slowly with time. 130

Less than seven hours of sleep may impair memory 135

Mirror Neurons

Neurons create the same pattern when we see some action being taken as when we do it. | 195

Mirror neurons are a form of cognitive mimicry that transfers active behavior and other cultural norms. 27, 196

What we see we become ready to do. | 194

Mirror neurons facilitate neural resonance between observed actions and executing actions. | 196

Rapid transfer of tacit knowledge (bypasses cognition). | 196, 215

Through [mental] reliving we recreate the feelings, perspectives and other phenomena that we observe. | 197

We may understand other people's behavior by mentally simulating it. | 197

Plasticity

Plasticity is a result of the connections between neural patterns in the mind and the physical world—what we think and believe impacts our physical bodies. | 7

Thoughts change the structure of the brain, and brain structure influences the creation of new thoughts. | 88

About Mountain Quest Institute

The view from the mountain!

MQI is a research, retreat, and learning center dedicated to helping individuals achieve personal and professional growth, and helping organizations create and sustain high performance in a rapidly changing, uncertain, and increasingly complex world. MQI has three quests: The Quest for Knowledge, The Quest for Consciousness, and The Quest for Meaning. *MQI is scientific, humanistic and spiritual and finds no contradiction in this combination.*

The Institute, situated on a farm in the Allegheny Mountains of West Virginia—where the mountains meet the library—is part of the Mountain Quest Inn and Retreat Center. The Retreat Center is designed to provide full learning experiences, including hosting training, workshops, retreats and business meetings for professional and executive groups of 25 people or less. The Center includes a 40,000-volume research library, conference room, community center, 12 themed bedrooms, and a four-story tower with a glass ceiling for enjoying the magnificent view of the valley during the day and the stars at night. The farm boasts a mountain terrain and farmland, a labyrinth, creeks, farm animals and a myriad of wild neighbors. Amidst parks, lakes and mountains, close neighbors include the Snowshoe Ski Resort, the Green Bank Radio Telescope and the CASS Railroad.

About the Authors

Drs. David and Alex Bennet are co-founders of the Mountain Quest Institute. They may be contacted at alex@mountainquestinstitute.com

David Bennet's experience spans many years of service in the Military, Civil Service and Private Industry, including fundamental research in underwater acoustics and nuclear physics, frequent design and facilitation of organizational interventions, designing and developing the Naval Sea Systems Command training program for acquisition management, and serving as technical director/chief engineer of two ACAT I DoD Acquisition programs. Prior to co-founding the Mountain Quest Institute with his partner Alex (achieving a dream he set out to accomplish a number of years earlier), Dr. Bennet was CEO, then Chairman of the Board and Chief Knowledge Officer of a professional services firm located in Alexandria, Virginia. He is a Phi Beta Kappa, Sigma Pi Sigma, and Suma Cum Laude graduate of the University of Texas, and holds degrees in Mathematics, Physics, Nuclear Physics, Liberal Arts, Human and Organizational Systems, and Human Development, with, more recently, a Ph.D. in Neuroscience and adult learning. His passion is researching the nexus of Science, the Humanities and Spirituality.

Alex Bennet, Professor on the faculty of Bangkok University's Institute for Knowledge and Innovation Southeast Asia (IKI-SEA), is internationally recognized as an expert in knowledge management and agent for organizational change. Alex is also Co-founder and Director of the Mountain Quest Institute. She served as the Chief Knowledge Officer and Deputy Chief Information Officer for Enterprise Integration for the U.S. Department of the Navy, and was co-chair of the Federal Knowledge Management Working Group. Prior to that she served as Acquisition Reform Executive and Standards Improvement Executive for the DoN Acquisition Workforce. Dr. Bennet is the recipient of the Distinguished and Superior Public Service Awards from the U.S. government for her work in the Federal Sector. She is a Delta Epsilon Sigma and Golden Key National Honor Society graduate with a Ph.D. in Human and Organizational Systems; degrees in Management for Organizational Effectiveness, Human Development, English and Marketing; and is certified in Total Quality Management, System Dynamics and Defense Acquisition Management, and a Reiki Master, among other energy modalities. Alex believes in the *multidimensionality and interconnectedness of humanity as we move out of infancy into full consciousness.*

Robert Turner served in the military in Army Intelligence and Organizational Development, where he founded and co-developed the U.S. Army Fusion Center, an advanced decision support center. He subsequently founded and directed the Federal Aviation Administration Team Technology Center and managed programs in support of FAA leadership development. His work at the FAA included representing the FAA at the Institute for the Future in Menlo Park and at the IBM Institute for Knowledge Management (KM). He served for four years as the Chairman of the government-wide Federal KM Network. In 2003, he received the first government-wide award for service in KM. In 2006, as co-developer of the FAA Knowledge Services Network (KSN) for virtual work, he received an acclaimed government-wide award for innovation excellence. He is a member of Phi Kappa Phi whose motto is "*Let the love of learning rule humanity.*" He graduated magna cum laude and received a special academic achievement medal from the University of Maryland in psychology and business. He completed his master's degree in education with Boston University. Bob co-led with his wife Jane a cohort in the joint BYU-Idaho Pathway & LDS Institute Program, an innovative global university endeavor. His research interests include individual and organizational high performance and accelerated learning. He has been an associate with the Mountain Quest Institute since its inception over 20 years ago.

Other Books Available from MQIPress

Expanding the Self: The Intelligent Complex Adaptive Learning System
by David Bennet, Alex Bennet and Robert Turner (2015)
The foundational study underlying *Unleashing the Human Mind*, this book is a treasure for those interested in how recent findings in neuroscience impact learning. The result of this work is an expanded experiential learning model called the Intelligent Complex Adaptive Learning System (ICALS), reflecting the five modes of concrete experience, reflective observation, abstract conceptualization, active experimentation, and social engagement, all undergirded by the power of Self. A significant conclusion is that adults have much more control over their learning than they may realize.

The Course of Knowledge: A 21st Century Theory
by Alex Bennet and David Bennet with Joyce Avedisian (2015)
Knowledge is at the core of what it is to be human, the substance which informs our thoughts and determines the course of our actions. **From a 21st century viewpoint,** we explore a theory of knowledge that is both pragmatic and biological. In a continuous journey towards intelligent activity, context-sensitive and situation-dependent knowledge, imperfect and incomplete, experientially engages a changing landscape in a continuous cycle of learning and expanding. *We are in a continuous cycle of knowledge creation such that every moment offers the opportunity for the emergence of new and exciting ideas, all waiting to be put in service to an interconnected world.*

Leading with the Future in Mind: Knowledge and Emergent Leadership
by Alex Bennet and David Bennet with John Lewis (2015)
This book provides a research-based *tour de force* for the future of leadership. Moving from the leadership of the past, for the few at the top, using authority as the explanation, we now find leadership emerging from all levels of the organization, with knowledge as the explanation. The future will be owned by the organizations that can master the relationships between knowledge and leadership. As the key ingredient, collaboration is much more than "getting along"; it embraces and engages. Wrapped in the mantle of collaboration and engaging our full resources—physical, mental, emotional, and spiritual—we open the door to possibilities. We are dreaming the future together.

Decision-Making in The New Reality: Complexity, Knowledge and Knowing
by Alex Bennet and David Bennet (2013)
We live in a world that offers many possible futures. The ever-expanding complexity of information and knowledge provide many choices for decision-makers, and we are all making decisions every single day! As the problems and messes of the world become more complex, our decision consequences are more and more difficult to anticipate, and our decision-making processes must change to keep up with this world complexification. This book takes a consilience approach to explore decision-making in The New Reality, fully engaging systems and complexity theory, knowledge research, and recent neuroscience findings. It also presents methodologies for decision-makers to tap into their unconscious, accessing tacit knowledge resources and increasingly relying on the sense of knowing that is available to each of us.

The QUEST: Where the Mountains meet the Library
by the Drs. David and Alex Bennet with Afterword by Robert Turner (2021)
"So here it is. This is a book of big ideas, the very ideas that have continuously filled our minds and hearts over the past 20 years, bubbling up and down as we traveled the world, then settling into printed text when we returned home to Mountain Quest where the mountains meet the library, a beautiful valley set in the Allegheny Mountains of West Virginia. Each chapter is, it seems, a chapter of our lives, yet all of it blends together in a magnificent unfolding of thought and learning. And while we do not have a whole lot of answers, we are eager to share with you that which we have learned, as one point of view in an infinite number of possibilities."

With Passion, We Live and Love: Research, Prose, Verse and Music
by Alex Bennet and Cindy Lee Scott (2021)
Alex, now in her 70's, and Cindy, 15 months younger, are sisters who finally met in 2016. Alex, who was given for adoption as an infant with an older sister, never knew Cindy existed. This gift provided Alex with the opportunity to personally explore her life-long questions regarding nurture versus nature. Cindy, a flower child of the 60's, is a veracious poet grown through a wide diversity of life experiences. Verse gave direction to and expression for her passion, and coupled with music and faith, served as the saving grace of her life. During the same time period, Alex was forging ahead with a hunger for knowledge that led her to take a consilience approach in her studies. She is widely published. All three of the sisters had a passionate love for, and life involvement in, music. The older sister (deceased) was a violinist, pianist, vocalist and guitarist; Alex was a flautist and opera singer; and Cindy was a pianist, vocalist, guitarist, harpist and composer. This book shares comparative glimpses of their lives.

Remembrance: Pathways to Expanded Learning with Music and Meta-Music® *by Barbara Bullard and Alex Bennet* (eBook 2013; softback 2020)
Authors Bullard and Bennet have a simple premise: your genes are not your destiny. They believe that you can significantly influence your ability to learn and what you remember. While there are a number of tips, tricks and exercises to help you reshape and focus your mind, one of the most surprising revelations is the power that music has over your brain. Music and sound are primal stimuli that affect the brain at the deepest levels, even altering dominant brainwave patterns. Designer music then, music created to stimulate specific brainwave states, can be used in a wide variety of ways, including maximizing your learning potential and your memory recall. In this book, particular attention is given to Hemi-Sync®, and injects some compelling insights into the "Mozart Effect" controversy, providing strong arguments for effectively using music to boost mental prowess.

The Profundity and Bifurcation of Change series (2020)
by Alex Bennet, David Bennet, Arthur Shelley, Theresa Bullard and John Lewis

Part I: Laying the Groundwork
Part II: Learning from the Past
Part III: Learning in the Present
Part IV: Co-Creating the Future
Part V: Living the Future (with Donna Panucci)

This supports progression of the *Intelligent Social Change Journey* (ISCJ), a developmental journey of the body, mind and heart, moving from the heaviness of cause-and-effect linear extrapolation, to the fluidity of co-evolving with our environment, to the lightness of breathing our thought and feelings into reality. Grounded in development of our mental faculties, these are phase changes, each building and expanding previous learning in our movement toward intelligent activity.

    Part I lays the groundwork, moving through the concepts of change, knowledge, forces, self and consciousness, recognizing that change comes from within. **Part II** explores Phase 1 of the ISCJ, focusing on the linear cause-and-effect relationships of logical thinking. **Part III** explore Phase 2 of the ISCJ, co-evolving with the environment, when we are able to pull together patterns from the past and think conceptually to extrapolate future behaviors. **Part IV** explores Phase 3 of the ISCJ, recognizing the power of thought and the role of attention and intention in an ever-expanding search for truth, fully embracing our role as co-creator. **Part V** delves into the ancient art of Alchemy to explore the larger shift underway for humanity, then looks at balancing and sensing, the harmony of beauty, and virtues for living the future.

And foundational work still relative:

Organizational Survival in the New World: The Intelligent Complex Adaptive System *by Alex Bennet and David Bennet* (2004)
A new model of the firm that enables organizations to react quickly and fluidly to today's dynamic business environment. Based on research in complexity and neuroscience—incorporating networking and knowledge theory to turn the living system metaphor into a reality for organizations.

Knowledge Mobilization in the Social Sciences and Humanities: Moving from Research to Action *by Alex Bennet and David Bennet with Katherine Fafard, Marc Fonda, Ted Lomond, Laurent Messier and Nicole Vaugeois in cooperation with The Social Sciences and Humanities Research Council of Canada (SSHRC).* (2007)
A powerful methodology that can unlock hidden value creation potential in a community or organization. Essential to increase competitiveness in a sustainable way.

<u>MQIPress Conscious Look Books:</u>

The Possibilities that are YOU!

by Alex Bennet

All Things in Balance
The Art of Thought Adjusting
Associative Patterning and Attracting
Beyond Action
Connections as Patterns
Conscious Compassion
The Creative Leap
The Emerging Self
The Emoting Guidance System
Engaging Forces
The ERC's of Intuition
Grounding
The Humanness of Humility
Intention and Attention
Knowing
The Living Virtues of Today
Me as Co-Creator
Seeking Wisdom
Staying on the Path
Transcendent Beauty
Truth in Context

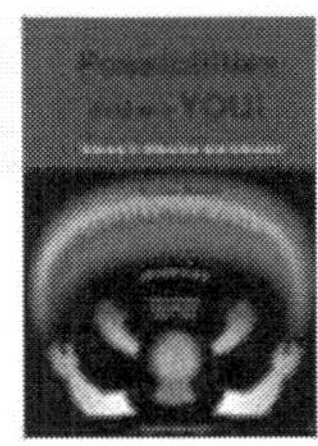

Of the 22 books in this series, only 11 covers are shown here.
The entire series with descriptions and "Look Inside" glimpses
is available at Amazon.com as individual copies with free kindle version.
Once you start reading or hearing them, you will find it difficult to decide
which are your favorites and which ones to share.

Consilience
Systems
ICALS
Epistemology
Constructivism

Appreciation to all those who dare to unleash their minds
to create an unknown future.
So many to thank for their contributions to this work.
Still, a few need to be called out,
specifically, Charlie Seashore, mentor to both David and Alex; and
Jane Turner, who helped edit this book and who's always been there
supporting the authors through many years of discovery.

There are so many friends, family, co-authors, and colleagues,
the network is vast and far reaching.
Not to forget, the stream of those who traveled
to quest with us at the mountain, who shared
their stories, their wisdom, and their passions
for knowing life in the deepest ways.